The Conservation Handbook:

Research, Management and Policy

2006

This book is being sent free to those practising conservationists outside Western Europe, North America, Australia, New Zealand and Japan who are otherwise unlikely to obtain a copy. These copies are provided at cost price by Blackwell Science, the publisher, and paid for with the author's royalties. Each book sold means another one donated. Administration and distribution of donated copies is handled free of charge by the nhbs.com bookstore. The Christensen Fund has generously provided a grant to cover the cost of postage.

We welcome names of people who live in the area outlined above and would benefit from this book. Please send your name and address, the name of the suggested recipient, their address and a sentence or two explaining why they should be sent this book, to NHBS, 2–3 Wills Road, Totnes, Devon TQ9 5XN, England. Email: gratis@nhbs.co.uk. The number of books donated can be followed on http://www.nhbs.com/info/sutherland/conservation_handbook.html

The Conservation Handbook: Research, Management and Policy

William J. Sutherland
Professor of Biological Sciences
School of Biological Sciences
University of East Anglia
Norwich
United Kingdom

Blackwell
Science

© 2000
Blackwell Science Ltd
Editorial Offices:
Osney Mead, Oxford OX2 0EL
25 John Street, London WC1N 2BS
23 Ainslie Place, Edinburgh EH3 6AJ
350 Main Street, Malden
  MA 02148-5018, USA
54 University Street, Carlton
  Victoria 3053, Australia
10 rue Casimir Delavigne
  75006 Paris, France

Other Editorial Offices:
Blackwell Wissenschafts-Verlag GmbH
Kurfürstendamm 57
10707 Berlin, Germany

Blackwell Science KK
MG Kodenmacho Building
7–10 Kodenmacho Nihombashi
Chuo-ku, Tokyo 104, Japan

First published 2000

Set by Keyword Publishing Services Ltd
Printed and bound in Great Britain at the
University Press, Cambridge

The Blackwell Science logo is a
trade mark of Blackwell Science Ltd,
registered at the United Kingdom
Trade Marks Registry

DISTRIBUTORS

Marston Book Services Ltd
PO Box 269
Abingdon, Oxon OX14 4YN
(*Orders*: Tel: 01235 465500
          Fax: 01235 465555)

USA
Blackwell Science, Inc.
Commerce Place
350 Main Street
Malden, MA 02148-5018
(*Orders*: Tel: 800 759 6102
          781 388 8250
          Fax: 781 388 8255)

Canada
Login Brothers Book Company
324 Saulteaux Crescent
Winnipeg, Manitoba R3J 3T2
(*Orders*: Tel: 204 837 2987)

Australia
Blackwell Science Pty Ltd
54 University Street
Carlton, Victoria 3053
(*Orders*: Tel: 3 9347 0300
          Fax: 3 9347 5001)

A catalogue record for this title
is available from the British Library

ISBN 0-632-05344-5

Library of Congress
Cataloging-in-publication Data

Sutherland, William J.
  The conservation handbook:
  research, management and policy/
  William Sutherland
      p. cm.
  ISBN 0-03-205344-5
    1. Biological diversity conservation.
  I. Title.
  QH75.A3 S88 2000
  333.95′16—dc21   00-028905

For further information on
Blackwell Science, visit our website:
www.blackwell-science.com

Contents

Contents

Contents

Contents

Contents

Foreword

The Conservation Handbook is a very welcome addition to the literature and armamentarium of conservation practice. It fills an important niche; we too easily forget, in the swirl of theory and global strategies, that the salvaging and management of biodiversity is eventually to be won on the ground, much like a war (which in many respects it is), by dedicated people who know how to proceed day to day in particular places and times, carrying with them the tools required. In this regard *The Handbook* is much like a field guide for the identification of species, not to be omitted from one's luggage or research station. It will be especially useful for conservation workers in developing countries, and I applaud the plan of the author and publishers to distribute as many free copies as possible to residents there.

Professor Edward O. Wilson
University Research Professor and Honorary
Curator in Entomology
Harvard University, USA
2000

Acknowledgements

I am very grateful to the following who made suggestions, commented on parts or answered questions: Petri Ahlroth, Graham Appleton, Diana Bell, Tim Benton, Colin Bibby, Tim Birkhead, Gerard Boere, Nigel Collar, Will Cresswell, Nicola Crockford, Séan Doolan, Bart Ebbinge, Chief Emil, Rob Freckleton, Alistair Gammell, Wenceslas Gatarabirwa, Jenny Gill, Jeremy Greenwood, Tim Halliday, James Harrison, Joh Henschel, Robert Kenwood, Hanna Kokko, Brian Lewis, Georgina Mace, Duncan McNiven, Bartshe Miller, Oliver Nasirwa, John Oates, Carlos Peres, Dave Pritchard, Edmunds Racinius, John Reynolds, Andreas Shilombaleni, Rick Shine, Guy Shorrocks, Carl Smith, Philip Stephens, Maris Strazds, Ron Summers, Alison Surridge, Peter Thomas, David Thomas, Jeremy Thomas, Hazell Thompson, Des Thompson, Charles Vatu, Martijn Weterings and Gerald Winegrad.

Discussions with students on our MSc course in Applied Ecology and Conservation clarified my thinking.

Ian Sherman showed his usual editorial enthusiasm and arranged for Blackwells to provide books instead of royalties.

Bernard Mercer at NHBS kindly organised the collation of addresses and the distribution of the extra copies.

The Christensen Fund extremely generously offered to pay for the postage to distribute the free copies. I thank the director, Keyt Fischer, for encouragement and guidance.

Special thanks to Nicola who read each chapter, checked the references and tolerated this book dominating our lives.

I would be very pleased to receive comments and corrections and would also be interested to hear whether the book has been useful (email: w.sutherland@uea.ac.uk; address: School of Biological Sciences, University of East Anglia, Norwich NR4 7TJ, UK).

1 Introduction

I take it for granted that the reader recognises the widespread and accelerating loss of biodiversity, realises the enormous cultural, economic and biological importance of this loss and is convinced of the need to do something about it. There are many books outlining the main concepts of conservation biology (for example Meffe & Carroll 1994, Noss & Cooperrider 1994, Caughley & Gunn 1996, Dobson 1996, Hunter 1996, Reaka-Kundla *et al.* 1997, Primack 1998, Sutherland 1998a). The aim of this book is to concentrate on what individuals can actually do to tackle some of the world's problems.

I believe that many of the conservation problems and solutions are similar everywhere, regardless of whether in polar, temperate or tropical regions, regardless of how affluent the region is and regardless of whether considering international conservation or conserving a small area. The universal problems are: habitat destruction, intensive agriculture, overgrazing, undergrazing, nutrient enrichment, pollution, hydrological changes, changes in fire regimes, overexploitation, introduced predators and introduced competitors. Similarly the universal solutions are: set priorities, plan, monitor, detect problems, diagnose problems, then bring about change through the main techniques of species management, habitat management, legislation, education, public awareness and integrating development and conservation (Fig. 1.1).

The global similarities in problems and solutions were evident when selecting case examples. To illustrate the principles of using education for conservation, should I use the campaign to reduce overexploitation of turtles on Pacific Islands or the programme to encourage gardeners not to use peat in north western Europe? To demonstrate how diagnosis of conservation problems should be undertaken, should I describe how biologists detected a decline in wandering albatrosses *Diomedea*

exulans in Antarctica and traced it to the long-line fishing off South America (Croxall *et al.* 1990) or how Cropper *et al.* (1989) discovered that if the habitat of the metallic sun-orchid *Thelymitra epipactoides* in Australia is never burnt it remains dormant because the competitors flourish?

I believe that it is a mistake to over-compartmentalise conservation biology. The conservation of plants and animals need similar approaches. Science theory, science practice and policy all need to be considered together. As examples, it is often useful for biologists to discuss with policy makers to determine what information is required before carrying out research and for practitioners to discuss with researchers so that their actions can improve understanding.

The case studies I have chosen tend to describe successes, although I have often outlined the problems encountered. This choice is because I am an optimist, because it is more enjoyable to read about successes and because people are more forthcoming about their successes. It does, however, have to be said that for each success story there are probably numerous failures. I hope this book will help increase the ratio of successes to failures.

Some themes run through this book. A major one is that we are throwing away the opportunity to learn from our actions. The lack of experimentation, monitoring and documentation means that we know far less about how to carry out conservation than we should. This criticism applies to every aspect including habitat management, species management, fund raising and education. Another major theme is the importance of determining objectives and how these will be achieved. This again applies to every aspect from selecting areas for conservation to determining priorities within an organisation.

I enjoy cooking and sometimes read recipe books looking for ideas and techniques which I modify

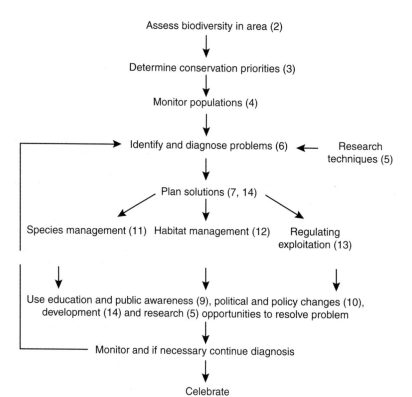

Assess biodiversity in area (2)

Determine conservation priorities (3)

Monitor populations (4)

Identify and diagnose problems (6) ← Research techniques (5)

Plan solutions (7, 14)

Species management (11)   Habitat management (12)   Regulating exploitation (13)

Use education and public awareness (9), political and policy changes (10), development (14) and research (5) opportunities to resolve problem

Monitor and if necessary continue diagnosis

Celebrate

Fig. 1.1 The logic for organising conservation work. Chapter numbers are given.

according to the ingredients I have and my own ideas. Similarly, I do not expect anyone to slavishly follow the procedures described here: they are presented to be modified and improved according to the particular problem and situation.

The successful conservationist often uses a myriad range of skills, and should ideally have some knowledge of planning, field skills, scientific methodo-logy, statistical analysis, advocacy, policy and education. Although they require a great range of skills, conservationists also tend to be underfunded and overworked. I wish to salute the remarkable achievements of many conservationists I have met, some of whom were working under very difficult conditions.

2 Assessing biodiversity

2.1 Why assess biodiversity?

Resources for conservation are always limited and thus, to maximise the benefits of any actions, it is advisable to focus on the highest conservation priorities (see Chapter 3). Much of the work of practising conservationists entails making judgements about the relative importance of different areas at a range of different scales. The objective may be, for example, to determine which countries or regions should be priorities for conservation funding, to determine which areas within a country are the most important, to determine which part of an area is most important (perhaps in order to create reserves), or may even be at a very local scale, say in deciding which part of an area should be cleared to create a camp-site or car park.

The conservation importance of an area is typically determined by assessing its biodiversity and, as the basic units of biodiversity may be considered to be species, this is done mainly by which species are present and by the abundance of key species (see Chapter 3). Creating a species list is not, however, straightforward and may be expensive. For example, it has been estimated that creating a list of higher plants and vertebrates of a single $40 \, km^2$ National Park in Australia costs about US$60 000 (Balmford *et al.* 1996a). The experience from a listing of hymenoptera and other taxa is that inventories cost about US$1–2000 per species (Gauld 1999). The major issue is thus how to use the available information and resources to provide the best assessment of the conservation priorities between areas and this needs to be considered before any survey starts.

Chapter 3 considers a range of ways of measuring biodiversity such as species richness and diversity and discusses whether it is sufficient to survey just a few taxonomic groups.

A key decision is the balance between estimating population sizes and discovering a wide range of species. The most rigorous approach is to carry out population estimates of all species using the methods of Chapter 4. For example, carrying out distance estimates from point counts or transects has much to recommend it, but it will not reveal as many rare species as the less rigorous methods described in this chapter. The other extreme is to simply produce a species list so that all the time is spent looking for additional species and this is best for discovering rarities. However it is also usually useful to know the abundance of the species. Some approaches described in this chapter can provide information on both diversity and abundance. Another solution is to concentrate on listing species richness whilst also collecting information on the abundance of a few species of particular concern.

An important issue is the availability of data. Data can be a valuable source of income for organisations, for example by selling to consultancies carrying out environmental impact assessments. One concern with commercial use of data is that some voluntary recorders become unwilling to provide records, another is that data centres can lose sight of their conservation objectives. There is an increasing shift towards free availability of data.

2.2 Biodiversity assessment techniques

Whatever technique is used, it is important to quantify the amount of effort put into sampling as otherwise it is difficult to determine the extent to which a long species list reflects the importance of the site or the enthusiasm and skill of the observers (Block *et al.* 1987, Gaston 1996). This is less important for well-studied groups that are reasonably easy to survey but is always a source of some bias.

Many biodiversity assessment methods involve sampling by selecting sample areas or transect routes. Will their locations be determined at random, selected to provide the highest total number of species or located where convenient, for example, near to trails from the camp? There are occasions when each is sensible but the method used must be stated. It is possible to analyse both the mean number of species per sample, as a measure of local richness, and the total number of species from all samples combined, as a measure of total richness. Most of these methods can be used to gain information on the abundance of each species as well as diversity.

Time-based observation methods, such as time-restricted search and timed species counts (see Section 2.2.10), are only practical if field identification is rapid and reliable. These are thus only appropriate for readily identified groups and for experienced naturalists.

A measure of biodiversity on its own is pretty meaningless. The objective of assessing biodiversity is usually to compare sites or to provide the data that can be used by others for comparing sites. It is thus important that the methodology is consistent and clearly stated so that it can be repeated. It is often a good idea for members of a survey team to start by going in the field together to ensure that their identifications and methodologies are consistent.

For each method the results will vary with the weather, time of day, season and habitat structure. It is important, when comparing surveys, to ensure this has not greatly affected the comparison. Key habitats, such as fruiting trees or drinking pools, bias results but are very effective at increasing species lists: they thus need to be dealt with consistently.

2.2.1 Total species list

A list of all the species within a group ever recorded is by far the most common technique. Species are simply added to the list as detected. Contributions to the list may come from a variety of sources and individuals. It has the enormous problem that it is difficult to compare lists if variation in effort is a major contribution to the number of species recorded per site. It is most useful for well-studied groups such as birds and large mammals where variation in effort is likely to be less critical. In its simplest form of a cumulative list it may include species that only exceptionally visit the site and species that have been extinct for some time in the site. An annotated species list, with information on the number of sightings or impressions of abundance greatly improves the accuracy and ease of interpretation. One option is to give a series of columns with headings such as, 1 'recorded by Smith July 1997', 2 'recorded by Smith September 1998', 3 'recorded by local people but not by Smith 1997 or 1998,' with ticks for each.

If simply creating a species list or inventory, it is sensible to use as wide a range of techniques as possible. Use a diversity of searching techniques such as searching, trapping, listening for calls, playback of taped calls, looking for droppings or looking for different life stages (such as eggs, larvae and pupae). If trapping, it is sensible to vary the methodology, for example, the species of invertebrates caught in water traps may depend upon the colour of the trap and its height above ground. Search or trap at different altitudes, in each of the habitats and in the centre and edge of habitat blocks. Animal species differ in their behaviour, so that being present at different times of day and night, under different weather conditions or at different seasons, will all increase the number of species found.

If identifying in the field, one approach for some taxa is to start with a list of possible species and cross off each species found and concentrate on methods and locations likely to detect those remaining.

For easily identified species some useful information can be gained by talking to local people and showing pictures. It is also useful to include some pictures of species that are definitely not present to evaluate the accuracy of the information. It can be useful to talk to hunters and fishers and inspect their catch. A new species (and genus), the Udzungwa Forest Partridge *Xenoperdix udzungwensis* (Dinesen *et al.* 1994), was discovered in 1991 when a pair of ornithologists in Tanzania could not identify the bird's feet in their supper!

Discovery and conservation of the Saola *Pseudoryx nghetinhensis* in Vietnam

In 1992 a joint expedition by the Ministry of Forestry in Vietnam and the World Wide Fund for Nature of the Vu Quang Nature Reserve in Vietnam discovered three sets of bovid horns that were clearly new to science (Dung *et al.* 1993). Since then over 20 specimens have been located in Vietnam from a range of evergreen forest along the Annamite Mountains with another 23 localities known across the border in Laos (Schaller & Rabinowitz 1995). Remarkably, subsequent surveys in the area found a second new species of large mammal, a giant muntjac deer *Megamuntiacus vaguangenis*, evidence for another small undetermined species of muntjac (Schaller & Vrba 1996) and a striped rabbit (Surridge *et al.* 1999). Although biologists spent considerable time surveying in the field, all the information on these three mammals came from talking to local people and especially hunters.

After these discoveries, and the considerable worldwide interest, the Ministry of Forestry enlarged the Vu Quang Nature Reserve, cancelled logging in the area and strengthened its protection laws. The Saola has been added to Appendix 1 of CITES.

Automatic cameras have photographed some species never seen alive by biologists such as this recently discovered Saola (photo: EU/SFNC).

2.2.2 Total genus or family list

This is the same method as the species list except a higher taxonomic level is used (Balmford *et al.* 1996a,b). Balmford *et al.* estimated that limiting a higher plant survey of a forest in Sri Lanka to the genus level reduces the cost by over 60% and that simply recording the number of families reduces the cost by over 85%. They suggest that the family level is too crude but that surveys of genera provide a reasonable estimate of richness and correlate well with the number of species. However some studies, for example on ants in Australia, show that even generic information is a poor surrogate for species richness (Andersen 1995). Furthermore, information on the distribution of species has so many other uses, such as for determining threat status or creating Red Data books that it is probably usually worth the extra expenditure to obtain the full species list. With the current practice of assessing conservation priorities, a list of genera from a site would usually be of little use.

2.2.3 Parallel-line searches

A parallel-line search is the best method for assessing the presence of visible and fairly sedentary species in reasonably small areas (Nelson 1987). It has been frequently used for plants and also for groups such as amphibians. Divide the area into blocks (not exceeding 10 hectares for higher plants). Systematically cross each block across the shortest width in a series of parallel paths, recording all species and marking the locations of rarities.

Return to record the populations of rarities. Closer parallel lines give greater accuracy but obviously take more time. Document the precise method used, the species list for each block and the precise location of rarities.

2.2.4 Habitat subsampling

A survey may have to be very incomplete as there is only time to visit a small fraction of the total area. Some groups, especially many invertebrates, are so abundant, inconspicuous or difficult to identify that it is almost always necessary to restrict the survey to small areas. The samples could be as small as soil cores for sampling soil invertebrates or as large as quadrats or transects for plants or amphibians. Many traps will only sample a very small proportion of the total area, for example a pitfall trap can only catch those invertebrates within walking distance. By standardising the methodology (e.g. diameters and depth of soil core or size of pitfall trap, bait used and time left) it is possible to compare sites.

The samples may be located in different micro-habitats to produce the greatest species list or at random locations (see Section 4.3) to allow comparisons between sites. Some researchers suggest long thin quadrats are better than square ones as they cross a greater range of habitats. Condit *et al.* (1996) found rectangular quadrats contained 10% more species than square ones and a very narrow quadrat (100×1 m) contained 18% more species than a square quadrat of the same area. Rectangular quadrats have a longer edge so are more inaccurate if there is a problem in deciding whether individuals are inside the quadrat.

2.2.5 Uniform effort

Uniform effort entails standardising sampling effort in each site, for example, by counting the catch per trap per day or the catch per 20 sweeps of a sweep net. Even the most straightforward techniques will be carried out differently by different people so that any difference between sites may be an artefact. It is best to use one person (who must also be consistent!) but otherwise give precise instructions and ideally ensure the samplers meet, watch each other and compare their sampling method and efficiency in the same location.

2.2.6 Time-restricted search

In time-restricted search, also known as rapid inventories or rapid biodiversity assessments (Crump & Scott 1994), a habitat is searched for a set period of time (e.g. an hour or a day) and the species recorded. It requires experienced field naturalists. A range of techniques can be used (but they should be consistent between sites) and the observer or team is free to search where they think most species will be found. As a result of this freedom, time-restricted search is less consistent than most other methods described here. It is, however, good for adding species at a high rate and discovering the most exciting species. This is thus often a good way to compare sites rapidly where little is known. Either the entire search can be repeated a number of times within different parts of a site, which is statistically more elegant, or alternatively, a consistent total time, say three days, is devoted to the entire site which, as it avoids repetition, is better for creating the longest species list. There is likely to be considerable variation between observers or teams.

2.2.7 Encounter rates

Estimating encounter rates is the most elementary way of incorporating effort into abundance estimates. The total number of records are divided by the time surveyed (confusingly the reciprocal, minutes per individual seen, is sometimes used). Thus 300 person hours and 60 sightings gives a value of 0.2 observations per hour. Two problems need to be resolved. What will the policy be for dealing with individuals that are encountered repeatedly, such as one next to the camp? A solution is just to state in the report when this is clearly a problem and also calculate the encounter rate excluding this individual. Secondly, it is necessary to decide what counts as observation time. For example, is time included when spent in other activities such as eating but when some records will be made? More than one person together conventionally counts as one person but detection often changes with group size. The rules used must be stated in the paper/report.

Measuring encounter rates is not a very good method but is a great improvement on a simple species list without any attempt to estimate effort or abundance. The site is often subdivided into areas or habitats so the encounter rate in each can be calculated.

2.2.8 Species discovery curves

One important question is whether further field-work will increase the species list. One method is to record the time and date of each new species seen along with a measure of the time spent in the field. Count the total numbers of species seen and plot against the number of days or number of hours spent in the field. This species discovery curve shows the point at which further effort is unlikely to reveal further species. Figure 2.1 shows a species discovery curve from West Java (Robertson & Liley 1998). After 25 days the rate of discovering new species had declined but new species were still being found after 70 days. This method cannot say whether you have found all the species, just whether further searches are likely to reveal more. In theory it is possible to estimate the asymptote to compare sites but a problem can be that some curves show no obvious asymptote within a reasonable time. A curve can be fitted directly by eye or a statistical package, using a logarithmic or exponential curve (Robertson & Liley 1998). Alternatively, the number of new species in each survey (e.g. day) can be

plotted against the $\log_{10}$ number already discovered; where the linear regression hits the x-axis is the total number of species (Pomeroy & Tengecho 1986) (Fig. 2.2).

This approach can be used on a smaller scale. Goff *et al.* (1982) describe a similar approach for surveying plants in which they suggest the survey stops once 30 minutes have been spent searching without finding a new species.

In practice, the number of species recorded may just keep increasing with sampling effort, especially for mobile species, such as birds, as rare species and vagrants continue to be recorded (Harrison & Martinez 1995). This can lead to the problem of oversampling in which the importance of an area is overestimated simply because of greater observer time.

2.2.9 MacKinnon lists

MacKinnon lists (MacKinnon & Phillips 1993) involve a series of surveys, each listing the species until a certain number (say 20) is seen. The next survey starts a fresh list. The cumulative total number of species seen from all counts combined is then plotted against the number of counts (see Fig. 2.2). Sites with higher diversity will have a higher cumulative species total after a given number of counts.

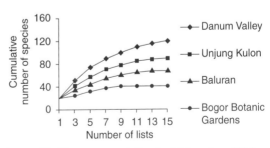

Fig. 2.2 MacKinnon curves plotting the total number of bird species seen against the number of surveys of 20 bird species for four sites in Indonesia (from Mackinnon & Phillips 1993). This shows that after seven surveys in Bogor Botanical gardens few further species were discovered but that even after 15 surveys in Danum Valley the rate of discovery of new species had barely begun to decline.

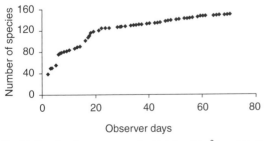

Fig. 2.1 Species discovery curve for birds in a 3 km² forest block in West Java, Indonesia (from Robertson & Liley 1998).

Again this does not indicate the maximum number that would be recorded but will show when further counts are unlikely to produce further species. This method is good for inexperienced observers as time can be spent identifying individuals without distorting the results. The fraction of lists in which each species occurs can be used as an index of relative abundance but it underestimates inconspicuous species and gregarious species because a group counts just once (Robertson & Liley 1998). More rigorous analyses of these curves is complex (see Soberón & Llorente 1993).

2.2.10 Timed species counts

This method takes advantage of the fact that common species are likely to be first seen soon after starting a survey while rare species, if seen, are almost as likely to be first seen at the end of the survey as at the start. This method involves dividing a one-hour observation period into six 10-minute blocks (an alarm function on a watch makes this easier). A list is then made of the species seen in each 10-minute period (or the 10-minute interval in which the species is first seen is noted on a checklist). Once a species has been recorded it is then ignored for the rest of the hour. Those seen in the first 10 minutes score 6, those in the next 10 minutes score 5 and so on with those not seen scoring 0. These values are averaged over a number of census periods

(usually 10–15) (Pomeroy & Tengecho 1986). As with MacKinnon lists the scores underestimate inconspicuous and gregarious species.

Timed species counts are a useful hybrid of the other methods. The point and transect count methods described in Section 4.7 provide reasonable estimates of population size of the commoner species but as so much time is spent documenting these and travelling between points or transects they tend to miss the rarer species. Simply compiling a species list documents the presence of rarer species but gives no idea of the abundance of each species. In each timed species count the commoner species are ignored once first located so most time is spent searching for further species. One disadvantage of this useful method is that timed species counts have been rarely used and thus there is little data for comparison.

2.2.11 Recording absence

Ironically, absence is much harder and much more time consuming to record than presence, as the searching should continue until the species must have been found had it been present. Paradoxically, it is thus even more important to describe the methodology when documenting species absence than presence. Reed (1996) used the concept of statistical power (see Section 5.16) to calculate in theory the necessary number of visits, N

$$N = \frac{\ln(\alpha \ level)}{\ln(1 - P)}$$

where P is the probability of detecting the species on a given visit and α level is the acceptable risk that the species is present but recorded as extinct. Of course, the ideal is to be absolutely positive that the species is extinct (α level = 0) but this is impossible! The value of α level used then depends upon the consequences of classifying a species as extinct when it is not. As is intuitively obvious, more visits are needed for elusive species (P just above 0) than for conspicuous ones (P just below 1). The detection probability can be assessed from studying the species elsewhere, from comparing other similar species, examining the

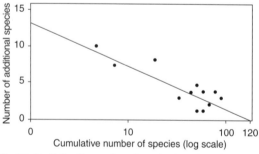

Fig. 2.3 Predicting the number of species in a site using timed species counts. This plots the number of additional species seen during a given time period against the log of the cumulative number of species. A regression predicts the total number of species (120 in this case) (from Pomeroy & Tengecho 1986).

probability of detecting known individuals or from the guidance of experienced field workers (Reed 1996). Another possibility is to use the number of visits and the detection probability to estimate the likelihood that it has been missed (α level).

There are considerable problems of deciding whether a species still exists within an area. For conspicuous and easily identified species it is often worth asking local people and showing them photographs or field guide illustrations. Including some pictures of species that cannot be present is useful for checking the observer accuracy.

2.2.12 Habitat feature assessment

Experts can often be very good at judging from a distance how good a site is likely to be. Habitat feature assessment involves bringing together a range of experts and asking what features they consider important for assessing quality. For example, for grasslands the characters may be physical (e.g. aspect, underlying rock, size and altitude), habitat structure (e.g. scrub abundance or grass height), management (e.g. is it grazed and by what), protection (e.g. has it been ploughed and reseeded) or the presence of certain conspicuous key species. These features can then be converted into a list which can provide a standardised response by non-experts. This method is clearly only suitable for an initial superficial survey, for example to identify which sites warrant a more detailed survey.

2.3 Documenting rarities

In too many cases the location of rare species is only known from vague descriptions and it is unknown whether the population has gone extinct or just cannot be relocated. Accurate descriptions prevent such uncertainty.

Cropper (1993) describes how to document rare plants and this advice applies to many other groups. Draw the boundaries of the population as accurately as possible on the largest scale topographic map available. Distribution notes should be added in the field, not afterwards. To ensure the population can be relocated, supplement with detailed

hand-drawn maps if necessary. It may also be useful to state areas surveyed or areas omitted from the survey. Final maps should include details of source of map such as publisher, series, name and code, a scale, arrow indicating north, name of the species, your name and date, and reference to any photographs taken.

Global Positioning Systems (GPS) are continually becoming cheaper and will presumably become the standard method of recording locations in areas which cannot be readily determined on a map.

2.4 Collecting

Collecting specimens is sometimes essential. Specimens are often needed for subsequent identification or in order that the record can be verified by others.

The conservation and legal consequences of collecting need to be considered. In many cases reliable identification is possible without collecting and collecting may be ethically unacceptable. Attitudes have changed since the 1906 expedition that rediscovered the Pinzon Island race of the Galapagos Giant Tortoise *Geochelone elephantopus* collected all remaining 86 for research (Thornton 1971). In countries where the natural history is well known, the collection of butterflies, dragonflies, amphibians, reptiles, mammals or birds is usually considered unacceptable. Many countries require permits for trapping, collecting, exporting and for importing. Furthermore there may be regulations that apply to firearms, alcohol, formalin, syringes and various killing agents. Local people may have been asked not to hunt or kill within an area, so it may be unacceptable for an outsider to collect specimens.

Make sure you do not collect more specimens than is necessary and than can be preserved and identified. If someone else is identifying the specimens then ask how they would like them to be preserved as experts usually have strong personal preferences. Chapter 4 describes a number of collecting methods.

2.4.1 Labelling

If collecting a number of specimens, each specimen should be given a field number and a preliminary identification (e.g. small blue Nudibranch) as a check in case of confusion. Specimens are traditionally numbered sequentially starting with your initials (e.g. WJS 921). It is conventional not to repeat numbers, even in different field seasons or studies, so that each is unique. Printed tags with consecutive numbers are available (e.g. from the National Tag Company, 815 S Brownschool Road, Vandalia, OH 45377, USA) which prevents double numbering and saves time. Specimens from the same individual, such as flowers and leaves, are given the same number. If fruit, flowers and leaves are present at different times, and it is necessary to have all stages, then mark a specimen and return later.

The more information that is associated with a specimen, the more useful it will be. Each specimen should be labelled with location, date, habitat, altitude, collector's name, ecological or behavioural details, reference to photographs or tape recordings, fresh weight, sex and reproductive status (if determined by internal examination when preparing the specimen), measurements such as body length that cannot be made on the preserved specimen and those features that will be lost in the preservation, such as colour, smell or details of external parasites.

Photography often distorts colours so either include a colour chart in any photograph or describe separately. Colours are most easily described using water-colour paints (Knudsen 1972). Using a pencil, sketch the outline of the organism and draw in the edges of the blocks of colour. For each colour (e.g. green) mix an appropriate combination (yellow and blue) and add a dab to the edge of the paper. Hold this against the specimen and decide which paints should then be added to get the colour correct and make another dab on the edge. Keep experimenting and comparing until the colour is right. Then simply circle the correct dab of colour and join to the appropriate area or areas of the sketch with an arrow. It is unnecessary to paint in the sketch.

2.4.2 Preservatives

Alcohol evaporates quickly even through cork stoppers. Adding 5% glycerol ensures that the specimen will not dry out if all the alcohol evaporates and prevents specimens from becoming brittle. Specimens in tubes of alcohol may also be stored within bottles containing alcohol. Alcohol dehydrates the specimen and thus becomes more dilute. The alcohol should thus be replaced during the early stages of preservation if there are many specimens and particularly if they are soft bodied. Alternatively ensure that the specimens are only a small part (10%) of the total volume. Alcohol may cause the precipitation of salts from sea water and then should be replaced. Larger specimens, e.g. over 2 cm thick, may benefit from a slit in the gut cavity to allow the alcohol to penetrate further. Labels on the outside of bottles usually eventually fall off. Samples in alcohol should be labelled using pencil on card placed in the alcohol. Tubes should be sturdy and watertight with a clip-on plastic top. Alcohol is heavily taxed and expensive.

Industrial methylated spirit (96% alcohol, so add 25 ml water to each 75 ml to make 70% alcohol) includes wood naptha but is almost as good as alcohol for preserving specimens.

Distilled drinking alcohol (e.g. gin, brandy or whisky) can be used in emergencies and provides reasonable preservation but these still tend to be rather weak.

Formalin is a better preservative than alcohol and cheaper but is highly carcinogenic and irritates the skin, eyes and upper respiratory tracts and so should only be used in a fume cupboard with full protection. With increasing concerns over health and safety the use of formalin is likely to be greatly reduced.

Freeze-drying consists of specimens being frozen and then dehydrated by sublimation, the process by which water is converted straight from a frozen state to a gas without becoming a liquid. Specimens do not become distorted or shrink. Specimens are placed in a freeze dryer at below −10°C and the water is extracted with a vacuum pump.

Dried specimens should be stored in a dry, warm location with an insect repellent such as naphthalene. Collections are prone to attack by insects and mould if not fully dry.

Collecting increasingly involves preserving a range of material for analysis by parasitologists, for example for DNA analysis or for tissue analysis for substances such as heavy metals. Samples for DNA are best stored in pure (100%) alcohol. This will extract water from the specimen and thus dilute the alcohol, so either replace the alcohol occasionally or ensure the specimen is a small part (e.g. 10%) of the total volume. DNA extraction techniques are rapidly improving and dried specimens, hair or feathers are now often sufficient for techniques using PCR (see Section 5.15.1).

2.4.3 Collecting plants

Plants are usually collected in the field in plastic bags; sealable freezer bags are ideal. They are usually dried which is normally straightforward unless the plant is large or succulent or the drying site is very humid.

Plants are usually attached to heavy duty white paper onto which the details are written. Acid-free paper is better if the specimens are being preserved in the long term. Specimens can either be glued using plastic cement, taped on using gummed paper or cloth adhesive tape (Scotch tape becomes brittle after a couple of years) or sewn on if the specimen is thick. Include notes on other features such as size, bark and branching.

Many plants are too large to fit on a herbaceous sheet (usually 29 × 42 cm) so it is sensible to ensure that the specimen includes representative parts. Specimens may be cut or folded to fit. Laying each plant out so that the features can be observed when flattened takes time, so do not collect more plants than can be pressed. Plants are usually placed within a sheet of folded newspaper arranged as they are intended to look in the herbarium. Show both sides of leaves and the underside of flat flowers. A sheet of paper, felt or foam rubber (if the specimen is bulky) is placed on top before the next specimen. This proceeds to create a stack of specimens and paper. In damp locations occasional sheets of corrugated cardboard (cut so the corrugations run along the width, not the length) should be inserted within the stack to allow air flow. In damp areas, or if pressing succulent plants, the paper should be

replaced every few days. The press may be kept above a stove or above oil or gas lamps to aid drying, ensuring that it is not a fire hazard. Creating a skirt around the presses but with a gap at the base for air to enter will funnel the hot air upwards.

The pile of specimens can be compressed with weights such as books but a plant press made of hardwood or plywood (softwood tends to crack under the strain) is much better and essential for serious collecting. These are commercially available but can be easily made. They consist of a wooden grid (typically 30 × 45 cm) at each end of the stack of specimens. Kneel on the entire stack and tighten with straps running round the press. Adequate small presses can be made from wire grids, such as cake trays and strong string.

If drying is really impossible in the field, stacks of plants pressed within newspaper can be sprayed with alcohol or a litre of 70% alcohol can be poured over a 20 cm pile of plants and kept in a plastic bag. The resulting specimens tend to be blackened and brittle. This is also a fire hazard.

Succulents should be killed by submergence in boiling water for a few seconds (up to a minute for bulky cacti) as the tissue will then dry more quickly and it will also prevent them growing new shoots in the press!

Fruit may be dried or sliced and pressed, or preserved in 70% alcohol and stored separately. Cones and wood are dried.

Mosses are usually placed directly into a paper packet for drying and are not pressed. Liverworts tend to shrivel so some gentle pressing is sensible. Lichens are best dampened before pressing otherwise they break. Mosses, liverworts and lichens are usually stored in paper packets and well-pressed material can be rehydrated for examination by placing in boiling water or water with a drop of detergent. Macroscopic algae can be pressed and dried, freeze dried or stored in 40% alcohol (although they lose their pigments in alcohol). Flimsy algae are best placed on a herbarium sheet under water and then gently lifted. Dry by pressing gently with a cloth.

Forman & Bridson (1992) and Brayshaw (1996) give further details on preserving plants.

2.4.4 Collecting fungi

Fungi should ideally be collected in open containers and kept separately from each other before transferring to rigid containers (not a plastic bag). Many key identification features such as colour, smell and structure may be lost during drying and must be recorded when collected. If necessary they may be stored in a refrigerator for a few days but once picked they may start changing appearance.

Permanent storage is best achieved through drying in a well-ventilated drying oven, preferably with a circulatory fan, for about 48 hours at 60°C. Larger fungi may need as long as a week to dry. They may also be preserved by freeze-drying (Hanlin 1972) and are then easier to section than are dried specimens. Dried fungi are stored in boxes or envelopes if small. Fungi on leaves can be preserved by pressing the leaf and drying as described in Section 2.4.1. Fungi can also be preserved in 70% alcohol.

Spore prints are made by cutting off the stem of a mature individual, placing the cap on white card or herbarium paper and covering to exclude draughts. Remove cap and spray with fixative used by artists for preserving charcoal drawings.

2.4.5 Collecting invertebrates

There are a wide range of methods for killing, relaxing, fixing and preserving different taxa. Knudson (1972) and Lincoln & Sheals (1979) give good reviews.

Most insects are easier to transport if alive than dead, except flies and other fragile species. Insects are often killed with ethyl acetate in a glass 'killing bottle' (ethyl acetate dissolves plastic) with a securely fitting bung. This usually has a base of set plaster of Paris onto which a few drops of ethyl acetate have been dripped to moisten the plaster of Paris but not leave surplus liquid. Alternatively add crumpled paper tissue with a few drops of ethyl acetate. If pieces of tissue are added then the dead insects are less likely to be damaged if carried in the jar. It is best not to mix large and small specimens in the same killing jar. Label the containers or killing jars in the field if visiting more than one location.

Use pencil for internal labels as ink runs in the presence of ethyl acetate vapour. Boiling water can also be used to kill tough insects such as beetles but is unsuitable for delicate species.

Many invertebrates are highly contractile and have to be relaxed by slowly anaesthetising prior to preservation. The anaesthetising period should be short to prevent the possibility of decay. Anaesthetics of general use include chloral hydrate (sprinkle on the water surface or add the animals to 2% fresh solution), propylene phenoxetol (add 1% by volume to water), 10% ethyl alcohol (make from absolute alcohol, not industrial methylated spirits, add small amounts in stages).

In practice almost any invertebrate, apart from butterflies or moths, can be preserved in alcohol. A concentration of 55–80% is usually used, and if in doubt any alcohol in the range 60–75% will be reasonable for preserving almost anything. Higher concentrations make many specimens excessively hard. Soft-bodied species are distorted more by high concentrations of alcohol as they lose more water. This can be overcome by placing in 30% alcohol for a few hours, then 50% for a further few hours, then 70% for a few days, followed by final storage in 70%. Insects with thin cuticles should be preserved in 70% alcohol and can be placed straight into full containers of alcohol in the field (they are more likely to be damaged during transit in partly full containers). Invertebrates without an exoskeleton are usually fixed in 70% alcohol and then stored in 40%. In general, use a high concentration if likely to be diluted by many specimens in a small volume of alcohol, specimens that will release much water or evaporation.

Small arthropods, including most insects can just be air dried but for some large species with thin cuticles it is best to remove the viscera and replace with cotton wool. Large specimens should be dried out.

These are general rules for some important groups.

Sponges. Either dried or preserved in 75–95% alcohol (formalin destroys the calcareous spicules).

Worms (Platyhelminthes and Annelids). Stored in 70% alcohol but anaesthetise first by adding small amounts of alcohol to the worms in water to ensure they are fully expanded.

Crustaceans. Preserve in 70% alcohol.

Echinoderms. Dry if hard bodied or preserve in alcohol usually after anaesthetising. They may be relaxed first by adding Epsom salts gradually until they do not respond to touch.

Snail shells. Either boil or deep freeze and remove the body then wash and dry or store in 70% alcohol.

Most aquatic molluscs. Add alcohol gradually to make a 10% solution over a period of hours or days and then store in 70% alcohol.

Insects. If small, specimens may be glued to card using clear fingernail polish (which may then be removed with fingernail polish remover if necessary) and then pinned to a board with an entomological pin (ordinary pins are too short and rust). Butterflies, moths, dragonflies, mayflies and lacewings are usually stored pinned through the thorax (usually to one side so no features are hidden) with the wings set at right angles. This is easiest on a cork or polystyrene setting board with a groove cut in it. The body is placed in the groove and the insect wings arranged and then pinned into position with strips of paper (glossy magazine covers work well). Most insects will be dry within a month unless it is cold, damp or the specimens are especially large. Butterflies are often brought back from the field in folded triangles of paper but if left too long they become brittle.

Other arthropods, such as spiders, scorpions, ticks, mites, centipedes and millipedes. Kill and preserve in 70–80% alcohol.

2.4.6 Collecting fish

Fish may be frozen in the short term as this preserves the colours, but this is unsatisfactory for long-term storage.

Fish can be killed by using an anaesthetic such as Benzonocaine, fixed in 70% alcohol and then stored in 40%. The colours are soon lost and so should be noted before preservation by description, sketching or photography. The colours are often particularly vivid a couple of minutes after applying the anaesthetic, making this an ideal time for photography.

2.4.7 Collecting amphibians

Amphibians may be killed by immersing in chloretene solution made by dissolving a teaspoon of hydrous chlorobutanal crystals in a litre of water. The solution remains effective for 1–2 weeks if used regularly and longer if used infrequently. Gels containing Benzocaine (sold to reduce toothache) kill amphibians within a few minutes when smeared on the head (Altig 1980) although this is an anaesthetic and so at low concentrations may not be lethal. Amphibians may also be killed by drowning in 15–25% alcohol solution or warm (43–47°C) water, but freezing is considered by many to be the most humane technique. They are usually fixed in preservative of 70% alcohol and stored in 40% alcohol. For a more detailed account see McDiarmid (1994).

Eggs and larvae are readily damaged and are best placed straight into preservative in the field. As they contain more water they initially need a stronger concentration or a relatively larger volume of alcohol than do adults.

2.4.8 Collecting reptiles

Reptiles are usually killed by freezing which simply reduces the metabolic rate and so is considered reasonably humane. Alternatively they may be injected with 10% nembutal solution. Drowning in warm water is slow and probably causes considerable pain.

Reptiles can be fixed in 70% alcohol and stored in 40% alcohol. Preservatives cannot enter specimens greater than about a centimetre thick so those should first be injected with 80% alcohol or cut to allow the preservative to enter the skin. Snakes are injected every 2–3 cm or cut along the left ventral surface. For larger reptiles, alcohol is unsatisfactory and some still use 10% formalin to fix and after a day or two, once they start to become rigid, transfer to 70% alcohol. As described earlier, formalin has considerable health problems and should be used in a fume cupboard.

2.4.9 Collecting birds

Bird identification is sufficiently straightforward that there are almost no conditions where killing specimens is justified. Injured birds can be killed in a few seconds by pressing hard on the breast bone. For large birds this may entail placing them on their backs and kneeling on them. Dead birds are best carried in a paper cone, slipped in head first as otherwise the feathers become displaced.

Birds are usually skinned by cutting with fine scissors from mid-breast to vent (Knudsen 1972). Ease the skin away from the body on each side using the back of the scalpel. Sever each knee at the joint. Care should be taken in detaching the tail. If cut too close to the tail, the feathers will detach, if too close to the body the body cavity will be ruptured. Invert the skin until the shoulder is visible and cut through with strong scissors. Continue inverting the skin. It may be necessary to push the skin over the skull. Pull the skin out of ears using fine forceps. Cut optic nerve behind eye. The skin should now be completely inverted but attached to the base of the bill. The skull is cleaned by cutting a hole in the cranium and removing the brain and then left attached to the skin. Remove fat and flesh attached to skin and on skull, legs, wings and tail stump. It is important to remove as much subcutaneous fat as possible. Sprinkle on corn meal, magnesium carbonate or fine hardwood sawdust to absorb fluids. It is usual to note the sex and size and development of the gonads on the label (see Section 5.8.6).

Apply preservative (usually borax) over inner surface of skin. Salt has been widely used but it tends to discolour plumage, soften the bill and destroy the writing on the labels.

Take a thin stick of eye to tail tip length and sharpen one end. Cover with cotton wool to create a 'body' of appropriate size. Insert point of stick into base of skull until it can just be felt under crown. Fill eye orbits with cotton wool. Dampen skin of head to soften. Reinvert skin over the skull and cotton wool body and sew together. For large species it is necessary to remove the wing muscles by cutting along the bones on the underside of the wing, cutting out the muscles and sewing the skin together. Tie the label to the stick. Dry in a tube of stiff paper which will take at least a week. Test if dry by feeling feet to see if they are hard.

2.4.10 Collecting mammals

Small mammals can be killed using chloroform or a similar chemical such as sodium pentobarbiturate (sold as Nembutal). Place the mammal in a container with an air-tight lid containing cotton wool moistened with a few drops of the chemical, or place the trap in a plastic bag containing cotton wool soaked in chloroform. The chloroform should not be breathed by the biologists or used near an open flame. Chloroform is intensely irritating to the skin and mucous membranes and should touch neither the mammal nor the biologist. Corpses should be kept cool and skinned within 12 hours.

Mammals are usually skinned by cutting with fine scissors from the back of one knee to just behind the anus and round to the other knee. Corn meal, magnesium carbonate powder or fine hardwood sawdust should be applied liberally to absorb fluid and aid drying. Use the back of the scalpel to tease apart the skin from the body. Push one knee through the incision and cut at the joint and then repeat with the other knee. Pull the tail out as far as possible and then cut. The skin is then inverted over the head as if removing a sweater. Free the eyes by cutting inside the eye sockets and pull the skin out of the ear sockets. By cutting the inside of the lips the body and skin are separated. Remove fat and flesh attached to the skin, limbs and tail leaving as little as possible.

Rub powdered borax thoroughly over the inverted skin. Salt can be used and is more readily available but is a poorer preservative. Small specimens (squirrel or smaller) are often preserved flat on a stiff white card cut to the correct shape inserted inside the skin with the details written on the end of the card. Larger specimens are usually prepared as a round skin which resembles the body shape around an imitation body of cotton wool. The details are written on a card label and tied to a foot. The skin may need to be dampened slightly before mounting if it is too inflexible.

Traditionally the skull is preserved along with the skin. The jaw muscles, tongue and brain should be removed. The skull may then be dried and kept before cleaning. Cleaning is possible using ants or by placing in cold water and heating till simmering (do not boil vigorously) for a few hours until the remaining flesh can easily be scraped off.

For further details of preparing mammals see Yates *et al.* (1996) and Brown & Stoddart (1977).

2.5 Ethnobotany

Traditional knowledge is often astonishing, for example, in Britain there were traditions of applying mouldy bread to wounds before the discovery of penicillin and of consuming willows *Salix* leaves to cure headaches before acetyl-salicylic acid (aspirin) was discovered. Six billion dollars are spent a year on the 20 most frequently used pharmaceuticals in the United States and all are based on chemicals in natural products (Primack 1998). It is clear that conserving biodiversity, alongside a knowledge of traditional use, can have considerable medical and economic benefits and ethnobotanical surveys can help evaluate the economic importance of natural habitats (e.g. Peters *et al.* 1989). Ethnobotany can also provide the basic data for widening the use of sustainable products, creating markets and leading to the development of novel compounds. It is no longer considered acceptable to gain profits from resources and knowledge in developing countries without returning any of the profits. One approach is to acknowledge the intellectual property rights of native people (Posey 1990).

The local classification of plants and illnesses (called emic categories) is likely to show some differences from those used by a botanist or doctor (etic categories). The main objectives of an ethnobotanical field survey are usually to document the emic categories, relate them to etic categories and evaluate quantitatively the use and management of the plants in the region (Martin 1995).

One possibility is to ask about local uses and names of each plant while collecting it and recording this information. Such information is rarely sufficient to be of great use. The usual approach for ethnobotanical surveys is to create a card index or file for each local name using a local writing system or a widely accepted phonetic alphabet. Record the meaning of each name (although of course many will have no meaning) to clarify the name and reduce misunderstandings, the collection number of each specimen of that name, the scientific name and the informant's name. Another card index or file is created for each informant, giving the age, sex and marital status, culture and other information. Each species the informant mentioned is listed along with information on the parts used, preparation, function, distribution and seasonality. Interviewees should ideally be selected randomly but stratified (see Section 4.3) by age, sex and culture. Although labour intensive and repetitive, such systematic data collection allows far more rigorous analysis than simply collecting lists of uses and local name for each specimen. Cross-verification, checking whether information is consistent across informants, is invaluable. Consistent use of a plant, especially across communities, is likely to indicate effectiveness. A tape recording of a number of local people's pronunciations can also be useful. Tapes are then labelled and stored like any other reference material and should be cross-referenced to the local name file and the informant file. Tapes are ideally linked to specimens and give local name, date, location and informants' names.

Bulk collections of specimens for detailed chemical or biological testing may be fresh, frozen, dried or preserved in alcohol depending on the requirements of the laboratory. They are usually accompanied by a voucher specimen for confirming the identification if necessary. Large-scale collecting needs to consider the consequences for conservation and local use.

2.6 Atlases

The importance of a particular site is often best placed in context using an atlas of the distribution of the different species. The conventional atlas

CASE STUDY

Southern African frog atlas project

Following the success of the southern African bird atlas, it was decided to map the distribution of all frog species in South Africa, Lesotho and Swaziland using a quarter-degree grid. There is a good field guide (Passmore & Carruthers 1995) and also a compact disk of calls.

The fact that all 110 frog species in the region can be identified from their calls is used as the basis of the survey. Members of the public are invited to post tapes of frog calls to a local organiser who identifies the species and replies with a list of the species recorded. Thus field work can be carried out by people with no knowledge of frog identification. To ensure accuracy and consistency the organisers insist that even those that believe they can identify the species must initially include tapes, to confirm that their identifications are accurate. All tapes are stored for future reference.

Observers were requested to start each tape recording with the time, date and name of locality so these can be matched with the recording form which asks for time, date, observer's name, address and telephone, grid cell, locality name, name of farm, nearest town (to check coordinates), locality description (optional), weather conditions and ambient temperature (optional). It was requested that each locality is given full coordinates (degrees, minutes and seconds) in order to pinpoint the exact location of rarities.

The initial plan was to record data for each locality but it became apparent that it was often difficult to distinguish locations in the dark and they often merge with each other so that now all records for a grid square are amalgamated for a visit. Localities may seem to be more useful for re-finding species but it creates extra paperwork, it does not take into consideration the itinerant manner in which naturalists tend to work and it creates huge problems in defining coverage in a simple and communicable manner (J. Harrison personal communication).

In practice there has been enthusiasm by professionals but a low rate of submission of records by the public. This may well be because the best times are wet nights! Successes so far include records for several species from localities hundreds of kilometres beyond known ranges. This project has also made a significant contribution towards raising public awareness of issues related to frog conservation.

The Southern African frog atlas project found a new and relatively large population of the rare Long-toed Tree Frog *Leptopelis xenodactylus*. This discovery has been passed on to conservation authorities so that the necessary protective measures can be implemented (photo: Marius Burger).

consists of plotting the distribution of each species on a map showing a standard grid and is extremely useful for conservation. Atlas data can be presented in two ways: distribution maps for species and species lists for particular areas. Both may be of value for conservation. An atlas project is usually either continuously ongoing or lasts for a specified period of a few years. In most cases atlas projects depend on volunteers for the contribution of most records.

Obviously the finer each grid, the more detailed and useful the map will be, but it is clearly necessary to balance this against the amount of effort that is available. Scales used include 100×100, 50×50, 10×10, 2×2 km and one degree, half degree and quarter degree grid cells. Before launching a large atlas project, it is strongly recommended to carry out a trial to test the methodology for bias, accuracy, time taken and popularity with volunteers.

A major problem with such atlases is that they confound the distribution with the effort. This is a particular problem if there are few recorders relative to the effort required. For species that require very specialised expertise to identify, the distribution is most closely linked to the location of specialists. At the start of the Diptera mapping scheme one recorder wrote his initials across England in Yellow Dung Fly *Scatophaga stercoraria* records! (K. Porter personal communication). For atlases in which the records come from largely haphazard sightings from naturalists or the general public then allowing for effort is likely to be difficult. However if the data is collected by an individual or team it should be possible to provide some information on coverage, such as areas visited, yet this is usually not provided.

Rich & Smith (1996) investigated the variation in recording between observers by sending a series of pairs of botanists to four tetrads (2×2 km). Twenty-two per cent of the variation in the number of species observed could be explained by the number of habitats visited by the pair of botanists. The most important factor was the ability of the botanists. Rich & Smith classified each botanist from previous experience on a scale from 1 (inexperienced) to 5 (experienced) and this explained 50% of the variation. Rich & Smith suggested that the best ways to improve atlas data collected by volunteers are:
• Improve the recording ability of the recorders, by training in identification skills and field craft, and through contact with other naturalists.
• Encourage recorders to visit many different squares rather than concentrate on one. Individuals differ in their expertise, thus visits from a different recorder tends to add significantly more plant species than further visits by the previous recorder (Woodell 1975).
• Try to achieve even coverage by recording for the same number of hours or having the same number of visits in each tetrad.

• Visit as many habitats as possible in each square and note which have been searched.
• Ensure adequate seasonal coverage.

As an example, when Rich *et al.* (1996) constructed a local plant atlas based on 2×2 km squares of the Ashdown Forest, they ensured that each tetrad was searched for about 10 hours.

The British Trust for Ornithology have co-ordinated two atlases of the breeding birds of Britain and Ireland and refined their methodology. The first atlas (Sharrock 1976) simply determined presence or absence in each 10×10 km square. Although very useful, it was realised that there were potential biases due to observer effort and the atlas gave no indication of abundance. The second atlas asked for two visits, each of 1 hour, to a minimum of eight tetrads (2×2 km) within each 10×10 km square (Gibbons *et al.* 1993). One visit was early in the season (April to May) and the other late (June to July) to ensure that both early and late nesting species are covered. The percentage of tetrads in which each species was observed was then calculated. Observers wishing to do more field-work could then visit more tetrads but were not allowed to spend more time in each tetrad. Colonial species were counted individually.

Such systematic surveys which standardise effort are better at providing good comparative information on the distribution and abundance of the commoner species. They have the disadvantage of not including other records such as casual records or those resulting from more extensive field work and thus exclude many records of rare and elusive species. To overcome this, Gibbons *et al.* (1993) also included a map of all records collected at any time.

In southern Africa birdwatchers were encouraged to send in species lists with one card per visit. Over 5000 volunteers submitted records over a 5-year period. These were used to record the abundance in terms of the proportion of cards for each quarter degree square of six countries (South Africa, Botswana, Namibia, Zimbabwe, Lesotho and Swaziland) in which the species is recorded (Harrison *et al.* 1997).

Atlas schemes usually require help from a wide range of people over an extended period.

It is necessary to balance accuracy and popularity and consider the number and nature of the volunteers. A complex and prescriptive methodology (such as the second British Trust for Ornithology atlas) may give more precise results but may be less likely to attract volunteers than the simpler southern Africa scheme. Providing feedback, such as thanking participants and producing a newsletter describing progress, exciting discoveries and anecdotes has proved to be very important in helping to maintain enthusiasm. For larger schemes it is often useful to create a network of local co-ordinators to organise and motivate local fieldworkers and ensure gaps are completed. It is also useful, if possible, to be able to employ fieldworkers to cover outlying areas not covered adequately by volunteers.

2.7 Habitat mapping

At a variety of scales it is important to know the abundance and distribution of different habitats. This may range from the distribution of the remaining forest within a country to the bamboo thickets within a reserve. Once the distribution and size of the major habitats are known it is much easier to plan priorities and it provides the information for quantifying past or future change. Comparisons with previous surveys, maps or photos can be useful ammunition when demanding conservation measures, for example showing that only 6% of raised peat bog remains in Britain, has been useful in improving the conservation status of the area remaining.

The first stage in habitat mapping is to create a precise definition of each habitat, which is not as straightforward as it seems. It often requires visiting a wide range of areas to describe the range of habitats. For example, if mapping the distribution of mangrove forest, define mangrove forest in terms of the density, tree size and percentage mangrove. Without such precise definitions, different field workers may give different classifications to similar areas, there will not be an accurate baseline for detecting changes, nor will it be possible to compare your study with another area. A habitat type, such as woodland, is often divided into a range of communities which can then subdivided into more detailed classifications. These can then also be mapped.

2.8 Remote sensing

Remote sensing is increasingly used to map habitats (Alaric 1994). This is an extremely powerful technique that will become even more useful with time. The simplest method is photography from the air. Ultraviolet and infra-red wavelengths can also be detected which give further information on water content and vegetation structure. Air photo interpretation is then necessary to assign different tones to habitat types. This has to be standardised and tested and validated in the field (known as 'ground truthing').

Satellite imagery is conceptually similar with data collected for each block ('pixel') which may now be as small as 10×10 m. The images are less distorted than aerial photographs and it is not necessary to overlap photographs to cover large areas. The reflectance within bands of given wavelengths is recorded and stored. This clearly involves enormous computing, storage and handling capacity, for example, a single Landsat scene (180×180 km) has 280 million numbers.

The results can be presented accurately using blue for the blue wavelengths, green for green and red for red giving a 'true colour' image. In practice it is often more useful to use 'false colours', for example, often for vegetation mapping the red is used for mid-infra-red, green for near infra-red and blue for the green, as this is easier for distinguishing habitats.

Statistical techniques can be used to map habitat types. The 'supervised approach' entails relating the data for each wavelength to sites of known habitat type. It is then possible to map the habitats elsewhere on the image. This is easiest for readily distinguishable habitats such as agricultural land.

In the 'unsupervised approach' to mapping, principal components analysis is used to distinguish classes with similar characteristics. The image then presents these different classes. The classes are then related to habitats on the ground and combined if a given habitat is assigned more than one class.

2.9 Databases

If data cannot be used then there is little point in collecting it. The usual approach for collating databases is to have a file for each record (for example a location for a species) (as in Table 2.1). With geographical coordinates it is then possible to transfer the data to a Geographical Information System (GIS) to produce species maps. This information can then be linked to other files describing sites or giving details of a species (as in Table 2.2). These can then be merged to give the distribution of each species or records per location. As well as data collected specifically and directly from the field, data can be obtained from published papers, books and reports, labels on museum specimens or field notebooks of naturalists and the origin of the data recorded, even if a letter, email or conversation. There is usually enormous variation between areas and taxonomic groups in the extent of the data available. In some

cases most records come from a few accessible or well-known locations.

Including inaccurate records in a database compromises the accuracy of the whole database. If possible the accuracy should be quantified, for example, by naming the identifier (a person's records can then be removed or reconsidered if they prove unreliable), or stating if there are supporting herbarium specimens, museum specimens or photographs and where they are. It is important to ensure records are accurate, but without offending observers. If someone submits a record it is reasonable to tactfully ask for details of the observation (was the bird flying away or seen clearly for some time, was the frog identified by its call or seen, or was the plant flowering or not), what their previous experience is of the species and similar species and how experienced an observer they are. In a survey of Pine Martens *Martes martes* in England and Wales, where they were once considered extinct, anyone reporting

Table 2.1 An example of a record database file structure (from Crosby 1994). This was created for the Endemic Bird Area project described in Section 3.4.1.

Field name	Description
Species	Scientific name of the species
Reference	Source reference for the record
Locality	Name and description of the recording locality
Country	International Standardisation Organisation (ISO) code for the recording country
Coordinates[a]	Geographical coordinates of the recording locality
Certainty[b]	Code for the certainty that the coordinates are for the correct locality
Accuracy[c]	Code for the accuracy of the coordinates
Record type	Year(s) of records at this locality, with a qualifier (e.g. circa) if required
Altitude	Minimum and maximum altitudes of records at this locality
Record type	Code for the type of record (e.g. collected specimen, sight record, etc.)
Status	Information on abundance at this locality. Information on the breeding status at this locality (e.g. month(s) of records, etc.)
Notes	Names of observers, etc.

[a]Represented in the database by six fields (latitude and longitude degrees, minutes, and bearing), or two fields for decimal coordinates.
[b]Recorded using the following codes: A, certain, there is an exact match for the locality name in the Gazetteer, and no possibility of ambiguity; B, probable, there is a close match for the locality in the Gazetteer, and/or ambiguity is possible, but thought unlikely; C, possible, there is a reasonable match in the Gazetteer, but a worrying ambiguity; D, unrealistic, there is only a poor match in the Gazetteer, and/or unresolved ambiguity.
[c]Recorded using the following codes: A, coordinates believed to be accurate to within 5 km; B, coordinates believed to be accurate to within 20 km; C, coordinates not believed to be accurate within 20 km of the recording locality.

Table 2.2 An example of a species database file structure (from Crosby 1994). This was created for the Endemic Bird Area project described in Section 3.4.1.

Field name	Description
Species	Scientific name of the species
Synonyms	A list of alternative scientific names used for this species
English	English name of the species
Family code	The family number, to enable indexing in taxonomic order
Countries	International Standardisation Organisation codes for all countries where the species breeds
Habitat	Descriptions of breeding habitats, with references to the sources of data
Altitude	Altitudinal range during the breeding season, with references to the sources of data
Threat	Threatened or near-threatened classification, including relevant amendments
Taxonomy	Explanation for the few cases where the taxonomy used differs from the standard reference (Sibley & Monroe 1990)
EBA	Codes of the Endemic Bird Areas where the species breeds

one was subject to a standard interview at the end of which a confidence score (on scale 1–10) was assigned to each report on the basis of the quality of the information given and the experience of the observer (Messenger *et al.* 1997).

The species name needs to be understandable and to a set standard because if they are not stored systematically then they can be difficult to extract. For some groups, such as conifers, there can be many more names than species. Site names are often ambiguous and grid coordinates are notorious for being misread. It is useful to have a double check such as giving both the coordinates and the nearest town.

Security can sometimes be a problem for very sensitive species and the precise locations of species vulnerable to disturbance, persecution or collection should not be irresponsibly publicised. It is however now generally accepted that more damage is usually done as a result of locations not being widely known than from disturbance and collecting. The trend is thus towards greater openness.

3 Setting conservation priorities

3.1 Why set conservation priorities?

There are far too many conservation problems but there is nothing like enough time or money to tackle them. Hence we need to prioritise. Conservation has been compared with triage, the policy of sorting wartime casualties into three categories: those with wounds so severe that the patient is unlikely to survive even with treatment, those that do not require immediate treatment and those with serious but treatable wounds who are then the priority for medical care. Whilst not approving of the need of conservationists to make such divisions, Bibby (1995) points out that it is better for conservationists to decide priorities rationally rather than having reserves and species conservation programmes being selected in a somewhat arbitrary fashion, often depending largely upon the preferences of key influential individuals. Another strength of logical decisions is that others are likely to take conservationists more seriously if their priorities are based on transparent principles and data as it is then much easier to participate in strategic, political and economic debate than if conservation is based on whims and personal preferences. Finally, identifying priority species and areas for conservation can be useful for generating money by emphasising their importance.

The current approach is to quantify aspects, such as rarity, extent of decline and rate of decline as rigorously as possible and then use this data to determine priorities. Difficulties include deciding how to combine these different aspects and insufficient information. Non-ecological aspects are sometimes included, such as the popularity of a species or the beauty of an area. It is still useful to quantify these, for example by questionnaires. The final decision on priorities usually requires value judgements and considerations of the political realities.

3.2 Prioritising species

It is clearly important to decide those species towards which conservation effort should be directed most urgently. The main criteria used explicitly or implicitly are vulnerability to extinction, evolutionary distinctiveness, popular appeal, likelihood of recovery and local status.

3.2.1 Vulnerability to extinction

Species can be classified according to their vulnerability to extinction. On a global scale, IUCN (1994) provides rigorous definitions known as 'the IUCN Red list criteria' (Fig. 3.1) which attempts to classify species according to the likelihood of extinction within a given period. The aims were to provide a method that can be consistently applied by different people and for different taxa and which improves the objectivity and clarity in classifying species.

'Critically endangered' refers to a 50% probability of extinction in 5 years, 'endangered' as a 20% probability in 20 years and 'vulnerable' as a 10% probability in 100 years. Table 3.1 gives precise definitions of these categories. The likelihood of a species going extinct depends upon factors such as population size, rate of decline and threats. Table 3.2 summarises how such information is used for classifying species as critically endangered, endangered or vulnerable. For the precise definitions it is necessary to refer to Table 3.3 for rapid decline, Table 3.4 for small range, Table 3.5 for small and declining populations, and Table 3.6 for very small populations. It is sensible to consult the full instructions before finally classifying a species (see IUCN 1994 or www.iucn.org/species).

Each species is classified according to the criteria summarised in Table 3.2 and the highest threat

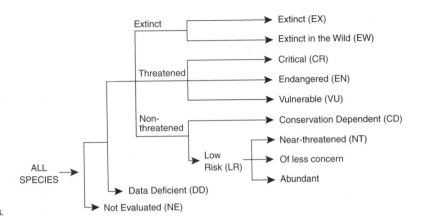

Fig. 3.1 Relationships between the IUCN threatened species categories.

category is used for the final listing. One approach, used by Birdlife International, is to record the threat category and code. For example the Lesser Florican *Sypheotides indica* is classified as 'Critical A1a; C1, C2b; D1' (Collar *et al.* 1994). This species of bustard has been in decline for over a century, initially probably due to hunting, but now as a result of conversion of grassland habitat. Once widespread, it is now restricted to tiny habitat patches in western India. The population is estimated to have plummeted from 4374 birds in 1982 to 750 in 1989. Due to the directly observed decline of over 80% in 10 years this species is classified as critical (see Table 3.3) and given the code A1a. Referring to the other tables gives the following codes and classifications: C1 endangered (less than 2500 individuals and a 20% decline in 5 years), C2b endangered (less than 2500 individuals all in one single subpopulation), D1 vulnerable (less than 1000 individuals). The highest of these threat categories for this species is critical and this is thus the final status. With this format used by Birdlife International the code linked to that threat category (A1a) is shown in bold. IUCN just gives the highest threat category.

This method seems to have worked reasonably well for classifying all birds, although the data were often incomplete. However, for many groups the majority of species would have to be classified as data deficient and for some groups of invertebrates it is inconceivable that there will ever be sufficient data for a sensible classification.

3.2.2 Taxonomic isolation

Species can be prioritised according to the extent of taxonomic isolation. Thus the Kagu *Rhynochetos jubatus*, Oilbird *Steatornis caripensis* and Aardvark *Orycteropus afer* are all extraordinary, distinctive and sole members of their families which most people would allocate greater priority than their simple measure of vulnerability to extinction.

The usual definitions of biodiversity include the variation within species, such as subspecies or races, thus the Tiger *Panthera tigris* is divided into eight subspecies, of which 1–3 are extinct, and there is concern to protect each of the remaining 5–7 subspecies. The next level is to consider evolutionary significant units (Moritz 1994, 1995), which are historically isolated populations that show some divergence.

There may even be demands for the conservation of variants. Ten per cent of a population of North American Black Bears *Ursus americanus* on two islands of British Columbia are pale ('spirit bears') and there are suggestions that this population of about 100 individuals should therefore be conserved (Anon 1997). Environmental groups are demanding that the government decision to allow logging in this area should be reversed. British Columbia is considering whether to create a 265 000 hectare 'wilderness park' to protect these bears. The spirit bears are beautiful and I hope they and the forest are saved but they make a negligible contribution to global genetic diversity.

Table 3.1 Definitions of IUCN categories (as given in IUCN 1994).

Category	Definition
Extinct (EX)	A taxon is Extinct when there is no reasonable doubt that the last individual has died
Extinct in the wild (EW)	A taxon is Extinct in the wild when it is known only to survive in cultivation, in captivity, or as a naturalised population (or populations) well outside the past range. A taxon is presumed extinct in the wild when exhaustive surveys in known and/or expected habitat, at appropriate times (diurnal, seasonal, annual), throughout its historic range, have failed to record an individual. Surveys should be over a time frame appropriate to the taxon's life cycle and life form
Critically endangered (CR)	A taxon is Critically endangered when it is facing an extremely high risk of extinction in the wild in the immediate future, as defined in any of the criteria (A–E) (see Table 3.2)
Endangered (EN)	A taxon is Endangered when it is not Critical but is facing a very high risk of extinction in the wild in the near future, as defined in any of the criteria (A–E) (see Table 3.2)
Vulnerable (VU)	A taxon is Vulnerable when it is not Critical or Endangered but is facing a high risk of extinction in the wild in the medium-term future, as defined in any of the criteria (A–E) (see Table 3.2)
Conservation dependent (CD)	Taxa which do not currently qualify as Critical, Endangered or Vulnerable may be classified as Conservation dependent. To be considered Conservation dependent, a taxon must be the focus of a continuing taxon-specific or habitat-specific conservation programme which directly affects the taxon in question. The cessation of this conservation programme would result in the taxon qualifying for one of the threatened categories above
Low risk (LR)	A taxon is at Low risk when it has been evaluated and does not qualify for any of the categories Critical, Endangered, Vulnerable, Conservation dependent or Data deficient. It is clear that a range of forms will be included in this category, including: (i) those that are close to qualifying for the threatened categories; (ii) those that are of less concern, and (iii) those that are presently abundant and unlikely to face extinction in the foreseeable future. It may be appropriate to indicate into which of these three classes taxa in Low Risk seem to fall. It is especially recommended to indicate an appropriate interval, or circumstance, before re-evaluation is necessary for taxa in the Low risk class, especially for those indicated in (i) above
Data deficient (DD)	A taxon is Data deficient when there is inadequate information to make a direct, or indirect, assessment of its risk of extinction based on its distribution and/or population status. A taxon in this category may be well studied, and its biology well known, but appropriate data on abundance and/or distribution are lacking. DD is therefore not a category of threat or Low risk. Listing of taxa in this category indicates that more information is required. Listing a taxon as DD acknowledges the possibility that future research will show that threatened classification is appropriate. It is important to make positive use of whatever data are available. In many cases great care should be exercised in choosing between DD and threatened status. If the range of a taxon is suspected to be relatively circumscribed, if a considerable period of time has elapsed since the last record of a taxon, or if there are reasonable chances of unreported surveys in which the taxon has not been found, or that habitat loss has had an unfavourable impact, threatened status may well be justified
Not evaluated (NE)	A taxon is Not evaluated when it has not yet been assessed against the criteria

Although these different levels are all part of biodiversity and worth protecting, it seems sensible that the conservation effort should relate to how deeply rooted a group is in the evolutionary tree, but the Black Bear example shows how other factors often override. If we took the idea of prioritising genetic separation really seriously, then we should put most of our conservation effort into protecting the diversity of archaebacteria and eubacteria and microscopic eucaryotes, as these show

Table 3.2 A summary of new IUCN threatened category thresholds. 'Extent of occurrence' and 'area of occupancy' are defined in the caption to Table 3.4. More precise definitions are given in the following tables: rapid decline Table 3.3, small range Table 3.4, small population Table 3.5, very small population Table 3.6.

	Main numerical thresholds		
Criteria	Critical	Endangered	Vulnerable
A Rapid decline	> 80% over 10 years or 3 generations*	> 50% over 10 years or 3 generations*	> 50% over 20 years or 5 generations*
B Small range fragmented, declining or fluctuating	Extent of occurrence $< 100 \, km^2$ or area of occupancy $< 10 \, km^2$	Extent of occurrence $< 5000 \, km^2$ or area of occupancy $< 500 \, km^2$	Extent of occurrence $< 20\,000 \, km^2$ or area of occupancy $< 2000 \, km^2$
C Small population declining	< 250 mature individuals	< 2500 mature individuals	< 10 000 mature individuals
D1 Very small population	< 50 mature individuals	< 250 mature individuals	< 1000 mature individuals
D2 Very small range	—	—	$< 100 \, km^2$ or < 5 locations
E Unfavourable population viability analysis	Probability of extinction > 50% within 10 years or 3 generations*	Probability of extinction > 20% within 20 years or 5 generations*	Probability of extinction > 10% within 100 years

*Whichever is longer.

Table 3.3 Criteria for the IUCN threatened categories: species undergoing rapid decline. 'Extent of occurrence' and 'area of occupancy' are defined in the caption of Table 3.4.

Main criteria	Sub-criteria	Qualifiers	Codes
A Decline of: > 80% over 10 years **(CR)** or 3 generations* > 50% over 10 years **(EN)** or 3 generations* > 50% over 20 years **(VU)** or 5 generations*	**1** Decline which has happened, observed, estimated, or suspected, based on:	**(a)** Direct observation **(b)** Decline in area of occupancy, extent of occurrence, and/or quality of habitat **(c)** Actual or potential levels of exploitation **(d)** Effects of introduced taxa, hybridisation, pathogens, pollutants, competitors or parasites	A1a A1b A1c A1d
	2 Decline likely in near future based on:	**(b)** As above **(c)** As above **(d)** As above	A2b A2c A2d

*Whichever is longer.

considerable genetic divergence while the multi-cellular animals, plants and fungi combined show comparatively little genetic divergence (Hugenholz & Pace 1996)!

3.2.3 What is a species?

This seems an innocent and straightforward question, but the definition of what is a species is currently of considerable scientific debate. At the time of writing (1999) most conservationists seem to have little appreciation that this issue could radically change conservation programmes and cause chaos.

The traditional definition of a species, the biological species concept, is groups of actually or potentially interbreeding natural populations which are reproductively isolated from other such groups (Mayr 1942). It has its roots in Darwinian thought

Table 3.4 Criteria for the IUCN threatened categories: species with a small range and declining. 'Extent of occurrence' is the area contained within the shortest continuous imaginary boundary which encompasses all known, inferred or projected sites of present occurrence. 'Area of occupancy' is the area within the extent of occurrence which is occupied by a taxon (this measure is often applicable to species with highly specific habitats).

Main criteria	Sub-criteria	Qualifiers	Codes
B Extent of occurrence estimated: $< 100\,km^2$ **(CR)** $< 5000\,km^2$ **(EN)** $< 20\,000\,km^2$ **(VU)**	**1** Severe fragmentation or At 1 location **(CR)** < 5 locations **(EN)** < 10 locations **(VU)**	None	**B1**
or Area of occupancy estimated: $< 10\,km^2$ **(CR)** $< 500\,km^2$ **(EN)** $< 2000\,km^2$ **(VU)** in *either* case with *any two* of:	**2** Continuing decline observed, inferred or projected in any of:	**(a)** Extent of occurrence **(b)** Area of occupancy **(c)** Area, extent and/or quality of habitat **(d)** Number of locations or subpopulations **(e)** Number of mature individuals	**B2a** **B2b** **B2c** **B2d** **B2e**
	3 Extreme fluctuations in any of:	**(a)** Extent of occurrence **(b)** Area of occupancy **(c)** Number of locations or subpopulations **(d)** Number of mature individuals	**B3a** **B3b** **B3c** **B3d**

Table 3.5 Criteria for the IUCN threatened categories: species with a small population and declining.

Main criteria	Sub-criteria	Qualifiers	Codes
C Population: < 250 **(CR)** < 2500 **(EN)** $< 10\,000$ **(VU)** mature individuals and *either*:	**1** Continuing decline $> 25\%$ within 3 years or 1 generation **(CR)** $> 20\%$ within 5 years or 2 generations **(EN)** $> 20\%$ within 10 years or 3 generations **(VU)**	None	**C1**
	2 Continuing decline in numbers of mature individuals *and* population structure observed, inferred or projected in form of *either*:	**(a)** Severe fragmentation no population > 50 **(CR)** > 250 (EN) > 1000 **(VU)** mature individuals	**C2a**
		(b) All breeding individuals in single subpopulation	**C2b**

and for practical purposes usually works well. However there are problems, for example, some species regularly hybridise and produce fertile offspring yet are considered different to retain species status. A common problem is that too little is known about them to know whether they ever interbred or alternatively they are geographically isolated so that it is impossible to tell whether they would hybridise if

Table 3.6 Criteria for the IUCN threatened categories: species with a very small population and/or a very small range.

Main criteria	Sub-criteria	Qualifiers	Codes
D Population:	None	None	
< 50 (**CR**)			**D1**
< 250 (**EN**)			**D1**
< 1000 (**VU**)			**D1**
mature individuals			
and/or			
Area of occupancy			**D2**
< 100 km² (**VU** only)			
or			
at < 5 locations			**D2**
(**VU** only)			

brought together. Most taxonomy is based largely on dead specimens and in such cases it is the experience of the taxonomists that determines species status. This subjectivity is unsatisfactory.

According to an alternative definition, the phylogenetic species concept, species can be defined by unique characters shared by all individuals, this is formally the smallest identifiable cluster of individuals within which there is a parental pattern of ancestry and descent (Cracraft 1983). As a result many isolated subspecies and races can be classified as species. With the phylogenetic species concept any difference, however, tiny, could be used to justify a new species (Collar 1997). With sufficient effort, it is likely that almost any isolated population can be classified as a species. Such an approach could increase the number of bird species in the world by many times (Collar 1996) and if identified by molecular techniques these may not be identifiable by a field biologist. Many of these new species will have small populations (e.g. many island races) and thus there will be a considerable rise in the number of species classified as threatened.

The conservation concern is usually enormously greater for species than for subspecies or races, as shown by those that have been demoted from species to subspecies status and now attract little interest, or those that have been promoted from subspecies to species status and are now targets of conservation programmes. It has been argued (Hazevoet 1996) that using the phylogenetic species concepts raises the status of many subspecies or populations to species and thus increases their conservation status and their likelihood of attracting conservation effort. Similarly, Urban *et al.* (1997) describe the Usambara Hyliota *Hyliota usambarae* as a full species although based on 'morphological evidence that is admittedly slender' in order to 'bring to attention a population that is rare and possibly endangered'. Such eccentricity gives conservation and taxonomy a bad name. Not only will the addition of so many new species (and even more new threatened species) cause chaos in the short term while priorities are adjusted, but they will dilute conservation effort from clearly distinct species, which is where the effort should be directed.

The uneven use of different species definitions causes considerable problems. Taxonomists usually work by reviewing a group or an area and the current fashion for splitting derived largely from the phylogenetic species concept means groups that have been reviewed recently are more likely to be finely split. For example, Cracraft (1992) reviewed birds of paradise, doubling the number of species and greatly increasing the number that would be considered threatened with extinction. Robertson & Nunn (1997) used the phylogenetic species concept to suggest there were 24 species of albatross compared with the present 14 species. Hazevoet (1995) reviewed the birds of the Cape Verde Islands and increased the number of species. This then leads to the problem as to whether these groups and areas deserve greater attention than do those that have yet to attract the attention of a splitting taxonomist. For example, Peregrine Falcons *Falco peregrinus* and Barn Owls *Tyto albo* show considerable geographical variation. Hazevoet (1996) reclassifies the populations on the Cape Verdes as a separate species: *Falco madens* and *Tyto detorta* but it would be odd to then give these populations greater priority than other equally distinct populations that happen not to have been examined by a believer in the phylogenetic species concept. Although much of the debate has concerned birds there have been other taxa to which this has been applied, such as the proposal to recognize a separate species of tiger on Sumatra (Cracraft *et al.* 1998).

Although changing species concepts would be hugely inconvenient, that is an insufficient reason for retaining the biological species concept if it is actually inappropriate. A major criticism of the biological species concept is that it is subjective and this is undoubtedly a major problem. However, Snow (1997) worked through a number of examples and suggests that, in practice, applying the phylogenetic species concept is likely to require equal subjectivity.

Once most subspecies, races and many practically identical forms become species, the need will arise to create a level between the genus and phylogenetic species, and the term 'species group' has been used. If this becomes the norm then the main consequence of all the reclassification may be to promote races, subspecies to species and the old species to a higher rank (Snow 1997). The main distinction between these concepts is the belief of those following the phylogenetic species concept that species must be the finest possible division. There seems to be little biological basis for this assertion.

DNA studies have been extremely successful in determining the pattern of evolutionary divergence, often confirming the results of traditional taxonomists but with some spectacular surprises. If there is a contradiction between the results of a taxonomy based on morphology and one based on DNA, then, as long as the DNA analysis is robust, it always provides the real evolutionary history. Differences in DNA sequence between populations is good evidence for isolation and indicates differentiation.

There is an increasing belief that a given difference in DNA alone is evidence for separate species. Most of the DNA differences recorded either concern sections of DNA that are not expressed or will have a negligible consequence on the appearance, behaviour or physiology. DNA differences largely measure isolation and this alone is no measure of differentiation. Although different taxonomists concentrate on different approaches such as genetics, morphology, geography, ecology or behaviour, each

approach should be giving a similar picture. It is misleading to concentrate on one aspect, such as DNA sequences, especially if it produces very different conclusions.

A very different issue is the discovery of cryptic species, that is clearly different species that appear similar. Thus, despite the legions of naturalists in Britain it was only discovered a few years ago that the commonest bat species (the Pipistrelle *Pipistrellus pipistrellus*) was actually two species (Barratt *et al.* 1997). The ranges of the two bat species overlap but they differ in call frequencies, with one at 45 kHz and another at 55 kHz they occur in single-species colonies and individuals with different calls differ markedly in mitochondria DNA sequence while individuals with the same call frequencies had similar sequences.

Classifying asexual species causes problems. The offspring of sexual species carry genetic material from each parent and thus individuals may differ markedly from the parents and each other. In practice, a taxonomist will classify a group as a species if there are consistent differences between the group and the nearest species and if the differences between them is large compared with the variation within. Some other species may reproduce largely or entirely by asexual means. This is particularly common in plants which grow via stolons, rhizomes or bulbs, or are apomictic (i.e. produce seeds asexually). All members of the clone contain the same genetic material. If a mutation takes place all subsequent members of the clone will possess the mutation. It is thus possible to claim that all of the clone possessing the mutation is a species as they are consistent within the 'species', they differ consistently from other 'species' and, being asexual, they do not interbreed. With each mutation you have a new species! Large numbers of virtually indistinguishable 'species' can then be named and some taxonomists do so. With the current desperate shortage of taxonomists it is a pity such people cannot find something more useful to do. The sensible measure is to look for differences in asexual clones of a magnitude that would be sufficient to recognise species in sexually reproducing groups.

3.2.4 Flagship species

Some argue that all species should be considered equal regardless of their public profile. However, it is clear that there is enormous public support for ensuring the maintenance of the Giant Panda *Ailuropoda melanoleuca*, Koala Bear *Phascolarctos cinereus* and Gorilla *Gorilla gorilla* which thus warrant greater conservation effort and the conservation of such 'flagship species' usually has benefits for associated less spectacular species (Eisner *et al.* 1995). Some groups such as primates, carnivores, birds and butterflies attract much more concern than others, such as springtails or nematodes. Thus in attempts by the British statutory conservation agency to gain private funding for single-species conservation, species such as the Large Blue Butterfly *Maculinea arion* were readily supported while there was no support for species such as the aptly named Depressed River Mussel *Pseudanodonta complanata*. Most conservationists would probably accept that different species should have different importance weightings but that conservationists have tended to concentrate too much on the particularly charismatic species.

3.2.5 Introduced species

Most conservationists stress the distinction between native species, towards which all the conservation is usually directed, and introduced species which often cause considerable problems, and are often subject to control (see Section 10.3). This distinction is usually clear but may occasionally be ambiguous. For example, attractive and reasonably harmless species present for a long time may be considered by some as honourable natives. There may also be doubts as to whether a species is native or introduced. For example, the Pool Frog *Rana lessonae* was considered an unimportant but harmless introduction in Britain. By the time that archaeological evidence had suggested that it was a native species it had gone extinct and considerable effort is now going into re-establishing it. The pool frog example is, however, atypical and in general introductions are one of the major causes of global extinction.

The general rule of protecting natives and excluding introductions is exceedingly sensible.

3.2.6 Likelihood of species recovery

Species for which conservation is likely to result in population recovery will tend to be a greater priority than those for which conservation is likely to be ineffective. This will depend on the size and distribution of the remaining population, the ease with which conservation measures can be implemented and the likelihood of success. It seems that with sufficient commitment, money and knowledge practically any species can be saved. The Mauritius Kestrel *Falco punctatus* had declined to four individuals including one breeding pair in 1974 so that some considered it unsaveable and thus not worth conservation effort. However, after concerted effort (Jones *et al.* 1995a) there are now over 300 birds.

3.2.7 Prioritising species within areas

There may be conflict between the national and international importance of species. In 1997, I visited Lake Kisezers in Latvia which had breeding Corncrakes *Crex crex* and Bitterns *Botaurus stellaris*. In the United Kingdom these are two of the highest priorities for conservation and each had received considerable investment, in the order of tens of thousand pounds per Corncrake and hundreds of thousand pounds per Bittern, so I was intrigued to see that the Latvians were burning and cutting the reed to convert it to meadows for Lapwings *Vanellus vanellus*, which is quite common in Britain but rare and declining in Latvia. As another example, in Britain the Holly-leaved Naiid *Najas marina* is considered a conservation priority as this plant is restricted to five sites, whereas in Israel it is controlled as a weed! Although accepting it is not internationally important I still care about Holly-leaved Naiid in Britain and would be sad if it went extinct in this country. One reason for this is to retain the distribution of a species while another is that an area's biodiversity is part of its culture.

National conservation priorities need to strike a balance between the local and global perspectives. Avery *et al.* (1995) suggest one approach in their

Table 3.7 The rationale for placing species along each of three separate biological axes of conservation priority to determine national priorities. Thresholds for qualification under each criterion are also given.

	Conservation priority		
	High	**Medium**	**Low**
National status	Declining rapidly in numbers or range[a]	Rare breeder[b] Localised breeder[c] Localised non-breeder[d] Moderate decline in numbers[e] or range	None of these
International importance for species of European conservation concern, category 1	UK holds over 30% of European population	UK holds 15–29% of European population	UK holds less than 15% of European population
International threat	Globally threatened[f]	Unfavourable status in Europe[g]	Favourable status in Europe

[a] > 50% over 25 years.
[b] < 300 pairs.
[c] 50% in top 10 sites and < 300 pairs.
[d] 50% in top 10 sites.
[e] 25–49% over 25 years.
[f] Critical, endangered or vulnerable, according to IUCN criteria. See Table 3.2.
[g] Species of European Conservation Concern, categories 2 and 3.

classification of the birds of Britain. They classified each species as low, medium and high in relation to national threat, international importance and international threat (see Table 3.7). Due to a lack of data the international status is based on the European population; it would be more satisfactory to have estimates of the global population. There are $3 \times 3 \times 3 = 27$ combinations of these (see Fig. 3.2). Those species of either high national threat or international threat were given 'red' listing (36 species), those of low national threat, low international importance and low international threat were classified as 'green' listing (280 species). The remainder were given 'amber' listing (110 species). Note that species that are scarce in Britain but flourishing elsewhere are just given amber listing. The same approach of balancing local and global perspectives has to be adopted at each scale. Even within a reserve or an area it is sensible to consider which species are most important priorities for conservation.

Section 3.2.1 describes how global priorities may be set on the basis of degree of vulnerability to extinction. The production of Red Data Books is an essential component of modern conservation. These list the endangered species within a taxonomic group. However, for some groups, such as almost all invertebrates, there is currently far too little data to compile a comprehensive and accurate Red Data Book. Many countries have also produced national Red Data Books listing threatened species within the country. This is often a useful step in assessing priorities as well as attracting wider interest and opportunities for funding. A common problem with them is that they do not distinguish between local and global importance.

3.3 Prioritising habitats

The information on species priorities can be used to determine priorities for which habitats to protect. This can take into consideration the habitat preferences of high priority species, the local and global distribution of each habitat and the local and global threats to each habitat. Once again, the approach is to

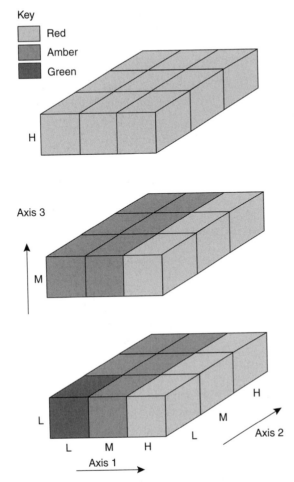

Key

■ Red

■ Amber

■ Green

Axis 3

M

L

L   M   H

Axis 1

L   M   H   Axis 2

Fig. 3.2 A suggested mechanism for prioritising species within an area. The three axes show (1) national threat, (2) international importance and (3) international threat (defined in Table 3.7). The combinations that are classified as red, amber or green are shown (from Avery *et al.* 1995).

quantify each aspect such as number of species, number of threatened species, rarity of the habitat and rate of decline. It is as a result of such analysis that there is so much concern for tropical rain forests and coral reefs: they are both of high global importance yet disappearing at a frightening rate. There is again the need to balance local and global importance, such as a habitat that is locally common but globally rare.

The great benefits to society from a range of services provided by natural and seminatural habitats are often underestimated. For example, natural habitats may reduce erosion, absorb pollutants and regulate water flow. The total annual benefits of natural ecosystems, i.e. what it would cost us to create systems that undertake the same functions, has been valued at US$16–54 000 000 000 000 (Costanza *et al.* 1997). Thus if we manage to destroy the functions of the world's ecosystems we simply do not have enough money to replace them. It is becoming increasingly accepted that the costs to individuals, organisations, governments and insurance companies from building on flood plains, polluting water catchments, destroying coastal defences or numerous other activities are unacceptable and conservation of natural processes is better value. Conservation of habitats is often considered a luxury when in reality it is an essential investment. Calling for sites to be protected because of the rare species they maintain may often attract little political or financial support but selecting areas on the basis of the ecosystem functioning may be more acceptable and so may also be more successful in protecting biodiversity.

3.4 Hot spots of global biodiversity

It is clear that some parts of the planet are of greater importance to biodiversity than others and identifying the most important areas is clearly a priority. One approach is to simply count the number of species in each area, but this does not necessarily identify the areas of greatest conservation need. A more sophisticated approach, as described here, is to incorporate information on the conservation priority of each species. The importance should be assessed independently of the threat. Of course in deciding where to act, the threats become important when considered alongside the importance.

3.4.1 Endemic Bird Areas

The most rigorous global attempt at categorising the areas of conservation concern is Birdlife International's Endemic Bird Area project (Bibby *et al.* 1992a, Stattersfield *et al.* 1998). This covers all bird species with restricted ranges (defined as less than 50 000 km^2). Considerable effort was spent in ensuring a standardised taxonomy and Sibley & Monroe (1990) was used. The data were extracted from the literature, an extensive network of contacts was used to obtain unpublished field records, and museum specimen labels were examined. In countries with little information data were extrapolated and interpolated from altitudinal and habitat information. All of these data are stored in computer databases allowing reinterpretation in the future.

An Endemic Bird Area is defined as an area with at least two restricted-range species confined entirely to it (Stattersfield *et al.* 1998). Two hundred and twenty-one of these have been identified (Fig. 3.3). Not surprisingly, the majority are in the tropics and half are on islands. A quarter of the bird species are entirely restricted to just 5% of the land surface and of course viable populations of a very large proportion of more widespread species occur within these areas. Some are well protected while many have a little protection. This work has provided a clear set of priorities for conservation and research and much of the subsequent ornithological fieldwork has been directed at these Endemic Bird Areas. The approach of prioritising using Endemic Bird Areas is,

however, unsuitable for some rare but dispersed species such as raptors.

3.4.2 Centres of plant diversity

Myers (1988, 1990) shows the location of 20 hotspots using the criteria of species abundance, endemism and high levels of threat of destruction. The top ten on the list cover only 0.2% of the land on Earth or 3.5% of the existing primary tropical forest but hold 13.8% of all plant species.

A series of three impressive books outlines 234 centres of plant diversity in the world (Davis *et al.* 1994, 1995, 1997). Sites were selected on the basis of either being clearly species rich (even if there was insufficient research to say exactly how many species are present) or because they contained a large number of species endemic to the area. Other criteria considered are whether sites include plants useful to humans, whether they include a diversity of habitats or whether they contain a high proportion adapted to particular edaphic conditions.

3.4.3 Important Taxon Areas

A recent priority has been to determine the most important areas for conservation of a taxon in a

Fig. 3.3 The Endemic Bird Areas (from ICBP 1992). Each point marks an area which holds at least two restricted-range species (defined as occupying less than 50 000 km^2).

geographical area using identical standards for inclusion in each country. The Planta Europa conference in 1995 called for the identification of Important Plant Areas in Europe as a top priority. These were defined as a natural or semi-natural site exhibiting exceptional botanical richness and/or supporting an outstanding assemblage of rare, threatened and/or endemic plant species and/or vegetation of high botanical value. The objective is to list the Important Plant Areas in each European country.

Important Bird Areas are similarly being created for each continent. Where possible the site should: (1) exist as an actual or potential protected area, with or without buffer zones, or be an area which can be appropriately managed for wildlife conservation; and (2) provide all the requirements of the birds that it is important for, alone or with other sites, at the appropriate season. In practice some sites such as seabird colonies tend not to meet the second requirement.

There are a number of criteria that apply globally in identifying Important Bird Areas (Heath & Evans 2000).

• The site regularly holds significant numbers of globally threatened species, or other species of global conservation concern.

• The site is known or thought to hold a significant component of the restricted-range species whose breeding distributions define an Endemic Bird Area or Secondary Area.

• The site is known or thought to hold a significant component of the group of species whose distributions are largely or wholly confined to one biome.

• The site is known or thought to hold, on a regular basis over 1% of a biogeographic population of a congregatory waterbird species.

• The site is known or thought to hold, on a regular basis, over 1% of the global population of a congregatory seabird or terrestrial species.

• The site is known or thought to hold, on a regular basis, over 20 000 waterbirds or over 10 000 pairs of seabirds of one or more species.

In addition to these global criteria (A categories) which apply universally and thus can be compared across regions, each area may also use criteria to describe sites of international importance but which relate to the local legislation. Thus for Europe, Important Bird Areas may also be recognised using criteria based on species of European conservation concern (B categories) and using criteria that relate to European Union law (C categories).

3.5 Prioritising areas and selecting reserves

Within a region, or a country, it is necessary to assess the importance of different areas. Chapter 2 describes how to evaluate the biodiversity. This section describes how to convert such information into priorities. The quality of the information varies greatly between groups. For example, information on birds and large mammals is usually better than for other groups. Whether areas selected to protect one taxonomic group will be satisfactory for another is a much-debated issue. Reserves selected for one taxon are likely to be high-quality examples of a particular habitat, such as primary forest, and will thus be suitable for a range of species. For example, in Britain, the reserves belonging to the Royal Society for the Protection of Birds are selected for their ornithological importance, yet a quarter would qualify for conservation status on the plants alone (Evans & Lambton 1992) and they contain important populations of the rarest reptiles and amphibians (Cadbury & Lambton 1996). One of the theories behind the endemic bird areas notion is that when birds have evolved into distinct species assemblages, it is likely other lifeforms will also have done so, in response to the same environmental pressures or historical events. So endemic bird areas become indicators of accumulations of biodiversity which other taxonomic specialists have yet to identify owing to their slower rate of taxonomic and distributional classification.

A comparison of the overlap of hotspots in South Africa of species richness, endemism and rarity between birds, snakes, freshwater fish, frogs, tortoises and terrapins, and various mammal orders showed that the overlap varied markedly between 0% and 72% (Lombard 1995). Other studies have similarly shown that congruence is too weak to be used (Lawton *et al.* 1994, Prendergast *et al.* 1994, K.D. Gaston 1996, Williams *et al.* 1996). The overlap between species diversity of different groups is higher at larger scales (Curnutt *et al.* 1994). For example the Endemic Bird Areas are reasonably similar to the centres of plant diversity (Bibby 1998).

In practice the four criteria most frequently used in site evaluation are naturalness, diversity, rarity/uniqueness and size (Smith & Theberge 1986, Usher 1986). Various criteria have been used for prioritising areas and these are listed in Table 3.8.

It is common to use diversity measures in those cases in which the number of individuals of each species found is routinely recorded, but many species are likely to be missed, for example for samples of invertebrates analysed in the laboratory. Diversity is a combination of richness (numbers of species) and evenness (whether the community consists of a few abundant species and many rare ones, or all species being equally frequent) (see Magurran 1988 for a good account of these). The same diversity index can thus be obtained for a community of many abundant

Table 3.8 Indices for scoring conservation priorities for comparing areas (modified from Turpie 1995).

Criterion	Index
Diversity	*Species Richness.* The total number of species in a site *Species Diversity.* Shannon's Index, $H' = -\sum p_i \ln p_i$, is the most often used measure of diversity where p_i is the proportion that species i contributes to the total. A high value of H' indicates a large number of species with similar abundances, a low value indicates domination by a few species
Abundance	*Total number of individuals* $= \sum n_i$ where n_i is the number of each species. This is just a count of all the individuals of all species *Abundance Index* $= \sum A_i$ where A_i is the abundance score of each species (e.g. 0 absent, 1 occasional, 2 common, 3 very common and 4 abundant). This is similar to the number of individuals except that it can be used for groups in which there is only relative estimates of abundance
Rarity	*Conservation Value Index* $= \sum 100 n_i / N_i$ where $n_i =$ number of individuals of species i within the site and N_i is the population size of that species in a wider area (e.g. a country or the world) (Nilsson & Nilsson 1976). A high value indicates the site contains high proportions of the country's or world's populations of many species *Site Endemism Index* $= \sum k/a_i$ where k is the total number of sites and a_i is the total number of sites at which species i occurs (Rebelo & Siegfried 1992). For a given site the calculation only includes those species found at that site. A high value indicates the site contains many localised species *Population Size Index* $= \sum p_i$ where p_i is a score of the abundance, e.g. 0:> 100 000; 1:50 000–99 999 etc. A high score means the site contains many localised or rare species (Williams 1980, Millsap *et al.* 1990)
Conservation status	*Conservation Status Index* $= \sum S_i$ where S_i is the conservation score of the ith species for example the score may depend upon the threat status as determined in Section 3.2. A high score indicates many species of high threat status are present
Multiple criteria indices	*Site Value Index* $= \sum (S_i + E_i) n_i / P_i$ where $S_i =$ conservation status score (see above), $E_i =$ site endemism score (see above), $n_i =$ number of individuals, and $P_i =$ wider population size (e.g. national or global) for the ith species. This gives high values to sites containing a large proportion of species that are high status and with restricted distribution *Priority Diversity Index* $= \sum (E_i \cdot A_i) R$ where $E_i =$ site endemism index, $A_i =$ abundance score of ith species and R is species diversity on the site. All of these are defined above This gives a high value to sites whose species are abundant within the site but occur in few other sites (Bolton & Specht 1983)

species as for a few rare species. The various criteria given in Table 3.8 often give different results. Calculating such measures can be an interesting step in evaluating areas, but in practice, diversity measures are rarely used by conservationists to determine the importance of areas (Götmark *et al.* 1986, Turpie 1995). Apart from other problems, the simple procedure of diversifying the habitat (for example by clearing some primary forest) will often increase local diversity although this would usually be considered detrimental.

How then can areas be compared on the basis of their importance for different groups? Howard *et al.* (1997) used a simple method to rank Uganda's forests after the government decided to dedicate a fifth of the forest to strict nature reserves. The initial selection of forests depended on size (above 50 km^2) or of unusual vegetation type. The groups studied were woody plants, birds, small mammals, butterflies and large moths (Saturniidae & Sphingidae only). Each species was scored by the reciprocal of the number of forests it was found in (i.e. score 1 if found in only one and 0.1 if found in ten). The total for each forest was calculated by combining the scores for all species. This method does not take into consideration any of the other aspects for prioritising species, such as global vulnerability (see Section 3.2), but is still a sensible initial approach to the problem.

Using criteria such as diversity or rarity may result in prioritising a number of similar areas which all contain a similar suite of species rather than protecting a range of habitats and species (Williams 1999). One suggestion is to select complementary sites (Vane-Wright *et al.* 1991), for example by ensuring each species is included a given number of times. Another solution is irreplaceability (Pressey *et al.* 1994). If just one site contains a particular species then it is irreplaceable, while a site in which all the species are present elsewhere is much more replaceable. In Ugandan forests there was only a weak consistency in the patterns of species richness of woody plants, large moths, butterflies, birds and small mammals but selecting reserves that complement each other using data from one taxonomic group, also protected the important forests for other groups (Howard *et al.* 1998). Similarly Hopkinson (1999) used data for liverworts, aquatic plants, non-marine molluscs, dragonflies, orthoptera, carabid beetles, butterflies and birds and showed that selecting complementary sites for one taxonomic group is also likely to include much of the diversity of other groups.

The major problem with much of the theory on selecting reserves is that where there is sufficient data for carrying out sophisticated analyses then the reserve network is usually well established. The areas in which it is possible to set up many new reserves, such as the Amazon Basin, are usually areas for which there is very little data. One approach suggested is to select areas of dissimilarity based on vegetation maps, climate and isolation due, for example, to river systems (Faith & Walker 1996, Schmidt 1996) and then use complementary methods to select a reserve network.

A major issue in selecting sites is whether the objective is to protect and enhance local or global biodiversity. There is a common tendency to protect habitats that are locally scarce even if abundant elsewhere. While this is understandable, as we all tend to wish to increase the biodiversity in our site, region and country, it is widely accepted that the emphasis should be on thinking globally not parochially.

Even in a fantasy world where all the species are known, their distribution and abundances has been determined and the conservation priorities of each species has been determined, the decisions as to where to locate reserves is far from straightforward. The reason is that such decisions depend only partially on conservation importance. In most cases other important considerations include cost, convenience, cultural importance, educational opportunities, aesthetics and political acceptability.

The other uses to which the reserve can be put are often important considerations. Sites may generate income in ways that are compatible with the conservation objectives, for example from appropriate tourism or agriculture. It may be considered important to choose sites with potential for education, for example by being near to cities. Extensions of existing reserves or sites near to existing reserves are usually easier and cheaper to run. When buying reserves it is important to think about the costs of maintaining and managing them.

The ability to maintain and enhance biodiversity may vary greatly between sites and this needs to be considered. For example, is the water quality and hydrology largely determined outside the site and does this matter? Is poaching or encroachment likely to be a problem? Is it practical to alter the grazing or fire regime?

Larger sites or those within dispersal distance of other suitable habitat will be more likely to retain viable populations. This is another reason for buying extensions to existing reserves. The scale will depend upon the species involved: a reserve intended largely to conserve large mammals will typically require a larger area than one to preserve a plant population. A reserve with a relatively short edge (e.g. round rather than long and thin) is usually easier to manage.

Sites will differ in their contribution to the economy such as the pollution-reducing function of wetlands, the role of mangroves and coral reefs in coastal protection and the role of estuaries as nursery areas for fisheries. In prioritising areas it is sensible to independently assess biological importance, threat and economic importance. The actual priorities will then be some combination of the three.

The main approaches for setting priorities for deciding which areas to protect or which areas to acquire as reserves are either top-down or bottom-up. The bottom-up approach is probably better if the information is available (it usually is not). In most cases with incomplete data the top-down approach is better. The bottom-up approach is a four-step process.

1 Decide on species priorities using methods such as those in Section 3.2.

2 Use the information on species priorities and the habitats they require to decide on priority habitats as in Section 3.3.

3 Make a list of objectives of the sort of sites required. Thus for each priority habitat type list the important criteria such as location, size, key species or educational potential. For example, from this stage it may be clear that the highest priority is one of the remaining mountain areas in the west of the country containing a particular vegetation type with particular associated invertebrates. The second priority could be one of the wetland sites near the coast containing various endangered fish, but must be a site where there is control over the hydrology.

4 Either carry out a case-by-case analysis of sites that it may be possible to protect or buy or alternatively use the objectives for determining which sites to buy from those that come on the market. In the absence of all the required information it is necessary to make best informed guesses with the available information for completing (1) and (2).

A top-down approach, modified slightly from that suggested by Margules (1986), is:

1 Decide upon a range of habitat classes and define the precise criteria for each.

2 For each habitat class, decide on minimum criteria for including a site, such as size, location or the extent of degradation. Thus it might be decided for saltmarshes that only those 1 km long, on the eastern coast and without artificial coast defences are included. List all the sites which fit these minimum criteria.

3 Rank each site within each habitat class using another set of criteria such as species diversity, presence of rare species, naturalness, education potential, threat or economic importance.

4 Take into account constraints such as finances or availability to convert (3) into a practical list suitable for determining acquisitions.

4 Monitoring

4.1 Why monitor?

People often ask me how to carry out a monitoring programme and I reply by asking what question they wish to answer. The answer is usually that they do not know! After unravelling the question, the census method usually becomes clear. Without a clear objective, monitoring may absorb considerable time without achieving anything useful.

Important conservation questions that monitoring could answer include: how are the populations of species of conservation interest changing nationally? How are the populations of a species changing on a site? How are the populations of pest species changing? Where are the most important areas for a species? What are the habitat requirements of a species? How do populations respond to changes in management? The techniques described here are also essential in determining why species have declined (Chapter 6).

It is depressing how often major changes have taken place without being quantified. Understanding the causes of declines is much harder without information on the timing and location of declines. Even estimates of rough numbers e.g. '100–1000' or 'over 1000' are much more useful than simply recording presence or stating 'many' or 'common'. Quantifying changes (i.e. the species has declined by 55% since 1970) also makes it much easier to attract public support or action from politicians than having to rely on vague statements ('a resident remembers seeing lots of the species in his childhood'). Conservation issues increasingly involve legal action and if this is a possibility the methodology should be defensible in court.

Performance indicators are widely used to monitor success, for example in economics and they are now beginning to be developed for conservation (Bibby 1999). By signing the Biodiversity Convention (see Section 10.8.1) most of the world's governments have committed themselves to create indicators of biodiversity and to monitor them. The UK government has been among the first to publish annual indices of various environmental measures. These include an aggregate measure of bird populations along with those for air pollution, river water quality and others. It can be useful for conservationists to press governments to establish such biodiversity indicators and help suggest effective means of monitoring them. It can also be useful to suggest that monitoring is incorporated in potentially damaging projects, such as changes in land use, both to document the effects and provide motivation for those involved to minimise the impacts.

Understanding the behaviour and ecology of the species is very useful in attempting to produce sensible census results. For example radiotracking (Section 5.6) or studying marked individuals can be extremely illuminating in interpreting data.

Measuring environmental variables, such as salinity or water depth, is often as important as monitoring species as it enables interpretation of biological changes. It can also be useful to compare environmental conditions between sites. Monitoring programmes on reserves should be capable of detecting gradual gross changes (such as the spread of scrub). Photomonitoring (Section 4.20) is one good way of achieving this.

Monitoring often risks causing disruption, both to the species under study and others present, as a result of disturbance or trapping. As well as considering how to minimise any disruption, it needs to be decided if the benefits are likely to be sufficient to warrant the study.

In this chapter I first review the main general techniques for monitoring populations and then describe the methods most commonly used to census each taxonomic group. I assume the section on the taxonomic group will be read in conjunction with the section on the relevant technique.

4.2 Bias and accuracy

Bias is the consistent overestimate or underestimate of numbers, for example a common bias is that certain individuals, such as juveniles or males, are missed. Inaccuracy means that the estimate is likely to be wrong but is as likely to be overestimated as underestimated. The importance of bias and inaccuracy depends upon the question. For comparing sites or different time periods, it may not be critical if each count is consistently biased. The project may, however, be worthless if the biases differ consistently, for example, if the expert observer goes to the primary forest blocks and the secondary forest blocks are counted by someone less experienced, as it will be impossible to distinguish whether the differences are due to observers or habitat.

One year I asked each member of my undergraduate seminar group to visit the lake on the University of East Anglia campus during the forthcoming week and count the number of ducks, gulls and willow trees. We then calculated the coefficient of variation as a measure of the variation between their individual counts. The value for ducks was 18%, the numbers of gulls vary markedly through the day and, not surprisingly, the value was higher at 37%, while for willow trees the value was a remarkable 137%. The variation in the number of willow trees (from 12 to 260) could be due to uncertainty in the area covered, tree identification or what size trees should be counted. I repeated the exercise the next year with similar results. Variations in estimates between individual recorders have been shown for a range of taxa (Inglis & Lincoln Smith 1995). The clear message is that even the most apparently straightforward census may be inaccurate.

As shown in the willow exercise, even the most simple instructions can be interpreted in different ways. If told the last census was 260 but the current census recorded only 12, most would deduce a catastrophic decline rather than suggest census error.

There are a number of ways of overcoming errors due to bias and inaccuracies and a combination of these approaches is often used.

- Acknowledge that there will be errors and try to reduce them rather than naively believing the results.
- Create and write down precise instructions.
- So that the methods can be repeated, create rules that deal with ambiguous cases such as the plants that overlap the quadrat edge, identifiable fragments of invertebrates in the trap, birds that are present but do not nest, or fish that are seen escaping as the net is pulled in.
- Standardising the method can greatly reduce the variation, for example by using the same observer, times and weather conditions. This can greatly reduce the data collected but is worthwhile if the variation would otherwise be large compared with other differences.
- Test any methodology first on yourself and then on others before making it widely available, both to see if it is practical and to check that the instructions are interpreted properly.
- Record the observer, time, weather and other variables, analyse their consequences and try and correct for these. This is only possible if there is overlap between methods and sites. Thus if the data consists of morning visits to site A and afternoon visits to site B then it is impossible to distinguish differences due to time of day from site differences.
- Equalise as far as possible the errors by balancing across sites, thus ensuring sites do not differ in observers, times or weather conditions during surveys.
- If it is essential to use a number of observers then agree on precise instructions and check in the field that the technique is consistent, for example by getting all observers to collect data from the same quadrat or to simultaneously collect distance sampling data from the same point.
- Rather than everyone collecting all types of data, consider allocating tasks, such as collecting census data, collecting habitat data or taking notes, to different individuals in such a way as to ensure the most consistent results.
- Quantify the errors. Carrying out independent studies in the same site by repeating the census

using a number of observers, or comparing the results with that from another method, can reveal inconsistencies in methods and indicate the accuracy of the census. This is useful both for improving the methods and interpreting the results.

4.2.1 Long-term data sets

It is particularly useful to be able to compare the current abundance and distribution of populations with the past. Most nature reserves should base their habitat management on such data. However, in practice, long-term data is usually difficult to interpret. As described in the previous section, the exact methods used are critical but the methods are usually not documented and it can be impossible to distinguish population changes from methodological changes. If long-term data is to be useful, it is essential that the methods are consistent or, if altered, that the consequences of changes are assessed, for example by carrying out the census for a few years using both the old and new methodology. Similarly it may be possible to improve the way the site is subdivided during censuses but this may not be sensible if it is then impossible to compare current and past distributions. Changes in methodology often occur when there is a change in land manager.

4.3 Sampling

Sampling is essential unless the entire population is counted. The importance of sensible sampling is greatly underestimated, and as a result many population estimates are almost worthless. It is much better to have a large number of small samples, as the samples are more likely to be representative, than a few large ones.

It is important to avoid bias in selecting the sampling areas. For example, if only the areas that seem most promising are counted, it is impossible to estimate the numbers in less promising areas. One sensible solution is to select areas at random. It is not sufficient to pick locations that look random or guess random numbers as both methods usually deviate markedly from randomness.

Random locations are usually determined using random numbers. Table 4.1 gives a thousand random numbers. They may also be generated by some computer packages, some pocket calculators, by using final digits in a telephone directory, randomly stopping a digital watch and using the value of the hundredths of seconds or, if desperate, by writing numbers on card, shuffling them and removing one at random. Repeat to obtain the next digit.

If the random number tables are always started from the top left then all surveys will start at the same point. Some statisticians get very excited about this issue and have suggested elaborate ways of entering such tables. Solutions include continuing using the tables from where last used or entering them at random. In reality, pointing a finger at random is likely to be sufficient. To obtain two random numbers between say 1 and 67, pick a random point, such as the top of the bottom left block (51832). The first number is 51, the second 83 is outside the range and

Table 4.1 A thousand random numbers.

11556	67225	02217	94283	99606	28751	77249	11193
41834	70057	14462	63533	20908	04812	79265	44683
29221	92881	45441	93867	05961	73770	72777	34468
12771	07865	70209	21115	87423	84502	08981	04694
07649	93484	94944	80805	64438	62430	79949	98690
49941	75664	56648	98016	89391	62389	79908	80599
95951	93070	70695	70741	93906	10534	75542	33623
74662	54062	35266	16631	02579	37087	63286	02564
07963	22264	14511	58299	46013	11248	44341	76829
27064	03329	06101	84194	88265	37283	96447	15318
41604	82536	70655	41871	03137	38734	00351	85900
37238	60466	63227	05959	49263	06893	85484	55558
46463	66466	54094	88336	88703	74166	54941	15648
58869	77281	98310	60239	18903	21900	03142	81707
75741	63864	86905	24722	99370	86118	93078	02690
49767	49637	09051	83491	42724	47303	82880	40679
62856	77792	70352	44607	51200	75732	05862	52702
89912	17075	02690	74069	95364	91286	33846	90096
31754	25647	33988	98123	91555	98777	76056	81241
09317	52702	38420	70444	92157	80707	87642	37754
51832	55723	57709	59597	42228	16671	87180	76938
89477	78699	91228	80775	04336	92048	70966	17704
81200	66232	52086	20371	41286	58113	73861	38611
98316	80845	63334	64464	63268	52401	80949	14248
55112	69853	13587	37813	54710	01715	05180	52085

ignored, the next is 28 and is included. Repeat numbers, such as a second 51, are ignored.

Sampling often needs to be done on a range of scales. Suppose the aim is to estimate the population size of a plant found only in wet grasslands and the locations of all wet grasslands are known. Firstly sites are chosen at random. Suppose there are 93 grasslands and 12 will be picked for study, then number each grassland, and pick pairs of digits from Table 4.1 until you have twelve between 1 and 93. The location of transects or quadrats within each chosen site is then chosen at random.

Random locations are most easily selected by overlaying a grid within which the site fits and then picking random x and y coordinates using Table 4.1. If the coordinates are outside the area to be sampled then another pair of random coordinates are chosen (see Fig. 4.1). It may be sensible to exclude random points that are so close to an already selected point that the same individual may be counted but this should be done systematically using a set distance.

Random transects are chosen in a similar manner. A constant transect direction is selected and the starting locations for each transect chosen at random. If all the transect lies outside the study plot it is ignored, while if only part of the transect is within the plot then this length is still surveyed (see Fig. 4.2).

The alternative to random sampling is to sample regularly, for example at intersections of a grid. It is much quicker to sample regularly than randomly and it gives a better coverage of the variation within a site (Mueller-Dombois & Ellenberg 1974). Sampling at precise random locations is so time consuming that many conservationists just wander around collecting samples yet state that samples were collected at random: this is deeply unsatisfactory. The main problem of random sampling is that there will often be unknown trends in abundance, for example due to variation in soil type or previous land use. Regular samples are more likely to result in each being sampled in proportion to abundance. It is also easier to analyse a regular distribution and so detect and present patterns in density as well as providing distribution data. The main problem with regular sampling is that it will be criticised by those obsessed with randomness.

There are occasions when for practical reasons random or regular sampling is impossible, for example when it is difficult to leave paths. One solution is to try and estimate the error caused by this biased sampling, for example by comparing sites on the path with those some distance away (Jones 1998). An alternative is to consider how much biased sampling is likely to matter, for example, if comparing the butterflies in two flat lowland forests then sampling near the tracks in each may be acceptable for comparing the forests but if the sites are mountainous and the track in one forest goes up the mountain but follows the valley floor in the other, then any

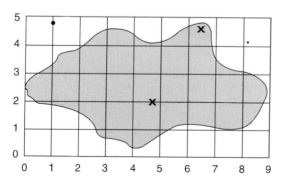

Fig. 4.1 Selecting random sites from an area. If the random coordinates lie within the area (e.g. 46, 20 or 65, 46) it is included and those outside the area (e.g. 82, 42 or 10, 48) are excluded.

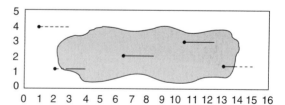

Fig. 4.2 Selecting random transects. First select a direction for all the transects (in this case east). This direction should ideally be the long axis of the site. Points are then randomly selected, as in Fig. 4.1, which provide one end of the transect. If the transect runs outside the study area then the length within (solid line) is still surveyed but the length outside (dotted line) is not. Note that the area within which random numbers can be selected extends to the west of the study site.

comparison is likely to be deeply affected by the sampling bias.

4.3.1 Stratified sampling

Species are often much more abundant in one habitat than another. If census points are selected at random, then the number of points falling in each habitat will greatly influence the mean and thus the estimate of total population. It is usually preferable to sample each habitat independently to give a population estimate for each habitat. These are combined to estimate the total population. Such stratified sampling is almost always preferable to random sampling.

If there is some evidence that the species is more abundant in one habitat then a sensible solution is to adjust the sampling so that it is proportional to the likely proportion of the total population in that habitat. Thus if preliminary studies suggest 60% of the population is in primary forest then 60% of the samples should be there. However, this is obviously circular as if the densities are precisely known in each habitat there is no need for the census, while if the densities are poorly known the survey design may turn out to be highly misrepresentative. The solution is to base the number of samples in each habitat on area if there is no information on population density (so that if 15% of the area is primary forest then 15% of the samples should be in that habitat) and on expected numbers if there is accurate information and an intermediate strategy for intermediate information.

A common mistake is to only visit the few areas where the species is known to be common. Without estimating density elsewhere, or confirming absence, it is impossible to estimate total population size.

4.3.2 Monitoring plots

It is often useful to record changes in abundance over time. Permanent plots are better than randomly selecting new plots each year as it removes one source of variation from the observed population changes. There are, however, practical problems with permanent plots. They are often difficult to relocate, the markers may be lost or vandalised, they may become damaged by repeated sampling or, if conspicuous, by attracting the attention of others. One way of reducing attention is to mark the exact site with buried metal pegs and use a metal detector to locate them, but even this is not infallible as one 57-year project ended unexpectedly after the metal posts were somehow located and removed!

Although it seems reasonable to place monitoring plots where the species is present and ideally abundant, over time the data may suggest a population decline when the population has merely shifted. The solutions are either to select sites at random or to divide into areas of high and low initial density (or areas where the species is present and areas where absent) and select random plots in each.

4.4 Indices and censuses

A census is an attempt to find out the actual population size. This is often too difficult and an index of abundance is used instead which is assumed to be proportional to the actual population. Examples of indices include the number of moths attracted to a light trap or the density of deer footprints. Indices are useful for comparing sites or changes between years. In reality, some censuses are actually indices, for example because only a fraction of individuals are seen.

One problem with indices is that they may be influenced by environmental conditions, for example if it is raining there may be few moths attracted to the trap but more deer footprints. Another concern is that a proportional change in the population may not be reflected by the same change in the index, for example, at high densities moths may hinder each other's entry into a trap or deer may interact more and so move further. Where possible, check the relationship of the index with environmental conditions and population size.

4.5 Counting recognisable individuals

This technique is used when individuals are distinctive and long lived and the population is small.

Fig. 4.3 Individual Bottlenose Dolphins *Tursiops truncatus* are identified from photographs using a combination of fin nicks, scratches, non-pigmented areas; areas of active or healed epidermal disease, unusually wide, tall or leaning dorsal fin or deformities of the normal body contour (Wilson *et al.* 1999). Each individual is shown twice. Dorsal fin nicks, deformities and unusual fin shapes persisted over the 3 years of the study but other marks were less persistent. The best estimate of this species in the Moray Firth, the only known population remaining in the North Sea, was of a minimum of 62 from a co-ordinated land-based survey (Hammond & Thompson 1991). Analysis using mark–release–recapture resulted in an estimated population size of 129 ± 15 (Wilson *et al.* 1999) (photos: Kate Grellier and Ben Wilson/Aberdeen University/SMRU).

It is regularly used to survey striking mammals such as African Elephants *Loxodanta africana* and Wild Dogs *Lycaon pictus* and has also been used for birds, fish, amphibians and snakes. Often only part of the animal is used, such as the tail flukes of Humpback Whales *Megaptera novaeangliae* or the belly pattern of a newt (Hagström 1973).

It is necessary to create a file (a file card system works well) listing the identification features of each individual or giving photos. Changes in features over time need to be noted. It can also be useful to add important details of the individual's history. Describing the identification features may be easier if the card includes a blank outline of the species or an outline of the part used for identification, such as face or tail.

Recognising individuals is less easy than it seems in producing sensible population estimates. Unless the recording is rigorous, two similar individuals can be considered as one or, conversely, by acquiring a new feature such as a scar, one individual can be recorded as two. Individual recognition can be used in one of two main ways. One technique is to attempt to identify every individual. However, even if individuals are equally likely to be seen, a large number of sightings are needed before being confident none are missed. In many cases some individuals are relatively conspicuous and frequently recorded, giving a false sense of confidence in the accuracy, while other inconspicuous individuals are missed. An alternative approach is to use mark–release–recapture methods in which the first 'capture' is the identification and the 'recapture' is a resighting (see Fig. 4.3).

Individuals can be identified from samples of hair, skin, feathers or faeces using DNA techniques (Section 5.15). It may also be possible to determine the sex of each individual. For example, to identify Brown Bears *Urso arctos* in the Pyrenees, DNA from hair and faeces was used and when combined with using paw size to separate adult and young showed that the population consisted of at least one adult female, three adult males and one yearling male (Taberlet *et al.* 1997). This approach can be useful for determining the minimum number of individuals but again a large number of samples is likely to

be necessary before there is any confidence that all have been detected. It can also be used to determine the home range of individuals; the bear study showed individuals outside the core range were wandering known individuals, rather than additional individuals, as had been assumed.

4.6 Quadrats and strip transects

Quadrats may vary enormously in size from being used to count diatoms to counting trees in savannahs. They are traditionally square but need not be. Circular quadrats, consisting of a peg (or weight if underwater) and string/rope of known length, can provide a simple portable quadrat of known area and have the advantage of the smallest possible edge for a given area. Strip transects, in which the observer moves a set distance by foot, vehicle, boat or plane and records the number of individuals that can be seen within a given distance of the transect, are really long rectangular quadrats. If individuals close to the transect line are more likely to be detected then distance sampling methods (Section 4.7) should be used instead.

Quadrats and strip transects are typically used when the species can be easily counted and is not disturbed by the observer. These are the commonest techniques for counting plants. They may also be used for counting signs such as dung or breeding sites. They can be used for mobile species as long as they can be counted before they move out of range.

Some individuals will usually be missed. It is thus useful to attempt to estimate this bias by other means such as seeing how many of a known number are detected. Caughley (1974) reviewed 17 studies in which known numbers of large terrestrial mammals were counted in aerial transects and between only 23% and 89% were seen from the air.

Strip transects have two advantages over square quadrats. It is often more practical to create a single transect in thick vegetation than to ensure all of a large quadrat is visited. Secondly, strip transects are likely to include a range of habitats and so reduce the variation between quadrats and thus increase the precision (Bormann 1953). Strip transects have the considerable disadvantage of a long perimeter so increasing inaccuracies due to deciding whether a particular individual is inside or not.

4.7 Distance sampling: line transects and point counts

A line transect consists of counting individuals seen from a transect. A point count consists of counting individuals from a fixed point. The theories of point counts and line transects are very similar with point counts simply considered as infinitely short transects. Table 4.2 compares line transects and point counts. Line transects should be carried out at a standardised speed, balancing the risk of double counting if too slow and missing species if too fast.

The easiest point count or line transect is to simply count all individuals that can be seen regardless of distance. This gives a relative index but cannot be converted to density. If individuals are counted within a given distance, and it is thought that the probability of detection is constant across that width, then the method being used is actually a strip transect or quadrat (Section 4.6).

Distance sampling (Buckland *et al.* 1993) estimates densities, allowing for the fact that more distant individuals are less likely to be seen (see Fig. 4.4). One approach, variable distance line transects or variable distance point counts (also called variable circular plots), entails collecting data on the distance of each individual from the transect or point when first detected; this can be used to estimate the detectability function, the relationship between the probability of detection and distance (Burnham *et al.* 1980).

If an animal moves then its initial position is recorded. If an animal enters the area, such as a bird flying over, then it is ignored. The distance can be determined by comparing with a marker set at a particular distance, calibrating the binocular focusing wheel, using a range finder, practising distance estimating with inanimate objects and checking the estimates or using the recently developed binoculars that record distance. Some researchers estimate distance from calls but this should be avoided because the volume of a bird's call, and therefore the estimate of distance, depends upon whether it is facing you, the type of habitat and the height of its perch.

In an ideal world, detectability curves should not contain data from repeated uses of the same transect line or point count locations. Nests, fruiting trees or vantage points may mean that a species is repeatedly seen at the same distance which will bias the results. In practice this problem is probably not critically

Table 4.2 Comparison of line transects and point counts.

Line transects	Point counts
Usually cover a much greater area in a given time so far more individuals are seen, which improves the precision	Fewer individuals usually seen
Transects tend to be more accurate as an error in estimating distance will change linearly with distance from a transect	Error changes with the square of distances from a point (see the equations in Box 4.1) so less accurate
Transects usually cover a larger area especially within the nearest band so that density estimates are likely to be more accurate (Lloyd *et al.* 1998). For example, for a transect 300 m long the area within 20 m would cover an area of 12 000 m²	Cover a smaller area especially for the nearest band. For example, for point counts 300 m apart the area within 20 m is 1270 m² for each point. Thus tend to be inaccurate unless numerous counts are made
In dense or inaccessible habitat, it may be necessary to concentrate on the track rather than observe animals	Under difficult conditions it can be better to stop at a series of points than attempt to record continuously
Transects may be subdivided into sections and the data for each section recorded separately and related to habitat or management but it is less clear that each subsection is an independent data point	Point counts are reasonably independent data points and so are easy to relate to habitat or management

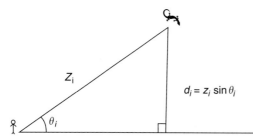

Fig. 4.4 Distance sampling entails estimating the distance of each animal to the transect line or point. If the distance to the animal is z_i and the angle from the transect to the animal is θ_i then the perpendicular distance $d_i = \sin\theta_i$.

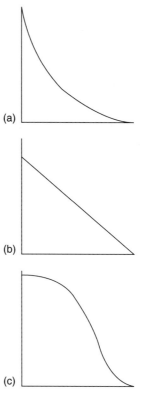

Fig. 4.5 Three possible relationships between detection probability and distance: (a) exponential; (b) linear; (c) half normal.

important and for rare species obtaining sufficient data for the detectability curves is often so difficult that it is sensible to carry out repeated visits to those transects or point counts where the species occurs to improve the curve. If the location is revisited but with different transects or point count locations then the results should not suffer from this source of bias.

Figure 4.5 shows three of the biologically plausible relationships between detectability and distance. The usual approach for fitting detectability curves is to count the numbers in a series of at least five different distance bands, such as 0–20 m, 20–40 m, 40–60 m, 60–80 m, 80–100 m and 100+ m. Software packages, such as DISTANCE (Laake *et al.* 1994), can be used to convert such data into density estimates. There is an excellent website (http://www.mbr.nbs.gov/software.html) that includes the software packages and can be downloaded for free. However, fitting detectability functions requires about 60 observations per species and the derived detectability function may not be applicable to other habitats and seasons. In practice, the analysis is often carried out for smaller data sets but the confidence limits are then large.

An alternative to determining detectability functions that is simpler, requires less data, but is less accurate, is to simply count the number of individuals seen within a given distance (near belt) and the number seen anywhere beyond that distance (far belt). Box 4.1 describes how to convert these data into densities. The distance should be selected so that

about half the records will be in each belt. Methods for determining whether individuals are in the near or far belt are the same as described in previous paragraphs.

In order to convert data on the numbers seen in the near and far belts into densities it is necessary to make an assumption about the shape of the detectability curve. However, a comparison of the three main models for the decline in detectability with distance (negative exponential, linear and normal) showed that all gave similar density estimates if there are similar numbers in each belt (Jarvinen & Vaisenen 1975).

Three crucial assumptions of distance sampling are that the sighting distances are unbiased, that individuals are recorded at their original locations and that all individuals on the line (for transects) or

BOX 4.1

Estimating populations from point counts and line transects

The simplest approach is to use two recording zones, one within distance r and one beyond that distance. Suppose that in m surveys, n_1 animals are counted in the nearest band and n_2 in the distant band. The following analysis assumes an exponential relationship between the probability of detection and distance.

If a point quadrat then:

$$\text{density} = \frac{n_1 + n_2}{\pi r^2 m} \log_e\left(\frac{n_1 + n_2}{n_2}\right)$$

while for a line transect of total length l metres

$$\text{density} = \frac{n_1 + n_2}{2rl} \log_e\left(\frac{n_1 + n_2}{n_2}\right)$$

(if only one side of the transect is counted then remove the 2 from the equation).

The equations give densities in terms of individuals per m^2 and need to be multiplied by 10 000 to give density per hectare or by a million to give density per km^2.

point (for point counts) are detected (Buckland *et al.* 1993). One way of providing a minimum estimate of the fraction missed on the line/point is on some occasions to have one recorder collecting the data as usual while others search for missed individuals along the line/point, for example by staring up in the canopy with binoculars to provide some estimate of the fraction missed.

It is common to collect data on the habitat (e.g. number of trees within 30 m, altitude, estimated percentage canopy cover) at the same time as collecting the transect data or point counts. Point counts are particularly suitable for this. Logistic regression can then be used to relate presence/absence to habitat variables. If the species being studied is rare and the habitat variables are time consuming to collect then it may be sensible to only collect habitat data when the species is observed and compare this data with a random subset of those locations where it is not observed.

4.8 Mapping

For territorial species, mapping the locations of individuals can provide good estimates of population size. It is labour intensive both in the field and in subsequent analysis and thus is usually only used for restricted studies such as researching a rare species or carrying out a census of a species within a reserve. The information obtained on territory location can also be useful for understanding habitat preferences. Compared with point counts or transects, mapping is more likely to detect inconspicuous species, produces more consistent results and is less affected by the timing of the visit (during the day or season) and weather.

The first stage is to decide on a study area, prepare a map which shows all obvious features and make copies. A major source of error is dealing with individuals whose territories overlap the edge of the study area. This problem can be reduced by minimising the edge (a large circular plot is ideal) and by selecting plots so that the edges run through poor quality habitat rather than through the middle of areas likely to hold many territories.

Each visit should use a separate map upon which all sightings are marked using a different code for each species. Figure 4.6 gives the standard codes for mapping birds. The information from each visit is then transferred to a master map for each species. All the sightings from the first visit are marked as 'A', those from the second as 'B' and so on. It is often useful to cross off or highlight each record on the visit map once transferred to ensure none are missed. The technique is then to draw territories around clusters of sightings. It is necessary to use the available ecological and behavioural knowledge to create rules for drawing territories using these maps. For example in the rules used by the British Trust for Ornithology, if there are 5–7 visits by the observer then two close registrations are considered a territory, while for 8–10 visits it is necessary to have at least three close registrations. Other information on breeding sites, observed movements, territorial fights and simultaneous observations are also incorporated. Simultaneous observations of adjacent individuals of the same sex (e.g. two singing males) are particularly useful in distinguishing territories, especially territorial fights as they often show territory boundaries. Sightings of the same individual watched moving is useful in showing the extent of an individual's territory. Further details on analysing such maps are given in Marchant et al. (1990). Even with strict rules the interpretation will still be partly subjective so all maps should ideally be analysed by one person. The maps should be kept so they can be reanalysed later if necessary.

Activity codes for use in mapping censuses in Finland

These activity codes have been developed from, and are very similar to, the mapping codes used by the British Trust for Ornithology. Most examples are for the chaffinch *Fringilla coelebs*. Some countries have standard codes for each species name (e.g. in the UK, CH = chaffinch). From Gibbons et al. (1996).

(F_{coe})	A chaffinch in song.
(F_{coe})	A chaffinch in song (exact location shown by the point).
(F_{coe}) ↗	A chaffinch in song (location is not exact; the point where the observation was made is shown by the cross).
$F_{coe}\male$	A male chaffinch repeatedly giving alarm calls or other vocalisations (not song) thought to have strong territorial significance.
$F_{coe}\male$	A male chaffinch calling.
$F_{coe}\male$, $F_{coe}\female$, F_{coe}, $F_{coe}\,2\male1\female$, $3\,F_{coe}\,juv$	Chaffinch sight records, with age, sex, or number of birds if appropriate. Use $F_{coe}\,\male\female$ to indicate one pair of chaffinches, i.e. $2\,F_{coe}\,\male\female$ means two pairs together.
$F_{coe}\male^{f}$	A male chaffinch carrying food (or faeces).
$F_{coe}\female^{m}$	A female chaffinch carrying nest material.
$F_{coe}^{*\,2E\,3N}$	An occupied nest of chaffinches, with 2 eggs (E) and 3 nestlings (N); * shows the location. Do not mark unoccupied nests, which are not of territorial significance by themselves.
$P_{maj}^{⊞\,10E}$	Great Tit *Parus major* nesting in a specially provided site. Such as a nest box.
F_{coe}^{*P}	Chaffinch nest with a parent bird incubating or warming young.
$F_{coe}\,juv$	A chaffinch fledgling.
$F_{coe}\,fam$	Juvenile chaffinches with parent(s) in attendance.

(Continued)

(Continued)

Movements of birds can be indicated by an arrow using the following conventions:

 A calling male chaffinch flying over (seen only in flight).

 A female chaffinch moving between perches. The solid line indicates it was definitely the same bird.

 A singing chaffinch perched, then flying away (not seen to land).

 A male chaffinch flying in and landing (first seen in flight).

 A Siskin *Carduelis spinus* circling above the forest.

The following conventions indicate which registrations relate to different, and which to the same, individual birds. Their proper use will be essential for the accurate assessment of clusters.

 Two chaffinches in song at the same time, i.e. definitely different birds. The hatched line indicates a simultaneous sighting/hearing of song and is of great value in separating territories.

The solid line indicates that the registrations definitely refer to the same bird.

The question-marked solid line indicates that the sightings/songs probably relate to the same bird. This convention is of particular use when your census route brings you back past an area already covered – it is possible to mark new positions of (probably the same) birds recorded before, without risk of double recording. If you record birds without using the question-marked solid line, over-estimation of territories will result.

No line joining the registrations – there is no assumption as to whether the records concern different birds, but depending on the pattern of other registrations they may be treated as if only one bird was involved (a question-marked dotted line indicates that the sightings/songs were almost certainly of different birds).

 Two chaffinch nests occupied simultaneously, and thus belonging to different pairs. Only adjacent nests need to be marked in this way. Where they are marked without a line, it will be assumed that they were first and second broods, or a replacement nest following an earlier failure.

 An aggressive encounter between two chaffinches; may be accompanied by notes on vocalisations.

4.9 Mark–release–recapture

Mark–release–recapture is typically used when individuals are difficult to see. The essential idea is that by marking and releasing a known number of individuals (say 100) and then determining the proportion of the population that is marked (say a fifth), the population can be estimated at 500. This idea is

BOX 4.2

Population estimates from mark–release–recapture

The Petersen method applies to the simple case in which n_1 individuals are marked, allowed to mix and then a sample n_2 are recaptured of which m will already have been marked. If n_1/m equals 3 then only a third of the population has been caught on the second occasion. It is then obvious that the total population $\tilde{N} = n_1 n_2/m$. In practice however a less biased estimate is

$$\tilde{N} = \frac{(n_1 + 1)(n_2 + 1)}{m + 1} - 1$$

For example on 21 February 1998 I marked 56 Greater Pond Snails *Lymnaea stagnatalis* in my garden pond ($n_1 = 56$). On 24 February I recaptured 86 (n_2) of which 10 (m) were marked. Ten marked snails retained in an adjacent water trough all survived and retained their marks for at least a month, showing that neither mortality nor loss of marks was important. Adding this data to the above equation gave a population estimate ($\tilde{N}$) of 449.8.

The approximate upper 95% confidence limit, U, can be estimated as

$$U = \frac{n_1}{\dfrac{m}{n_2} - \left\{\dfrac{1}{2n_2} + 1.96\sqrt{\dfrac{m}{n_2}\left(1 - \dfrac{m}{n_2}\right)\left(1 - \dfrac{m}{n_1}\right)} \Big/ (n_2 - 1)\right\}}$$

$$= \frac{56}{\dfrac{10}{86} - \left\{\dfrac{1}{2.86} + 1.96\sqrt{\dfrac{10}{86}\left(1 - \dfrac{10}{86}\right)\left(1 - \dfrac{10}{56}\right)} \Big/ (86 - 1)\right\}}$$

$$= \frac{56}{0.1163 - \left\{0.0058 + 1.96\sqrt{0.1163 \times 0.8837 \times 0.8214/85}\right\}}$$

$$= 1149.9$$

while the lower confidence limit, L, is estimated as

$$L = \frac{n_1}{\dfrac{m}{n_2} + \left\{\dfrac{1}{2n_2} + 1.96\sqrt{\dfrac{m}{n_2}\left(1 - \dfrac{m}{n_2}\right)\left(1 - \dfrac{m}{n_1}\right)} \Big/ (n_2 - 1)\right\}}$$

$$= 304.7$$

Thus in this case the population is estimated as 449.8 but with 95% confidence that it is between 304.7 and 1149.9.

To check this I caught and retained all visible individuals, $n_2 = 511$, over the next week and recaptured a total m of 45. This gave an estimate of 647.6 (confidence limits 560.3 to 735.1).

usually called the Lincoln Index (after Lincoln 1930) or the Petersen Index (after Petersen 1896) although Laplace invented this idea in 1783 to estimate the population of France. Box 4.2 outlines how to calculate population size and confidence limits using this method. This approach can also be used if individuals can be recognised by individual features (Section 4.5) but are not marked, or if they are marked and then observed rather than recaptured. It is sensible to use different capture methods for the marking and recapture as one method may only capture a subsample of animals (Seber 1982). For example, if a method only captured males then the population estimate would only be of males

and if the sex bias was undetected then the population estimate would be much lower than it should be. However if the method was used to mark a subset of males but a different method was used to capture them which was equally likely to catch each sex (but might be biased towards adults) then the population estimate would be much more accurate.

The Petersen/Lincoln Index only applies if the same population is present during marking and recapture. It is commonly used even when there is considerable immigration and emigration or births and deaths, thus producing results of dubious value (Linderman 1990). It is then necessary to use the Jolly–Seber method (Jolly 1965, Seber 1965) which

BOX 4.3

Population estimates from capture frequency

If there are a number of samples obtained in rapid succession, each of constant effort, and the animals are either given unique marks or marks for each sample, then the number of times each individual is caught is known. The underlying assumption of this approach is that each individual is equally likely to be caught. The number of individuals caught a total of 1, 2, 3, etc. times is then expected to follow the Poisson distribution.

A population of Bank Voles *Clethrionomys glareolus* was trapped on 6 successive nights (Blower *et al.* 1981) and the number of times each vole was caught was recorded.

Times caught (x)	1	2	3	4	5	6	Total
Number caught (fx)	21	13	12	1	3	1	$51 = \sum fx$
Product (xfx)	21	26	36	4	15	6	$108 = \sum xfx$

Calculate the average number of times each is captured x.

$$\bar{x} = \frac{\sum xfx}{\sum fx} = \frac{108}{51} = 2.12$$

To see if the data does fit a Poisson distribution calculate

$$x^2 = \frac{\sum x^2 fx - (\sum xfx)^2 / \sum fx}{\bar{x}} = 37.4$$

A significant deviation, as shown by a value of χ^2 greater than shown in the table below, means the data deviates from the assumption of random capture and this method cannot be used. The degrees of freedom are the total number of animals caught minus 1 (50 in this example).

Degrees of freedom	20	40	60	80	100
χ^2 ($P < 0.05$)	31.4	55.8	79.1	101.9	124.3

For larger numbers of individuals, v, the χ^2 values can be approximated from

$$x^2_{\alpha, \nu} \cong \nu \left(1 - \frac{2}{9\nu} + Z_{\alpha(1)} \sqrt{\frac{2}{9\nu}} \right)^3$$

where for a level of significance $\alpha = P < 0.05$ the value of $Z = -2.575$ (Zar 1974).

Find by trial and error the value of m that gives the result

$$\frac{m}{1 - e^{-m}} = \bar{x}$$

In this case m equals about 1.75. Then calculate the total population size P as

$$P = \frac{\sum xfx}{m} = 61.7$$

provides an estimate of population size even if there are births and deaths or immigration and emigration between catches. This involves catching on at least three occasions and marking with a batch-specific mark and requires quite a large data set. If individuals differ greatly in their probabilities of being captured then the Burnham & Overton (1979) method is better. Greenwood (1996), Begon (1979)

and Blower *et al.* (1981) provide good accounts of these methods.

4.9.1 Frequency of capture

The frequency of capture method involves making multiple captures over a short period in which there will be negligible births or deaths. Individuals are either individually marked or marked each capture and the number of times each is caught is recorded. This data is then fitted to a Poisson distribution to estimate the number that have never been caught, as described in Box 4.3 on page 49. The Poisson distribution strictly assumes all individuals are equally likely to be caught, which will rarely be true but the Poisson still seems a sufficient approximation.

4.10 Catch per unit effort

Catch per unit effort provides a relative estimate of abundance by recording the number caught in a given time period. It is best to standardise the catching period and intensity of sampling (such as density of traps set) as doubling the catching period or sampling intensity is usually unlikely to result in twice as many individuals being caught (e.g. Minns & Hurley 1988). If the catching is set for the best time, such as mist netting birds for 3 hours after dawn, then extending the period on some days will result in a proportionally smaller catch, thus distorting the catch per unit time. A solution is to ensure the data is always collected for a standard period and additional data from any extended period is clearly separated.

Capture rate usually declines with the number of days in which the catching takes place, as individuals learn where the traps are or avoid the area of disturbance. It is thus sensible to also standardise the number of days at a site or sample point.

4.11 Monitoring plants

Plants are usually a relatively straightforward group to monitor as they stay still. However individuals can vary enormously in size and in many species the genetic individual 'genet' grows to produce a number of individually rooted 'ramets' so that it can then be difficult to know what should be counted. The four main qualities that can be measured are:

Density (number of plants per unit area) is often the best measure but is difficult for species with clonal growth or may mean little if there are considerable differences in size. Some studies of trees only consider individuals above 10 cm diameter at breast height (1.3 m).

Cover (proportion of surface covered) avoids having to distinguish between individuals and can be useful for vegetation surveys. It is rarely suitable for monitoring populations of rare plants. Cover is probably the best measure if it is not possible to determine density.

Frequency (fraction of samples, such as quadrats or squares within a quadrat, in which a species occurs) is difficult to interpret and so is rarely used.

Biomass (total weight, usually dried weight, per unit area) is time consuming and destructive.

See Bullock (1996) for a more detailed account.

4.11.1 Total counts of plants

Total counts seem easier than they usually are. Unless carried out methodically it is easy to miss individuals or count them twice. One approach is to grid out the entire area and systematically search each grid square, marking each individual with a flag once found. This can be very accurate and provide excellent information on distribution but is time consuming.

4.11.2 Quadrats

Quadrats (see Section 4.6) are easily the most widely used technique for a plant census. The numerous designs are usually variants of four strips of wood, metal or plastic joined to make a square. Quadrats joined with bolts can be disassembled for carrying. Larger quadrats (over 4 m^2) are usually marked out with posts using a tape measure and set square or compass. Common areas are: bryophyte, lichen or algal communities 0.01–0.25 m^2; grassland, tall herb, short scrub or aquatic macrophyte communities 0.25–16 m^2; tall shrub communities 25–100 m^2 and trees in woods and forests 400–2500 m^2 (Kent & Coker 1992, Bullock 1996).

The number of individuals of the species of interest can be counted. It is usual to only count those rooting in the quadrat. Percentage cover may be used for mat-forming species or when it takes too long to count all individuals, but it is less accurate. Percentage cover is often used when the observer can stand above the vegetation or for estimating canopy cover above the observer, but it is difficult to estimate scrub or tall herbaceous vegetation at the observer's height.

4.11.3 Seed sorting

Determining seed densities is useful for studying potential regeneration, studying individual plant species and quantifying the food supply of granivorous species. Seed samples are usually obtained by a soil core inserted to a given depth. For sampling underwater it is usually more practical to use a grab but this is much less precise than a core. The best way of extracting seeds is usually to hand sort samples after washing through a series of graduated sieves. The smallest sieve usually has holes about 0.15–0.2 mm wide. A finer lower sieve can be used to prevent the sink from becoming blocked. For counting most seeds a binocular microscope is useful. Seeds tend to be highly clumped so that many samples are necessary to provide an accurate density estimate. In some areas it is possible to buy identification guides to the commoner seeds, otherwise make a reference collection by collecting and drying mature seeds from those plant species likely to be in the seed bank.

4.11.4 Measures of vegetation density

In many cases, particularly when collecting vegetation data relating to animal populations, it is useful to simply have measures of how tall or dense the vegetation is. Measuring maximum plant height as the height of the tallest, or five tallest individuals is usually highly subjective and a poor measure of overall sward height. The sward stick is a useful method for simply, but crudely, estimating the amount of vegetation present. Make a polystyrene disc with a hole in the centre through which a rod marked in cm (allowing for the thickness of the disc) can be inserted. Measure and document the disc's diameter and weight. Place end of rod on soil surface. Place disc over rod and lower gently so that it rests on the vegetation and read the height of the vegetation compressed under the disk. If the ground is rough and the sward short then most of the variation in readings will be due to surface irregularities, so this method may be unsuitable. The weight of the disc should be such that it pushes down to rest on the bulk of the vegetation. If it is too light it will be affected by single stems and if too heavy it will crush all vegetation to a constant height. The more irregular the vegetation the larger the disc should be.

A similar approach to the Secchi disc (Section 4.18.8) can be used for horizontal measures of vegetation density. One person stands at a point and another inserts a card (e.g. a 5 cm^2 bright orange card) at various distances into the vegetation; the point at which the card can no longer be seen is recorded. This can be repeated for different compass locations and the average taken.

4.12 Monitoring invertebrates

Invertebrates are often highly seasonal and often show considerable variability between years and vary in activity according to the weather, so that from a short series of censuses it is usually difficult to tell if the variation is part of normal fluctuations, due to sampling variation or part of a long-term population change. Ausden (1996a) and Southwood (1978) describe monitoring invertebrates in detail.

4.12.1 Direct searching for invertebrates

This involves systematically searching an area to find every individual. It can be used in a range of habitats using quadrats or transects (Section 4.6) or, less satisfactorily, catch per unit effort (Section 4.10). This is a good method for species that are easy to find and move little. If the efficiency of searching can be assessed, for example by releasing marked individuals in an enclosed area and recording the fraction relocated in a repeat survey, then the data can be converted to absolute densities.

4.12.2 Beating for invertebrates

Beating involves collecting and counting the individuals that drop out of hit vegetation. A standard size of tray is placed under the vegetation and then the branch or bush is hit sharply to dislodge individuals. Sudden hits with a stick are far more effective than shaking. The height of vegetation hit is estimated so that some index of abundance can be determined.

4.12.3 Water traps for invertebrates

Water traps are water-filled bowls usually either placed on plant pot bases screwed to the top of stakes or on wooden platforms with diagonal elastics attached by nails to the underside of each corner keeping the bowl in position. Adding a few drops of detergent makes escaping harder and can double the catch (Harper & Story 1962). Water traps should be emptied at least weekly. The contents may be poured through a sheet of muslin or a sieve into a bowl. The water is returned to the trap and the catch is placed in 70% alcohol.

The colour of the water trap has a considerable effect on the species caught, with yellow bowls attracting flies and Hymenoptera while white attracts flies but deters Hymenoptera. Neutral-coloured bowls, such as brown, reduces the selectivity. Height also influences the species composition with the greatest numbers usually caught if the trap is just above the top of the vegetation. The exact location, in relation to water, shelter and nearby vegetation, will also influence the catch.

4.12.4 Light traps for invertebrates

Any bright light will attract adult moths and some other nocturnal flying insects. Light traps are a major way of documenting their distribution and abundance. A high-pressure mercury vapour bulb is ideal as these also emit ultraviolet light. The catch will be increased if the light is next to a white wall or white sheet and if the light is visible from a distance and from above. Ultraviolet light is damaging to the eyes so sunglasses should be worn.

Light traps usually consist of a mercury vapour bulb suspended above a box. Insects drop through a cone into the box (usually filled with egg boxes as resting sites). The insects must not be regularly released in daylight or local birds will learn of this predictable food source. The number of insects attracted will vary greatly with the weather such that comparisons over short periods are often of little use.

4.12.5 Emergence traps for invertebrates

Emergence traps are placed above open water, soil, mud or dead wood to catch emerging adults. The usual design is an inverted plastic box submerged in the substrate or held at the water surface with stakes so that emerging individuals are trapped. If a funnel is fixed to a hole in the upper surface leading to a transparent collecting tube then insects seeking the light will collect inside the tube.

4.12.6 Pitfall traps for invertebrates

Pitfall traps are steep, smooth-sided glass, plastic (e.g. plastic cups) or metal containers buried so that the rim is at soil surface. Animals fall in and cannot escape. They are frequently used for studies of active surface-dwelling invertebrates such as spiders, beetles, ants, centipedes and collembola. One problem is that the numbers caught reflect both abundance and activity. Traps should be at least 2 m apart to ensure independence and further for widely dispersing species. Traps should be marked to enable them to be relocated.

A preservative prevents decay and stops the specimens eating each other. Alcohol is suitable but evaporates rapidly and it needs replenishing at least weekly. Ethylene glycol (antifreeze) does not evaporate but is toxic so gloves and protective glasses should be worn; it can also be difficult to obtain in warm seasons and countries and can poison animals if they can drink the contents (cows are particularly enthusiastic consumers of pitfalls).

Pitfall traps may be baited, e.g. with fermenting fruit or raw meat, to attract certain beetle species, although as the attractiveness will change over time, this may complicate population studies. Preservatives mask the smell of such bait and so should not be used, although this means the traps have to be visited daily.

Pitfall traps also catch other species. The risk of catching and perhaps killing other species including vertebrates should be considered. Pitfalls can be fitted with a mesh or raised cover to reduce the risk of such accidental captures.

4.12.7 Sweep, pond and tow nets

Sweep netting is a cheap and easy method of collecting a wide range of species from vegetation but is poor at catching those that are firmly attached, sensitive or highly mobile. Sweep netting simply involves firmly sweeping a net with a strong rim through the vegetation. After each sweep the handle is twisted so that the net rests on the frame and the captures cannot escape.

Different sweeping techniques will catch a different range of species. For example, rapid sweeps result in a larger catch and a higher proportion of active species. The numbers of sweeps and the method used need to be standardised, which is easiest if carried out by one person. One standard approach is to take sweeps of approximately 1 m long during every alternate step while walking at a steady pace (Ausden 1996a). Weather effects the vertical distribution and activity, both of which will influence the catch. Sweep netting is unsatisfactory in damp or short (< 15 cm) vegetation.

Pond nets need to be stronger than sweep nets. A similar sweeping method is used only underwater and it should also be standardised. The catch is usually tipped onto a white tray for sorting and counting.

A tow net is a fine funnel-shaped net, often with a collecting bottle attached to the end, which is pulled a fixed distance underwater, usually from a boat. This method is used for counting zooplankton and small invertebrates. The contents are washed off the inside of the net into the collecting bottle by splashing water against the outside of the net. For small organisms it may be necessary to subsample the bottle contents by shaking the contents and taking measured volumes.

4.12.8 Benthic cores for invertebrates

This is the main method for counting invertebrates in mud or sand. It consists of inserting cores of a known area. For wide cores the substrate inside is dug out to a given depth, such as the bottom of the core. Narrow cores are usually dug out with the contents in place and the substrate pushed out of the centre of the core; it can be useful to devise a plunger to push out the core. The sample can be sorted by hand in a tray or washed through a sieve. Adding 1% solution of rose bengal dye makes translucent invertebrates pink which can aid sorting.

4.13 Monitoring fish

Information is often available from commercial or recreational fisheries. Such fisheries can also be used for providing data on species present, abundance and the methods that are successful in catching particular fish. Further details of techniques for monitoring fish are given in Perrow *et al.* (1996) and Nielsen & Johnson (1983).

4.13.1 Fish traps

There is a huge range of fish traps available. The basic idea is that the entrance is easier to find than the exit, as for example with a funnel leading into a trap. One simple technique for catching small fish is to cut off about the top third of a plastic drinks bottle and invert it. Traps can be weighed down, fixed to the bottom or suspended in mid-water using weights or floats. All traps need to be removed at the end of the study as they will otherwise continue catching.

Fyke nets are traps consisting of a frame covered in netting with a funnel leading to a subcompartment. A funnel from here may lead either to the main chamber or the next subcompartment. The idea is that fish are unlikely to find the way out of a series of funnels.

If the effort is standardised these may be used to give catch per unit effort, an index of abundance. Alternatively fish can be marked (Section 5.10.3) and populations estimated using mark–release–recapture (Section 4.9).

4.13.2 Gill and dip nets

Gill nets are suspended vertical nets with the upper edge attached to floats and the lower surface attached to weights. They are usually used for sampling in deep water and in large water bodies. A fish attempting to swim through the net may get the mesh caught behind the gills. A particular mesh will only catch a narrow size range as small fish will swim straight through while large fish will not penetrate sufficiently to be caught by the gills. Some gill nets are designed to be unselective by incorporating a range of mesh sizes. Alternatively, a number of nets differing in mesh size can be used. Gill net catches are usually quantified as catch per unit effort but the netting period and density of nets should be standardised. Gill netting has the considerable disadvantage of often damaging fish and killing those caught for a long time.

Dip nets are strong pond nets used for catching small fish. The net is rapidly inserted either vertically or at an angle down the plane of the net and then pushed a set distance through the water. A jar may be attached to the end of the net to collect the fish. To estimate the volume of water sampled, and thus convert the counts to densities, a large plastic bag can be tied to the frame for a few trials and the water volume determined.

4.13.3 Electrofishing

Electrofishing entails applying a current to the water and if a sufficient voltage gradient is created along an animal it is stunned. Direct current (DC) has the advantages that fish swim towards the anode and it is safer for both the fish and operators than alternating current (AC). Electrofishing is potentially dangerous and requires training. Electrofishing data is usually analysed as catch per unit effort.

4.13.4 Transects and point counts for fish

Counts may sometimes be made from the shore. Polarising sunglasses, to reduce glare, and wide brimmed hats make this much easier. The fish may be sensitive to disturbance, in which case the technique is to wear inconspicuous clothing, move slowly and try to stay in front of cover.

Strip transects and point counts (Sections 4.6 and 4.7) can also be carried out underwater. This method is often used in coral reefs but may be used in other habitats where there is sufficient visibility. In shallow water snorkelling is sufficient but depths greater than 1–1.5 m in freshwater lakes or 3–4 m in the sea usually require scuba. Transects may be laid on the substrate using weighted ropes or chains. Point counts are usually made by staying still so that the fish become acclimatised before estimating the numbers at each distance. A single diver can count all fish while moving upstream in a stream but a line of divers is required to achieve this for larger rivers. The accuracy of counts is influenced by depth, temperature, clarity, fish behaviour, time of day and diver experience (Thompson *et al.* 1998). Data is best recorded underwater using soft pencil on a perspex sheet abraded with wire wool or sand paper.

4.14 Monitoring amphibians

Amphibians are usually counted when they are at the breeding area. However, some individuals, particularly females, may not move to the breeding area every season. The breeding season may be short in temperate areas, extended in the tropics and highly unpredictable in some desert species. Most species are most active just after dusk and looking for them during the day is often futile. See Halliday (1996) and Heyer *et al.* (1994) for more details.

4.14.1 Drift fencing and pitfall traps

Drift fencing entails building a fence around the breeding site of plastic cloth or aluminium sheeting buried at least 20 cm in the soil with at least 40 cm height above ground to catch all individuals entering or leaving the breeding site. Pitfall traps, such as buckets, are then dug alongside the inside and outside of the fence about every 4 m. Pitfall traps need drain holes, cover for the amphibians and regular checking (ideally twice a day), especially if predators may also be caught. Individuals are released on the opposite side of the fence. Drift fencing requires considerable effort to build, maintain and run. It is only appropriate for small water bodies and if the species migrate to or from the water body in a particular short period. Trough traps, which are rectangular pitfall traps, have the advantage of being less conspicuous or elaborate than drift fencing.

4.14.2 Direct counts of amphibians

The number of individuals at spawning sites is often used. For very small ponds and conspicuous species it is sometimes possible to count all adults present. Counts per unit time (Section 4.10) have been used for large ponds to give an index of abundance. The number of egg masses provides an index. Strip transects or quadrats (Section 4.6) can be searched either in water bodies or on land. Artificial cover, such as plywood sheets, may be placed in a grid and searched for amphibians (Fellers & Drost 1994). They should be painted or covered to reduce vandalism.

4.15 Monitoring reptiles

Reptiles are often difficult to see. As they are ectotherms, the weather will greatly influence their activity. Gloves should be used when hand searching or checking traps if there is a risk of encountering poisonous snakes, scorpions or poisonous spiders. Blomberg & Shine (1996) give a more extensive account.

4.15.1 Mark–release–recapture of reptiles

Many species can be caught by hand by searching likely habitats, for example by inverting rocks or logs (which need to be reverted to their initial state). Reading (1997) suggests using galvanised corrugated sheets, painted to camouflage the upper surface to reduce vandalism or theft, as a means of attracting reptiles.

Noosing is often an effective technique especially for lizards that are wary or inaccessible. A noose comprises a long pole with a loop of line (e.g. fishing line or dental floss) which can be pulled tight after manoeuvring over the lizards neck. They can also be caught using a grasshopper on a fishing line and when they bite it, flick them either into a bucket placed nearby or a butterfly net held in the other hand and swung into position at the same time (this needs practice).

Pitfall traps are often the best way of catching lizards. These should have steep smooth sides with drain holes and a cover. They need to be shaded in hot climates to prevent overheating. Drift fencing such as a 30 cm high wall of plastic or another material fixed to posts or aluminium sheeting can increase the catch markedly as individuals are diverted along it. Floating pitfall traps can be used to catch basking turtles (Petokas & Alexander 1979). Troughs, which are just wide pitfall traps, can be used and these have the advantage of being more inconspicuous than drift fencing. It is essential not to leave traps after the study ends as they will continue catching. Although mark–release–recapture (Section 4.9) is often used there is often too much population turnover to use a Peterson/Lincoln Index but insufficient data for a Jolly–Seber analysis. Means of marking reptiles are given in Section 5.10.5.

4.15.2 Direct observations of reptiles

Strip transects (Section 4.6) or line transects (Section 4.7) can be used for reasonably conspicuous species. Visibility usually varies markedly with temperature and should be undertaken under standardised weather and time periods. A grid of artificial cover such as corrugated sheets, camouflaged to avoid human attention, can be searched to provide an index of abundance (Fitch 1992).

4.16 Monitoring birds

Birds have the advantages that they are often reasonably conspicuous, have diagnostic calls or songs and many people have the expertise necessary to identify them in the field. As a result of the ease of counting them, birds are good for monitoring environmental change. See Gibbons *et al.* (1996) and Bibby *et al.* (1992b, 1998) for more detailed accounts.

4.16.1 Direct counts of birds

Conspicuous species may be directly counted. This is used especially for counting colonies and for birds on open water bodies. It is often better to subdivide the area and record the number in each. Ideally these subdivisions should be unambiguous and used for subsequent counts. If parts of the area cannot be seen then this should be recorded.

Nests on cliffs are easier to count from slightly above, or at the same level and perpendicular to the face than at an angle. The number of birds in a colony will vary over the season. It is necessary to decide exactly what to count (e.g. individuals, pairs or nests) and at what stage in the season and to document clearly what was done.

If roosting birds are packed too close to count easily it may be easier to count small groups arriving or leaving the roost. Most species tend to arrive in small groups but leave in larger flocks so it is usually better to count those arriving. Large flocks can be estimated by counting a number of individuals, say 20, and then counting the number of groups of that size that would fit into the total flock. There is a tendency to underestimate large flocks (Prater 1979).

4.16.2 Transects for birds

Line transects can be carried out by foot, from a vehicle for large species or from a boat. The movement rate should be standardised such as walking

1 km an hour in forest and 2 km an hour in open habitats. If belts are used for distance sampling (Section 4.7) they are often set at 25 m in forest and at 50 m in more open habitat.

Transects at sea or very large lakes can be carried out using the method of Tasker *et al.* (1984). All birds seen on the water within 300 m from one side of the boat are counted each ten-minute period. The area covered is calculated (1 knot = 0.32 km in 10 minutes); thus at ten knots, 3.2 km and 0.96 km² are covered. Birds in flight are recorded within the same width but as far ahead as it is thought possible to see all individuals. Thus if it is possible to see 400 m then eight (i.e. 3.2/0.4) counts at 1.25 minute intervals (i.e. 10/8) are required to count the 3.2 km transect. Flying birds are counted if present in the block 400 m ahead and 300 m wide at the instant the interval starts. Thus the transect can be viewed as a series of snapshot counts: any leaving the block are still counted and any entering subsequently are ignored. The flying and swimming birds can be added together. Birds following the boat are either ignored or counted separately.

4.16.3 Point counts for birds

It is often appropriate to count all individuals on either side of a circle such as of radius 25 m in forest and 50 m in open habitat. Each individual should be counted just once. It is often sensible to wait 5 minutes before counting so the birds are less disturbed. The count is for a fixed period of 3–10 minutes depending upon how conspicuous the birds are. The counts should be completed as quickly as possible to reduce the risk of double counting and allow more points to be visited. If many individuals of a species are present at a single count and are likely to be confused it may be sensible to draw a circle divided into quarters in a notebook and mark the position of each individual. Points should be at least 200 m apart to prevent double counting.

4.16.4 Territory mapping

This method is labour intensive and thus rarely used for comparing sites or monitoring national population changes. However, for an intensive study of a particular species, mapping is often the most effective and accurate way of counting the number of breeding pairs (see Section 4.8). Many species do not have clear territories and thus are inappropriate for this method. The method is to move through the study plot a series of times and plot the location of all sightings. Records of birds singing at the same time or birds of the same sex observed together are particularly useful as this helps identify territory boundaries. These maps are collated at the end of the season onto a master map (one per species) and the number of territorial pairs is determined by consistent rules as to what is considered evidence for a territory (see Section 4.8). It is reasonable to alter these rules according to what is known about the biology of the species.

4.17 Monitoring mammals

Some mammal species are obvious and can be readily counted. However, most species are difficult to see. Some of the species of greatest conservation interest are both secretive and occur at low densities. See Sutherland (1996a) and Wilson *et al.* (1996) for further details on how to census mammals.

For some species it is easiest to census the breeding or hibernating structures, such as beaver lodges or squirrel dreys. These are then surveyed using the same techniques used for counting individuals and can be used as a relative measure of abundance. Alternatively, by determining the number of beavers per lodge or the number of individual squirrels per drey the total population size can be estimated.

4.17.1 Direct counts of mammals

Direct counts are commonly used for counting colonies of bats and seals. Bats can be counted during the day using a torch, ideally with a deep red filter (e.g. Kodak Wratten Filter No. 29) to reduce disturbance. Waking hibernating bats reduces their fat supplies and can increase mortality. For large numbers of bats it is usually better to take photographs of clusters (this requires careful documentation) then count individuals in the photos. Caves and mines are

dangerous and a further risk is histoplasmosis, a fungal disease, caught by inhaling spores in bat roosts in warm humid regions, so either use a self-contained air supply unit or a respirator capable of filtering 2-micron diameter particles (Constantine 1988).

Fruit bats can be estimated by counting all visible individuals at roost, then disturbing them with a loud noise and counting them in the air to estimate the proportion missed. Repeating the exercise can improve the estimate of the fraction missed. This estimate can then be used to estimate the fraction missed in those sites where it is unacceptable or impractical to count the roost. The percentage visible may, however, vary between sites especially if there are considerable habitat differences.

Seals are usually counted when on shore at the breeding sites. Aerial photographs can also be used by placing a grid over the photograph and counting the number in each grid square. Crossing off individuals on an acetate sheet placed over the photograph is both more accurate and allows others to return to the photographs later and check the consistency.

4.17.2 Transects of mammals

Transects can be carried out by foot, vehicle, plane or boat. The methods described for birds (Section 4.16.2) apply here. It can be possible to census some nocturnal species by shining a strong light (some can be plugged into a car's cigarette lighter) and looking for the light reflection from the back of the retina, 'eye shine'. Aerial transects are often used for large mammals. The surveys are carried out at a fixed height. It is usually easiest to collect data into a tape recorder.

4.17.3 Mapping mammals

Mapping is best for those species that are clearly territorial such as many primates and carnivores. As with mapping birds (Section 4.8), the location of sightings and territorial calls are mapped. It is useful to be able to locate calls from different territories simultaneously. In many mammals the territory may be held by a group and thus mean group size has also to be determined.

4.17.4 Trapping mammals

Trapping is the standard way of determining the population of small mammals (Barnett & Dutton 1995). The two main types of traps are box traps (e.g. those sold as Longworths or Shermans) and pitfall traps. Pitfall traps are simply steep-sided containers (e.g. buckets) buried so the top is flush with the surface. They should have a cover if rain is likely and be positioned to avoid flooding. The traps are usually checked at dawn and dusk. If insectivores are likely to be caught then the traps should be checked every 6 hours. Food improves capture rate and is necessary to keep the captives alive. Dry food (e.g. sunflower seeds, wheat or corn) is easiest to use but insectivores need to be given live food such as fly pupae. A stiff mixture of porridge oats, peanut butter, water and cooking oil is often used. Dry bedding should be added if the nights are cool. Traps should be sheltered in hot climates. Adding bait to the area can increase trapping success but, by attracting outsiders, can severely bias the results. Placing a stick alongside the entrance of each trap to deflect them into the trap can increase catching success. The easiest handling technique is to empty the trap contents into a plastic bag. Active species can jump out between your arm and bag opening or even by running up your arm. To prevent this hold the top of the bag around your arm or have someone else do this for you.

It is usual to estimate densities using mark–release–recapture, within a regular grid. The standard technique has been to remove fingers and toes, but fur clipping to expose the underfur, which is usually a different colour, works reasonably well and is less barbaric.

As an index of abundance the number caught can be divided by the number of trapping nights (Erlinge 1983). Half a trap is subtracted for each trap that catches a mammal or is sprung accidentally. For comparisons over time or between sites it is necessary to use a standard trap density.

4.17.5 Dung counts

Many mammal species are too difficult to trap and too elusive to see. Counting dung is often the best remaining option as an index of abundance. This is usually accomplished by counting numbers in quadrats for abundant species or transects when the dung is less plentiful. Dung decays, especially in wet weather or when dung beetles are abundant. One way to overcome the problem of decay is to remove or mark all the dung in an area and after a time interval count the additional dung. An alternative that is less satisfactory, but more practical over large areas, is to find very fresh dung, mark it and leave in a typical location, revisit after a set period and only count dung that looks fresher than that dung. By measuring the rate at which animals produce dung (if necessary using captive animals on a similar quality diet) it is possible to convert the dung counted to population densities.

4.18 Monitoring environmental variables

It is often impossible to interpret population changes or the distribution of a species unless there has also been a programme of monitoring environmental variables.

Most environmental measures can be simply collected manually and I will concentrate here on such methods. The alternative is to use complex, automated equipment which can be superb in collecting and storing data automatically, every few minutes if required. The data may be automatically downloaded onto a computer. Manufacturers, brochures are the best sources for details on the latest sophisticated equipment. See Jones & Reynolds (1996) for a detailed description of environmental monitoring.

4.18.1 Temperature

Maximum–minimum thermometers can be used to give the daily temperature range. They are best located 1.25 m above ground as this is the standard height for meteorology (and thus best for comparison with other sites) and they are usually placed within standardised Stevenson screens, or other designs, for meteorological data collection.

For conservation studies, thermometers are often located in relation to the ecology of the species being studied. Temperature loggers which record the temperature at set intervals are very good for ecological measures of microclimate. For ectotherms ('cold blooded' species) in temperate areas, temperature is often critical; sward height, aspect, slope and colour may all have marked effects on the local temperature and detailed measurements can help interpret the ecology and behaviour.

4.18.2 Rainfall

Rain can be collected in open containers, but they are likely to be inaccurate due to evaporation. Rain gauges are cylinders 10–20 cm in diameter, usually buried in the ground, with a funnel at the top to collect the water into a removable collecting vessel. The rain is poured into a measuring funnel calibrated for the diameter of the gauge. Alternatively calculate

$$\text{rainfall} = \frac{\text{ml or cm}^3 \text{of rain}}{(\text{diameter of rim of gauge in cm})^2}$$

A rain gauge should be positioned so water cannot splash in and all vertical objects (e.g. trees) are four times their height away. Rain gauges are usually emptied daily at a fixed time. If visited irregularly then a little oil may be added to reduce evaporation.

Snow may be melted by taking the gauge indoors, if this does not miss collecting any precipitation, or by adding a known quantity of hot water. Roughly 12 mm of fresh snow is equal to 1 mm of rain but this is very approximate as the density of snow may vary about six-fold.

4.18.3 Water depth

Water depth is best measured in relation to stage boards which are water-tolerant boards marked in centimetres with a longer line every 5 cm and numbered every 10 cm (Gilman 1994). These are ham-

mered into the ground to read from somewhere accessible. When setting up consider whether extreme values can be recorded. It can be useful to have a discreet back-up post in case the conspicuous post is vandalised. Automatic water depth loggers can be bought which store water depth on computer chips and thus may be downloaded straight into a computer file for analysis. Clockwork chart recorders which record depth onto a chart are losing favour as they are difficult to maintain and the charts are fiddly to analyse.

Dipwells are used to measure water table depth. They can be made from a section of pipe (e.g. PVC), 6–50 mm internal diameter with holes (e.g. 4 mm diameter) drilled at least every 10 cm so the core may equilibrate with the water table. A perforated pipe can also be used. One leg of a pair of nylon tights can be pulled over the dipwell before inserting in the ground to stop soil entering. A flanged collar is useful to reduce vertical movement of the pipe. A cap should be placed on the top to prevent insects, leaves, soil or water from entering. A bung may be added to the base if the soil is likely to enter, which is important in mud or peat. The dip well should be dug to a depth that exceeds the expected minimum level of the water table. It is rarely necessary to dig deeper than 2 m. Narrower dip wells respond more quickly and so are more accurate but are harder to read. The main methods are to insert a ruler and measure when it touches the water (use a torch), insert a ruler and record the depth at which it becomes wet or insert an electronic dipmeter which responds to the water. Dipwells can also be bought and some designs record water level continuously. Temporary dipwells can be created using a soil auger (e.g. 2–5 cm width) and recording the water levels at half-hour intervals until it has stabilised. Water level range gauges are dipwells with a weighted float attached to a scale which records the maximum and minimum reading since the last visit by displacing foam blocks (Bragg *et al.* 1994) and are useful if the dipwell is only visited occasionally and extreme values are of particular interest.

Dipwells can be used to determine how ground water level varies seasonally, how it fluctuates in response to changes in weather, management, water levels in rivers or ditches, land use and whether there are any long-term changes in ground water levels. This is useful for managing the site and if planned sensibly can improve the understanding of the hydrology, for example, a series of dip wells perpendicular to a ditch indicates the degree of lateral movement, which is essential for understanding the importance of ditch level on ground wetness.

Piezometers are similar to dipwells but have perforations at just one depth and so measure the hydraulic head at that depth (soil water moves from the higher head to the lower). They should have the narrowest feasible internal diameter. It is frequent to use groups ('nests') of piezometers differing in perforation depth often located in the same hole to determine the hydraulic head at each depth. The data analysis requires a trained hydrologist.

For some sites it can be useful to make a simple map of soil wetness by creating a grid of points across the site and assessing the wetness at each (Brooks & Stoneman 1997). Water depth is obviously useful if it can be measured. It can be useful to dig a grid of dipwells and draw the contours of equal water depth. Visual measures can be used to map relative soil moisture, such as whether there is surface water at the point, whether the ground is sodden, whether there are pools within a fixed distance and the presence of species or communities associated with either wet or dry conditions. One possibility is to mark each point and repeat the survey at the seasonal extremes or under typical and flood conditions.

4.18.4 Water flow

The amount of water flowing through a ditch, stream or river can be estimated by calculating the cross-sectional area (mean depth × width) by the water speed. Speed is ideally calculated using a flow meter at a range of evenly spaced locations across the ditch/stream/river with depth measured at each. Hydrologists prefer at least 15 locations with at least 20 if the cross-section is irregular (Shaw 1994). Measuring water speed at 40% of the water column depth

(i.e. at 40 cm in a metre-deep stream) usually gives mean speed of the column (Francis & Minton 1984). A rougher, but sometimes sufficient, technique for estimating speed is to time the movement of an orange down a measured section. Oranges are used as they are too heavy to be affected by the wind.

A V-notch weir across a stream or ditch consists of a weir, which retains the water, with a notch, through which the water flows. If the discharge is low there will just be a trickle through while with high discharge it will gush through. The height of the water in the notch is obviously dependent on the flow and thus is a relatively simple way of measuring discharge (Brooks & Stoneman 1997). The notch is either cut into the weir or a separate notch plate is bolted and sealed with a waterproof sealant on the weir. The notch plate may then be made of a more expensive material (e.g. steel notch for a plywood weir) and replaced if damaged. It can be useful to place a mesh screen upstream of the notch to prevent debris from blocking the notch at low flows. It is important that water cannot move around the edge of the weir so the weir should be dug into a narrow groove into the base and sides of the stream.

The V-notch should be wide enough to allow high flows to pass through but sufficiently narrow for measures of low flows to be meaningful. A 90° notch is often used but is poor at measuring low water flows. A 20° notch requires a deeper weir but allows measuring of a greater range of flows. A curved notch that widens at the top allows both low and high flows to be recorded accurately but is harder to calibrate. The water depth in the notch is measured using a gauge attached to the weir. The relationship between water depth in the notch and flow is calibrated by visiting at different levels of discharge and measuring both the water level and collecting the water flowing through the weir over a set time period. The practical difficulties of collecting the water for calibration should be considered when designing the weir. Problems include being able to place the basin under the notch and being able to collect water trickling down the face of the weir at low flows (it may be necessary to use a funnel). The relationship between flow and water depth is then calculated (usually a regression of log discharge against log depth). After calibration the water depth is either measured manually or using an automated recorder and using the calibration can be converted to discharge.

4.18.5 Evapotranspiration

Measuring evapotranspiration is important for creating a water budget (Section 12.7.1) and for understanding the water loss from the system, but is very difficult to measure. The standard approach is to create a lysimeter, which is an area of natural vegetation enclosed by a waterproof barrier (Calder 1976). Rainfall and lateral water movement are measured, vertical movement is either prevented or a site selected where this will not affect the results and any unaccounted water loss, according to the water budget calculated for the lysimeter, is attributed to evapotranspiration. There are two main techniques for creating lysimeters.

Container lysimeters: dig out small blocks of vegetation and place in a container such as a plastic drum (giving time to revegetate if damaged in the transfer). Containers are easier to manage and may even be attached to balances to record changes in water content but if the vegetation of soil structure is disturbed then this may affect the results.

Natural lysimeters: surround an area with a waterproof barrier, for example, by digging a ditch, pushing in a double layer of thick polythene and backfilling with clay. For larger habitats, such as forests, natural lysimeters are the only practical possibility.

To determine the water balance equation (see Section 12.7.1) within the lysimeter, measurements are made of the rainfall (see Section 4.18.2), water level (see Section 4.18.3) and water exchange. There are two approaches to measuring water exchange. One approach is to record the outflow from pipes fixed to the lysimeter at the usual water table height. The second approach is to add or remove water to keep the lysimeter at the same water level as the surrounding areas (this also reduces problems with leakage) and measure the necessary import or export of water required to achieve the balance.

To convert changes in soil water levels to changes in amount of water stored, it is necessary to know the specific yield. This measure, which depends upon the soil type and the pore sizes, specifies the extent to which a change in soil water volume results in a change in water level. This can be estimated by adding a known quantity of water to the lysimeter, waiting for the level to equilibrate and recording the changes in dipwell reading.

$$\text{specific yield} = \frac{\text{water added (mm)} \cdot 100}{\text{change in water level (mm)}}$$

Thus if adding a volume of water equivalent to 10 mm depth raises soil water level by 40 mm then the specific yield is 25. There are also published values of specific yields for various soil types.

Lysimeters are difficult to use and in reality are rarely used by conservationists. However, land managers sometimes have continuous data on water levels in dipwells and in isolated sites (i.e. with no lateral water movement) the daily decline in water level during a period without rain will measure the evapotranspiration. However sites are rarely isolated and there is likely to be some lateral seepage into the site. The seepage into the site can be detected during the night when evapotranspiration is reduced as the water level may then increase by several millimetres (Goodwin 1931). The hourly rise of the water table during the night (e.g. midnight to 4 a.m.) can then be used to estimate r, the rate of lateral seepage (assuming that evapotranspiration is negligible during this period). By making the reasonable assumption that the lateral seepage is constant through the 24 hours the evapotranspiration, E, can be calculated from

$$E = \frac{S}{100}(24r + f)$$

where S is the specific yield (see above) and f is the fall in water level over the complete 24-hour period (Gilman 1994). The same approach could of course be used if there is seepage out of the site. The specific yield can also be measured using dipwell data by determining the rise in water table in response to a measured depth of rainfall. Strictly this should be measured as net rainfall (rainfall minus evapotranspiration), but if measured after heavy rain on days when evapotranspiration is likely to be small then the relative contribution of evapotranspiration is likely to be small and can be ignored.

There is a range of techniques for estimating evapotranspiration from published constants and weather data (described in Shaw 1994). These are however difficult to apply.

4.18.6 Wind speed

Wind speed is best recorded with a cup anemometer comprising three or four cups on horizontal spokes which spin with the wind and are conventionally placed 2 m above the ground. They can give either instantaneous readings or total distance moved since last read and are often automated. Hand-held gauges, which consist of a cylinder containing a float which rises as the wind enters the hole near the base, are less accurate ($\pm 3\,\mathrm{km\,h^{-1}}$) but cheaper and more portable. The Beaufort scale (Table 4.3) is the least accurate but free.

4.18.7 pH

The pH measures how acid ($<\mathrm{pH}\ 7$) or alkaline ($>\mathrm{pH}\ 7$) the soil or water is. Indicator paper is a simple but rather inaccurate method. Wide-range indicator papers can be used for giving the approximate pH to be followed by the appropriate narrow-range paper. pH meters and electrodes are more accurate. The electrode is placed in the water and the pH is read off the meter. Meters need to be regularly calibrated using buffer solutions of constant pH. Tablets of pH 4, 7 and 9 are often provided with the meter. Either pH is measured in the field or samples are collected in clean glass bottles ensuring there are no air bubbles, as the carbon dioxide in the air can lower the pH.

Soil pH is determined by mixing one volume of soil with twice the volume of distilled water (pH 7), waiting for 10 minutes and taking the reading.

Table 4.3 The Beaufort scale for assessing wind speed.

Beaufort number	Name of wind	Observable features	Velocity (km h^{-1})
0	Calm	Smoke rises vertically	<2
1	Light air	Smoke drifts downwind. Wind does not move wind vane	2–5
2	Light breeze	Wind felt on face; leaves rustle. Vane moved by wind	6–12
3	Gentle breeze	Leaves and twigs in constant motion; wind extends light flag	13–20
4	Moderate breeze	Raises dust and loose paper; small branches are moved	21–29
5	Fresh breeze	Small trees in leaf begin to sway; crested wavelets form on inland waters	30–39
6	Strong breeze	Large branches in motion; whistling heard in telegraph wires; umbrellas used with difficulty	40–50
7	Moderate gale	Whole trees in motion; inconvenience felt in walking against wind	51–61
8	Fresh gale	Twigs break off trees; progress generally impeded	62–74
9	Strong gale	Slight structural damage occurs (chimney pots and slates removed)	75–87
10	Whole gale	Seldom experienced inland; trees uprooted; considerable structural damage occurs	88–101
11	Storm	Very rarely experienced; accompanied by widespread destruction	102–121
12	Hurricane	At sea, visibility is badly affected by foam and spray and the sea surface is completely white	>121

4.18.8 Underwater light

The depth to which light can penetrate determines where photosynthesis, and thus plant growth, can take place. An Italian admiral submerged a dinner plate to assess turbidity, and a modified but less genteel version of his technique is still used today. The Secchi disc is a 30 cm diameter disc with alternating black and white painted quarters with a small hole in the centre through which a calibrated line is tied. The disc is lowered into the water and the depth at which it can no longer be seen is recorded. The disc is submerged further and then raised slowly. The depth at which it becomes visible again is recorded and the average of these two depths used. The depth at which photosynthetic species have problems persisting (the euphotic zone depth) is about 1.2–2.7 times greater than the Secchi disc depth (Moss 1988). Secchi disc measurements should be taken under similar light conditions to allow comparisons.

4.18.9 Salinity

Salinity is expressed as the weight in grams of total salts per kg of water expressed in parts per thousand (‰). Full strength sea water is 35‰. The best approach is to place a hydrometer in the water and read off the specific gravity. If the temperature is measured at the same time then specific gravity can be converted into salinity (e.g. see Jones & Reynolds 1996).

4.18.10 Water chemistry

Jones & Reynolds (1996) outline the basic techniques. There are numerous water testing kits which include ones for measuring pH, dissolved oxygen, ammonium, nitrate, nitrite, phosphate, calcium, iron and silicate. These are useful to give rough guidance or to check other methods but are not as accurate as the more elaborate methods described in Golterman et al. (1978) for fresh water and Parsons et al. (1984) for sea water.

4.18.11 Soil characteristics

Soil profiles can be determined by digging a hole with a straight vertical side and measuring the depth of each layer. Soil augers are quicker and easier for obtaining a soil sample at a given depth. The soil core is laid on a flat surface so the profile can be measured. Successive sampling will give a complete depth profile.

4.19 Monitoring human impact

It is often useful to be able to document human impacts such as the number of hunters, number of visitors, the area mined or the extent of logging or agriculture. The approach is the same as in monitoring populations or environmental variables. It is necessary to find a sensible sampling regime and a repeatable way of monitoring. Write down the exact method and criteria. For example, it is essential to have precise definitions for who counts as a hunter: does a hunter and unarmed companion count as one or two? Does someone hunting elsewhere who walks through the area count? Similarly if a single tree is removed is the area no longer primary forest and if so does this apply to the area the tree occupied or the entire forest? Without precise definitions it is difficult to distinguish variation in classification from actual changes.

4.20 Photographic monitoring

Photographs are a good way of documenting changes to sites. Photographs are not usually useful for documenting small-scale changes for which data from quadrats is usually preferred. Aerial photographs are invaluable for monitoring and documenting gross changes to sites, such as changes in the extent of woodland or open water. A series of photographs may show changes imperceptible to site managers, especially when managers change. Photographs may also be valuable for legal uses although the documentation must then be rigorous. Photographs can be a very dramatic way of illustrating change and problems and are likely to be of more widespread public interest than, say, presenting data on changes in species composition within a quadrat.

The following approach is based on that used by the Countryside Council for Wales (Jones 1994). First walk over the site and select potential fixed points which are likely to be recognisable well into the future. Global positioning satellites are useful for describing fixed points. A fixed lens is more satisfactory than a zoom lens, as it is easier to standardise. Photographs are taken using a camera on a tripod with a 35 mm lens in a panoramic arc from left to right with 10–30% overlap. This has the advantage that features such as pools can be included within a single shot. The alternative approach is to take photos at fixed compass directions. A 35 mm lens covers 62° and thus eight photographs can be taken at 0° North, 45°, 90° 135°, etc. Alternatively six photographs can be taken every 60° with negligible overlap. A 50 mm lens covers 46° and then nine photographs can be taken every 40° or eight every 45° with negligible overlap.

The camera is taken off the tripod and 1–3 shots are taken of the tripod in location to ease relocation. The first tripod location is marked A on the map and numbered arrows are used to show the direction of the shots (unnecessary if fixed compass directions are used). The date and focal length of the lens used is recorded. This procedure is repeated in a number of locations within a site. Where old photographs are available it is worth considering using the same locations. For very large sites aerial photographs are probably better in conjunction with ground photographs of specific areas. It may be sensible to label each film prior to processing to ensure they are not muddled.

The films can be processed to archival quality to ensure they do not deteriorate. Black and white prints last longer than colour but may show less detail. Careful documentation is essential so that the photographs can be linked to the locations. Archiving and documentation will take time and money and this needs to be considered before starting a programme. A set of prints and a copy of the documentation should also be sent to the site manager.

Jones (1994) suggests that repeat surveys are made every 15–20 years with a shorter interval for rapidly changing sites. Extra surveys can be added to answer specific questions or to monitor particular events.

5 Ecological research techniques

5.1 Why carry out research?

Good research helps reveal the truth, but the truth can be a real nuisance if it contradicts current beliefs, decisions and practices. It is important to create a climate in which researchers can question dogma rather than simply having to justify existing decisions. A student did some surveys on a nature reserve in England and made some management suggestions to the warden. The warden swore at him and told him not to return. Another student subsequently carried out research there and made useful discoveries with important management implications but I instructed him not to mention anything to the warden. Similarly Chase (1987) suggests that the lack of independent research and intolerance of information that conflicts with current beliefs has greatly hindered the management of some of America's National Parks. Wilkinson (1998) gives a series of examples of the difficulties of scientists whose research brings them into conflict with the government environmental agencies they work for. In my experience research conclusions are much more likely to be used if partly funded by the organisation that can implement them, as the results can be fed into their decision-making apparatus. By contrast, completely external research is easy to ignore. It is also my experience that independently derived research is more likely to be used either if the message is compatible with the land manager's beliefs or if the researcher has managed to gain the respect of the community that will implement the results. It is, of course, important to be tactful when presenting research that undermines current practices. Scientific reports similarly need to be objective and avoid politics.

There is, of course, a huge range of research techniques so I outline those that I consider particularly useful. For a newcomer to a subject, working alongside an experienced researcher can be invaluable. I review statistical problems in the final section.

5.2 Designing a research project

In designing experiments or observations it is best to think backwards. This method seems clumsy, but too many projects either could never answer the question or even do not have a question.

1 What is the question or questions you want to answer?

2 What analyses are necessary to answer this question? I recommend drawing the possible graphs and tables as this concentrates the mind on the exact analyses. Decide on exact statistical tests that could be carried out.

3 What data are needed for these analyses? For example how many data points are likely to be necessary? How many sites have to be visited?

4 How can this data be collected? It is often useful to design a protocol as described below.

5 Is it practical to collect the data necessary to answer the question? Estimate how much time (e.g. number of days) is available, allowing time for travelling, shopping, bad weather, illness, repair of equipment and inevitable unforeseen problems. Estimate how long each research activity takes, allowing time to travel between sites and sort samples, and then write out a realistic timetable. The ideal methods almost always would take longer than the available time and so it is usually necessary to make difficult decisions deciding what to exclude. It is much better to start with a realistic programme which can include a plan for further work if there is spare time than to have an overambitious programme which is only partly completed.

6 If the data that you can actually collect is likely to be inadequate to answer the question, then why are

you going to collect it? It may be more useful to do something else.

In practice, designing a project usually involves working up and down through the above list. For example there may not be time to carry out all of the work so the decision will have to be made to either exclude one of the analyses or restrict the study to one species or one area. Any such change will affect the protocol and timing. Once started projects continue to evolve, but changes should still be within the above framework.

It is often sensible to devise a protocol which makes best use of the available time to collect the required data in a systematic manner. A protocol might be to visit each site between 11 a.m. and 3 p.m. and count lizards along five 200 m transects, record the temperature at the start of each transect, describe the vegetation as percent cover using a 1 m^2 quadrat at each end of the transect, estimate the percentage of bare rock within 1 m^2 every 50 m along each transect, and take one 5 cm diameter $\times$ 10 cm deep soil sample from the centre of the site. At the end of the visit spend half an hour catching as many lizards as possible and record species, sex, snout to vent length and weight.

Many projects involve data sheets and their creation is often a critical stage in clarifying thinking. The creation of the data sheet and protocol often go together; following the protocol then involves filling in the data sheet in a systematic way. It is useful to create a range of data sheets, e.g. one for the transect and habitat data of each site, one for the laboratory soil analysis and a third for the details of the captured lizards. Data sheets make it much less likely that any data is missed and make it more likely that the data will be analysable rather than a set of unsystematic observations in a notebook. It is sensible to prepare a draft data sheet, test and then improve it before adopting. The initial testing can be by imagining being in the field and considering how to add data and the likely problems but a field test is also important. If the sheets have to be prepared before starting field work, they can still be roughly tested on a different species. If others will be using the data sheet then watch others filling in the form to discover ambiguities and problems.

The data should be easy to transfer from the data sheet to the computer file for analysis. It is thus useful to consider the expected analysis as part of the process of designing the data sheet. Using abbreviations saves time in the field but can lead to confusion and doubt over the accuracy afterwards. One useful technique is to confirm any unusual data such as extreme values, for example by underlining, so as to remove any concern that it might be a notation error.

Many biologists have a nightmare notebook story. Mine is remembering at high tide that my rucksack was still on the mudflat. Thankfully I discovered it in a creek the next day and the notes were still readable. Completed data sheets do not need to be taken in the field and so are less likely to become lost than notebooks. They can even be photocopied and stored in a separate building to reduce the risk of fire, flooding or theft. Creating data sheets ensures that you think about the project, how the data will be collected and how it will be analysed. Furthermore others can view the data sheets and make suggestions and comments.

Quantifying variation in the field may appear to be simple but actually requires great skill. For example, in a project to study the decline of a rare orchid, sites may be studied which differ in orchid abundance, cattle densities, slope, tree height, bee densities, soil type and amount of bare ground. The first decision is how each can be measured. One approach is to divide into classes such as high, medium and low cattle density, but this is a poor technique, as it is often difficult to repeat and is difficult to analyse statistically. It is thus much better to quantify these as continuous variables. The skill is to find simple ways of obtaining sufficient precision, balancing between methods that produce data that is too inaccurate to be useful and methods that are more time consuming than the results obtained justify. Once the time and cost of quantifying each variable is roughly known then decisions can be made as to which to include. Although it is tempting to quantify all variables, this takes time and reduces the sample size or increases the cost. Thus, in the earlier example, estimating bee density in a sufficiently accurate manner may be considered less important than using the time saved to visit more sites.

A frequent mistake is to have such a time-consuming protocol for each site that it is only possible to carry it out a few times so there is too little data to analyse. It is common to be told by experts that a huge amount of data is necessary to measure a population or a physical feature with sufficient accuracy and thus a proposal to compare say, 20 sites, is completely impractical. This is often a result of a common misunderstanding. The expert is often explaining how to obtain a precise measure for each site, say of the abundance of a predator. For comparing sites to see if declines in a target species can be linked to the spread of an introduced predator, simple measures may be sufficient. Thus counting faeces along a transect may distinguish between sites where the predator is abundant from those where it is scarce and this may be sufficient to answer the question.

5.3 Experiments

Experiments are a powerful but considerably under-used tool in conservation and considerably better than drawing conclusions from observed patterns. It is very useful to relate observed population changes to other factors, such as changes in salinity, predators or grazers but any correlation may be due to chance. Thus, a plant population may show a decline that correlates with an increase in the rabbit population, which may seem good evidence that the rabbits caused the decline. However, many other factors are also likely to have increased or decreased in that period and thus would also show a correlation. A second problem is that the correlation may be entirely due to a linked third factor, thus high rainfall may increase plant abundance, but reduce the rabbits. The implication of the observed correlation is that reducing the rabbits will benefit the plants but this is untrue if the correlation is due to coincidental trends or a linked factor. Experiments are a powerful technique in keeping everything constant except the factor of interest. Thus an experiment to exclude rabbits will reveal whether or not they have an impact on the plants.

An experiment should have a control, be replicated, randomised and monitored (see Section 5.16). Many use the term 'experiment' when they mean 'trial'. If it is just a change in management then it is not an experiment.

5.4 Hygienic fieldwork

In visiting a range of sites or individuals there is a real risk that diseases or unwanted species can be spread, thus increasing the conservation problems (see Section 11.3). For example, population declines of frogs in Panama and Queensland, Australia, have been linked to the same chytrid fungus which has also been identified from captive amphibians in zoos and aquaria in Australia and the USA (Berger et al. 1998). There is the worrying possibility that herpetologists are spreading the disease (Halliday 1998).

These measures outlined below may seem extreme and many conservation biologists will dismiss them. However, it is clear that many fieldworkers are far too lax about hygiene and there is a real risk of causing more harm than good. Not only do conservation biologists visit particularly sensitive habitats and species but they also are more likely to cause problems by activities such as touching rare species, taking equipment into the water or digging into the soil.

- Minimise the amount of materials taken from a site.
- Remove mud, seeds, vegetation and water from equipment, boots and vehicles before leaving. Rinse with boiled or treated water.
- Before reusing, clean the equipment and if necessary disinfect with bleach or 70% ethanol and rinse with sterilised water, especially those items likely to have close contact with soil (e.g. pegs), animals (e.g. traps) or plants (e.g. secateurs). Nets can be washed within a mesh bag using bleach on the delicate cycle of a washing machine.
- If there is a particular problem or risk then wear a new pair of disposable gloves for handling each individual. Even handling bird eggs can add bacteria which can pass through the shell to the developing embryo.

• Allocate one set of equipment for each site and clean and store separately.

• If the species could catch human diseases then consider wearing a face mask. Keep contact to a minimum. Do not eat on the site and especially do not leave partially eaten food.

• If it is thought that some areas are infected with a pathogen or pest species then start at the least infected site and finish at the most infected site.

5.5 Determining habitat use

The most useful techniques for determining habitat use are to compare areas where the species occurs with those where it is absent, the areas where it occurs with random locations or relate abundance to environmental conditions. This should take into consideration current knowledge. For example, if it is already definitely known that the species is restricted to calcareous grasslands at altitudes below 500 m then restrict the comparison to this habitat. This can be carried out on a range of scales (Sutherland & Crockford 1993), for example compare the sites where a species is present with those where it is absent, the patches within the site in which it occurs with those in which it does not and even the exact locations within a patch where it feeds or breeds with the rest of the patch.

Habitat use can be assessed using Jacobs (1974) preference index, D.

$$D = \frac{r - p}{r + p - 2rp}$$

where r is the proportional use of a habitat by the study species and p is the proportion in the environment. Both r and p thus have values between 0 and 1 while D has values between -1 (always ignored) and $+1$ (restricted to that habitat).

Habitat use, as shown by such methods, does not necessarily mean that the habitat is preferred, just that it is the habitat in which it has been able to persist. There are numerous examples of species restricted to suboptimal habitat where survival and breeding is reduced. For example, Takahe *Porhyrio mantelli* and Nene *Branta sandvicensis* (Black 1995, Clout & Craig 1995) were only known from the uplands (of New Zealand and Hawaii respectively), but once the source of persecution was removed have been shown to survive better in the lowlands.

5.6 Radio tracking

This is an invaluable technique for studying species which are active at night or live in habitat which makes them difficult to see. It is then possible to study habitat choice, movement patterns and home range. Radio tracking also makes it easier to find the individual and collect observational data. It also enables breeding and roosting sites to be located, often in a less biased manner than checking suitable habitat. For a little extra weight and size, sensors can be added in which the pulse changes according to activity, orientation or light intensity. Even without these sensors it is also usually possible to tell if an individual is active by shifts in signal strength as it moves. Mortality sensors, recording a lack of movement over a given time period, are useful for studying the cause of death as they distinguish dead from temporarily inactive animals thus making it easier to find corpses soon after death. Radios can be the only practical way to get reliable data on survival of rare species, especially during releases, and give the opportunity to relate that survival to movements and habitat use. Similarly, tracking by satellite can give unique insights into long distance movements of larger species.

Radio tracking is not a panacea. It is expensive and time consuming. The animals may be difficult to catch, the equipment may break and it may take time to collect sufficient accurate data. Furthermore, radio tracking usually provides data on only a few individuals (assume at least a quarter of the individuals will die, disperse, lose their radios or the radios will fail) and when the data is then divided into say age or sex classes the sample sizes can become very small. It is then difficult to know whether the behaviour of, say, the few juvenile males tracked are typical of all such individuals.

There is a wide range of means for attaching radios. For birds, if the tail feathers are strong, tail mounts are good for short-duration tracking, for example after releases. However, necklaces or

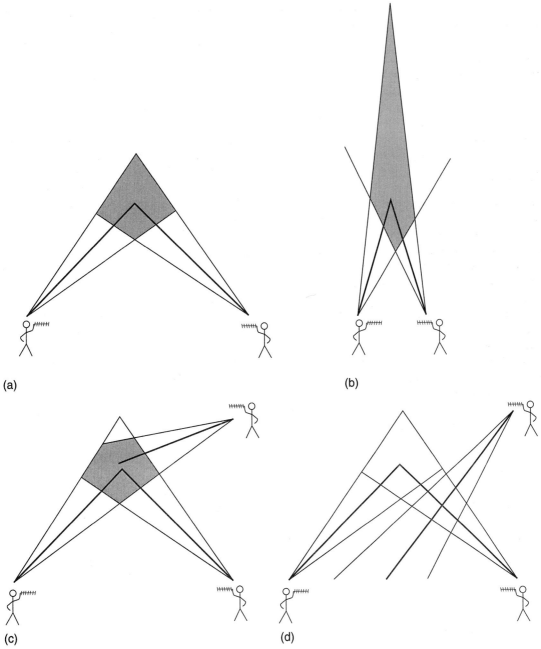

(a)

(b)

(c)

(d)

Fig. 5.1 Locating radio-tracked animals from signal directions. The lines are drawn from the signal directions from two locations. The area of error is shown, assuming the direction is accurate to within 10°. Note that the accuracy is much higher if the directions are roughly perpendicular (a) rather than almost parallel (b). A third direction can be used to confirm the accuracy (c). If the directions do not coincide (d) it shows at least one of the directions is inaccurate or the animal has moved during the measurements.

backpacks are necessary if tracking for several years but they need to be attached very carefully. Collars are usually used for mammals but should not be fitted to juveniles unless they are designed to allow growth. Other techniques include gluing with epoxy resins onto seals, wiring to pangolin scutes or attaching to horns or spines. The scutes of large reptiles may be drilled and the tag wired on. Small reptiles are more difficult to tag. The only practical solution for snakes, amphibia and fish is surgical implants which require expert veterinary advice. Implants need to be coated in a physiocompatible compound such as paraffin wax or non-acetic silicone.

It is also crucial to prepare all stages of data collection and analysis extremely carefully, including, if possible, a pilot study to maximise efficiency in the field (Harris *et al.* 1990). Analysis can become a nightmare through collecting inappropriate data for analyses of home range and sociality. Many projects have wasted a great deal of time gathering data without producing useful conclusions.

Radio tracking is easier from a height. For example, raising the antenna 4 m above a flat surface can double the reception range (Kenwood 2000). Height is particularly important for detecting a tag underwater as all radiowaves greater than 6° from the vertical will be reflected.

Interpreting the signal is often less straightforward than it initially seems. The signal direction can be distorted or reflected by hills and trees while the signal strength depends not only on how far away the animal is but also whether the signal is blocked by vegetation or rocks. It is thus very useful to know the study area well and how signals are modified there. Starting off by radio tracking with someone experienced is invaluable. Trying to locate transmitters placed by a colleague is a useful insight.

There are two main approaches to locating the animal. The first is to simply follow the signal until the animal can be seen; for nocturnal species this may entail using an image intensifier. The second is to use the signals to map where each individual is. This entails gathering the signal direction from two locations, drawing these on a map and assuming the animal is at the point where the lines cross.

The accuracy is highest if the two lines are approximately perpendicular (Fig. 5.1). The accuracy may be estimated by determining the error in detecting the direction of test transmitters placed by someone else and using this to draw an error polygon around the estimate (Litvaitis *et al.* 1986). If the animal moves rapidly then triangulation estimates from a single observer will be highly inaccurate. An alternative is to have pairs of observers recording directions simultaneously at preset times or ideally in contact using short wave radio. If locations are mapped immediately then any uncertainty (for example the animal appearing to be in unsuitable habitat) can then be resolved. If tracking at night, it is very useful to become familiar with the locations of patches of suitable habitat during the day as the signals are easier to interpret. Kenwood (2000) is a useful guide to radio tracking. White & Garrott (1990), Samuel & Fuller (1994) and Kenwood (2000) review data analysis methods.

5.7 Diet analysis

Determining the diet is often central to understanding an animal's habitat requirements. The usual technique is a combination of observation determining the prey present and faeces/pellet/dropping analysis.

It is often not straightforward immediately to identify the prey being eaten. Observations are particularly useful when prey diversity is low. When combined with sampling the habitat, and thus knowing which species are present, it is often possible to deduce what they are eating by considering what each possible prey would look like when being eaten. There are obvious biases, such as that individuals may be conspicuous when feeding on certain prey such as fruiting trees. Another problem is that only certain prey types can be identified so it is thus important to note the number of prey unidentified or identifiable to category (e.g. 'benthic invertebrate – not bivalve').

The traditional method of killing the animal and studying the stomach contents is now rarely acceptable, especially for rare species. Individuals may however be found dead, providing a good opportunity to analyse the stomach contents. These can be preserved in 70% alcohol although pigments may be lost in alcohol which can hinder identification. Alcohol can be injected into the digestive system to reduce digestion if there is a delay before the stomach can be removed. The whole animal can also be immersed in alcohol. Freezing can also be used but the stomach contents may disintegrate.

The faeces of different mammal species often differ sufficiently to be identified confidently. Pellets are regurgitated prey remains produced by birds consisting of hard parts such as bones or shells. Birds produce faeces and urine together which are usually referred to as droppings. The droppings of different bird species tend to be similar so it is often necessary to watch them and collect the droppings produced. One approach is to have one person watching the bird with a telescope on a tripod who after seeing a dropping produced keeps watching the spot and directs a second person using hand signals or handheld radios. Droppings and pellets may also be collected from nest sites or single-species roosts. Adult birds may be watched carrying away the faecal sacs from the nests and these may be located.

Pellets are usually stored dry while faeces and droppings can be preserved in 60–80% alcohol (higher strengths make the remains brittle). Adding water prior to examination softens the remains so they can be teased apart. They can be examined using a magnifying glass or dissecting microscope. Faeces may be preserved in formal acetic acid (FAA), then macerated in FAA, washed in fresh FAA and mounted on microscope slides and examined at ×100 or ×400.

It is necessary to create a reference collection of the likely prey items. Each faeces, dropping or pellet may contain the remains from more than one prey individual or only part of one individual so it is necessary to decide which items from each species in the diet should be counted to quantify the diet. Faeces, pellets or droppings should be analysed one at a time so the variation between them can be determined. The prey species are then recorded as the minimum number of individuals; thus three mandibles = 1.5 individuals. For counting abundant remains, such as earthworm chaetae, it may be useful to analyse a random subset of the sample. The size of the subsamples may be assessed by weight or by sampling a random section of a Petri dish. For plant fragments it is best to use a grid on the Petri dish to sample the species at each intersection. Diet can be analysed using Jacobs' preference index (Section 5.5).

The diet of vertebrate and invertebrate herbivores can be assessed through comparison of faecal fragments with identified epidermal fragments obtained by physical or chemical damage (e.g. Stewart 1967). One approach to assembling such a reference collection of fragments is to pass potential food species of the target species through slugs (Jennings 1979). Provide slugs with carrots (or another readily identifiable food) to empty the digestive system of other species and orange faeces will be produced within 24 hours. Plant species that could be within the diet are then saturated with distilled water, cut into small sections and placed on damp filter paper in Petri dishes, one species per dish. Add a slug to each dish. The slug faeces containing the plant being tested can be readily distinguished from those containing carrot (or filter paper!). The faeces can be fixed in 70% alcohol and then temporarily mounted, e.g. with gum choral, or dyed in an alcoholic solution of gentian violet or safranin, dehydrated using increasing concentrations of alcohol and then mounted using a permanent mountant. Those plants with hard exteriors are better analysed by scraping off fragments with a sharp scalpel (Metcalfe 1960) rather than the slug method.

The size of prey items taken can often be assessed by firstly determining the relationship between fragment size (e.g. mandible length) and body size for a number of prey and then converting the measured fragments to prey size.

Some remains may be digested or too fragmented to recognise. It is useful to know the digestibility of the selected prey fragments. If there is a captive or tame animal available for study then the usual technique is to feed it known numbers of each prey species (or biomass if using plants) and measure the abundance of fragments in the faeces. It is then possible to correct for bias due to differential digestibility. To assess the digestibility of different sized prey, carry out a series of experiments each providing a restricted size class of prey. Individuals raised on a highly artificial diet may, however, have markedly different digestive efficiencies from those in the wild.

5.8 Ageing and sexing

There are three main methods for ageing individuals.

1 Examine tissues that have recognisable annual layers. This method only works for species with annual cycles of fast and slow growth and thus often does not work in the tropics. It is important to remember that any stressful change is also likely to produce a growth ring, although these are often less pronounced. It is thus sensible to test this technique using individuals of known age.

2 Adults and juveniles may differ in morphology or pattern. Subtle differences are often detected by studying which features of young individuals are retained. This method is not always as straightforward as it seems. It is well accepted that different year classes of Herring Gulls *Larus argentatus* can be distinguished by plumage but Monaghan & Duncan (1979) examined the plumage of known aged birds, which had been marked as chicks, and some showed the supposedly incorrect plumage for their age.

3 Variation in size. Those species which continue growing through life can often be divided into size classes and assigned to ages. Lengths can be analysed by plotted on graph paper to reveal age cohorts or sex differences or, preferably, by using specialised packages such as those designed for fisheries data. It is necessary to consider whether more than one size cohort may be produced a year, whether sexes differ in size and whether there may only be cohorts from certain years as breeding, recruitment or survival may be irregular. It is easier to use sizes if the breeding season is short (as in seasonal climates) as the year classes are then more distinct. Once the within-year variation equals the growth between years this method no longer works. It is thus usually only suitable for the first few years and so may be of limited use for long-lived species. It is advisable to check the accuracy, for example by remeasuring

marked individuals, by following the length classes over a year to confirm they grow at the rate expected or by ageing some individuals using another technique, such as growth rings.

It can be useful to assign individuals to size categories even if it is not known how old each is. For example, baboons have been classified into adult male, adult female, subadult male, two juvenile stages and two infant stages (National Research Council Committee on Non-human Primates 1981). It is important to have a clear definition of each category. With experience of known aged individuals it may then be possible to state the usual age of each category although there will, of course, still be individual variation and variation depending on environmental conditions. Similarly plants may be categorised by height, diameter or number of leaves.

Sexing is obviously possible by examining the reproductive structures although in many species the differences will be internal. One conclusion from sexual selection theory is that the sex that spends the least effort (Trivers 1972) or time (Sutherland 1987) on producing and caring for the young will be the sex that competes most to gain further matings. Females produce the offspring and often undertake a disproportionate share in caring for the young. Selection then favours those males able to gain more matings through being better fighters, more attractive, more brightly coloured or better singers. Many species can be sexed by such morphological and behavioural differences. In some other species, such as the Bronze-winged Jacana *Metopidius indicus* the males undertake more parental care and it is the females that are larger and more aggressive (Butchart *et al.* 1999). Finally, and obviously, the males are usually on top when mating, but with some exceptions, for example, in many grebe species the females mount the males in a manner difficult to distinguish from actual copulation (del Hoyo *et al.* 1992).

5.8.1 Ageing plants

Young trees and shrubs with seasonal growth can be aged by counting the scars from the terminal buds. Most conifers and some other species produce a set of branches at the top of the trunk each year, making ageing young individuals simple. For older individuals there may be few signs of branches produced when very young at the base of the trunk so the number of years growth here has to be estimated. Very old individuals in which many sets of branches have disappeared, or those that have lost the top of the main stem, cannot be aged.

Trees in seasonal climates can be aged by annual rings. Ageing is more accurate using wedges or cross-sections (± 1 year) than cores (± 10 years) (Zackrisson 1977). However the great advantage of cores is that they may be extracted from living trees, usually without harming them, using a borer with a sharp cutting edge (Stokes & Smiley 1996). In the Swedish increment borer, which is better for large trees or hard wood, the narrow borer is surrounded by a screw thread which draws the borer into the tree. The borer is directed towards the pith near the centre, bearing in mind that for trees on steep slopes there is more growth on the downhill side. Cores are more accurate if taken near to the ground (Peterken 1996). The core is removed using an extractor spoon. The core is fragile and may be stored in folded paper in an envelope or corrugated cardboard. The pith looks like a vertical band. Rings can often be counted in the field. Otherwise air dry for a few days, glue into a slot in a block of wood and either sand down or cut with a razor and count bands with a microscope. Accuracy can be improved by preparing cores from 10 to 20 mature trees to create a baseline chronology against which cores from individual trees can be checked to see if some rings appear missing (Agee 1993). When a nearby tree dies then trees may show a marked growth pulse for a couple of years so allowing the year of the neighbour's death to be determined.

The relationship between diameter (usually at 1.3 m height) and age can then be used to roughly estimate the age of the other trees, but this will obviously vary with growing conditions. Diameter is determined by dividing the circumference by π (3.14) or reading from a forester's tape in which the conversion has already been done. Some very long-lived tropical species without rings have been aged by carbon dating of the central wood.

Heights can be estimated from a measured distance by determining the angle to the top (ϕ_{top}) and bottom (ϕ_{bottom}) of a tree using a clinometer and calculating:

$$\text{height} = \text{distance} \cdot \tan \phi_{top}$$
$$+ \text{distance} \cdot \tan \phi_{bottom}$$

With this method it is not necessary to be level with the tree. Cycads and palms have no secondary thickening, thus width gives no indication of age. Height or the number of leaf scars can be used instead. Cycads and palms grow laterally in the ground to gain sufficient width before vertical growth; this process can take a decade and needs to be added to the estimate of age.

Perennial herbaceous species may increase in size with age but are usually difficult to age with any accuracy. Some species with underground crowns may have annual rings on the upper surface. Some species with rhizomes may show distinct annual bulges in growth and it is even possible in some species, such as Yellow Flag *Iris pseudocorus* to distinguish flowering, branching or the number of leaves produced per year, sometimes for 10 years previously (Sutherland & Walton 1990).

It is easy to distinguish between monopodial species, in which growth usually continues through the apical bud with flowers developing from lateral buds, from sympodial species, in which the apical bud becomes the flower and growth continues from a lateral bud. Thus for sympodial species with annual flowering, each curved lateral branch is a year's growth.

5.8.2 Ageing and sexing invertebrates

Some invertebrates, and especially crustacea, continue growing for ever during a series of moults while others, especially insects, do not. Most arachnids do not moult as adults, but do swell markedly the more they eat or the longer they live, as the exoskeleton is often less sclerotised than in insects or crustacea. Many exopterygote insects and other arthropods have a very strong relationship between age and size, with a progression factor of about 1.25 from one moult to the next. Some molluscs in seasonal environments can be aged by their annual growth rings. Only adult earthworms have a ctilellum ('saddle'). Many species of centipede and millipedes accumulate extra eyes during moults.

Adult insects may look increasingly tatty as they age, especially butterflies and other big flyers, who may frequently have bite marks in their wings. Newly matured adult insects often have a clean waxy or shiny look whereas older ones look rather dull.

A general rule, with many exceptions, is that female invertebrates are larger than males. As a result of sexual selection males may possess weapons used in fights such as horns or big claws. Only males have claspers or hair pads on the tarsi to hold the mate. The presence of an ovipositor makes it female. In most moths the males have longer antennae while Hymenoptera can often be sexed by counting the antennae segments with the males having more. Most insects can be sexed by the shape of abdomen or visible external genitalia. Lepidoptera can usually be sexed by the shape of the abdomen tip: females typically taper to an ovipositor while males usually have a blunt tip which may possess hooks or spines. Most male spiders have enlarged palps as if wearing boxing gloves.

5.8.3 Ageing and sexing fish

Fish often show annual cycles of growth which can be counted from various body parts. Scales are often the easiest to read. Scales can be removed from dead fish using the back of a scalpel. For live fish use tweezers to remove two or three from each side below the lateral line and behind the pectoral fin. Wash hands and instruments after dealing with each fish or scales easily become attributed to the wrong individual. Look at the scales under a microscope and count annual rings. Scales can be lost and regrown which is why it is necessary to count more than one. Any stressful change is likely to produce an extra growth ring.

Otoliths are often used as, unlike scales, they cannot be replaced and thus are more accurate but they cannot be collected from live fish. For small fish cut off the top of the head just above the eye. There are usually six otoliths and it is conventional to use the largest, the sagittal otolith, situated at the base of the spine. For larger fish split the head down the middle. Clean and fix the otolith in clear nail polish which can be dissolved later in either alcohol or nail polish remover. It may be necessary to cut and polish the otolith to read the rings. Elasmobranchs do not have otoliths but many can be aged by sectioning the vertebrae and counting the rings.

Fish are conventionally classified as: egg (before fertilisation), embryo (after fertilisation), larva (once feeds independently), juvenile (once fins fully differentiated) and adult (starts with the maturation of gametes) (Balon 1975). The term fry should only be used in the kitchen.

Fish are usually difficult to sex. Some may show external differences, with males tending to be more colourful and females tending to be larger. Most species can only be sexed by dissection, and usually only after they become sexually mature. The ovary is much larger than the testes and developing eggs can usually be seen even in fish that have just spawned. Many fish change sex from male to female or vice versa with age and social status.

5.8.4 Ageing and sexing amphibians

Amphibians may show both age and sex differences in size, which is conventionally measured from the snout to vent, but there is considerable variation in growth rate which makes this technique fairly inaccurate (Halliday & Verrell 1988). Amphibians in seasonal environments can be aged by sectioning bones (e.g. toes) and counting the rings.

In some amphibians in which the males clasp the females during mating and mate guarding, the males may have dark rough 'nuptial pads' along the thumbs and inner fingers of the fore feet during the breeding season. Only male frogs and toads call. Sexing urodels is generally very difficult but in a few species of salamander and newt the males in the breeding season are more brightly coloured or have a larger crest. Adult newts can often be sexed in the breeding season by the cloaca which is darkly pigmented and bulbous in males but smaller and less conspicuous in females (Fig. 5.2); the difference is more subtle in salamanders.

Male

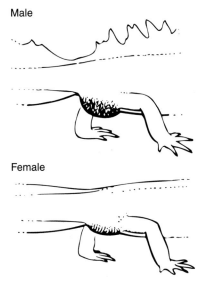

Female

Fig. 5.2 The cloacal region of male newts (above) is often darker and more protruding than in females (below) (from Buckley & Inns 1998).

5.8.5 Ageing and sexing reptiles

Reptiles are usually aged on the basis of size by measuring snout to vent length but this generally only works for the very youngest cohorts. The plates of the shell (scutes) of turtles and tortoises have growth rings in seasonal environments which give a reasonably accurate measure of age. Many fully aquatic species shed their scutes periodically so that the earlier growth rings cannot be read. Skeletochronology, sectioning toes and counting

the annual rings may work, but is currently insufficiently researched to used with confidence. They are also ethical and conservation implications of removing toes.

Many reptiles can be sexed in the breeding season by the greater activity of males, and this can then determine colour or morphological differences. Male lizards are often more brightly coloured. Many species can be sexed by the male's hemipenis which may be everted by pressing on each side of the vent. In some lizards (anolis, scolopids) the male has two large post-vent scales which the females lack. Many male lizards have a penial bulge, a swelling at the base of the tail and can often be sexed using a hand lens by the femoral pores on the posterior row of scales on the underside of the thigh (Fig. 5.3). Male turtles and tortoises typically have thicker, longer tails with a vent further back and the underneath, the plastron, of males is often concave to ease copulation. The only crocodilian that can be readily sexed is the Gharial *Gavialis gangeticus* (mature males have a protuberance at the end of the snout).

Female snakes tend to have a distinctly wider body narrowing abruptly to a narrow tail, while males often have a penial bulge (Fig. 5.4). Male snakes tend to have longer tails and in the hand the tail length (tail tip to vent) as a fraction of the total length is often used. A more precise distinction is that in most species males have more sub-caudal scales between the vent and tail tip (count the number of scales running along the shortest line between the two), although there may be some overlap. Dead specimens can be sexed if adult, although this takes

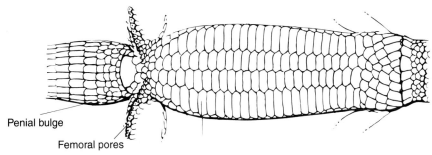

Penial bulge

Femoral pores

Fig. 5.3 In many lizard species the male has a penial bulge at the base of the tail and femoral pores in the scales along the underside of the thigh (usually has to be confirmed with a hand lens) (from Buckley & Inns 1998).

Male

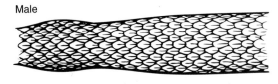

Female

Fig. 5.4 The underside of many male snakes (above) tapers gently from the body to tail but is interrupted by the penial bulge. In females (below) there is usually more abrupt tapering (from Buckley & Inns 1998).

some practice. The testes are obvious, spherical, white or yellow and positioned about two-thirds of the way down the body. They are joined to thick, tightly coiled efferent ducts (vas deferens) which are white from the sperm inside. The ovaries are large, with obvious elongate follicles if the female is reproducing.

5.8.6 Ageing and sexing birds

Adult birds are usually the same size as juveniles. Juveniles often have a distinct more buff-coloured plumage. In many species the first moult may always be incomplete, thus certain juvenile feathers are consistently retained for the first year. This is often most conspicuous in the coverts (the small feathers overlapping the flight feathers on the wing) and especially the greater coverts. This method is easiest in those species in which juveniles consistently moult some of the inner greater coverts but not the outer ones and thus show contrast. Beware of the pitfall that the adult plumage of some species includes this very contrast (Svensson 1992).

The structure of the plumage may differ between adults and juveniles. Juveniles often grow their underwing coverts last resulting in bare 'armpits'. Each period of poor conditions during feather growth results in a bar on the feather (a 'fault

bar'). As birds in the nest grow all tail feathers concurrently these fault bars will line up across feathers, while they are staggered in adults as these moult their tail in stages (see Fig. 5.5). Juveniles tend to have more pointed tail (see Fig. 5.5) and primary feathers. The growing tips of juvenile feathers often become broken, resulting in a castellated appearance.

Juvenile passerine birds fledge with a skull consisting of a single layer of bone while a second layer grows over a period of usually 3–7 months. The extent of skull ossification can thus be used for distinguishing adults from juveniles. Wet a small patch on the back of the skull and part the feathers. Gently make the skin taut in this patch using thumb and index finger. In bright light examine for the contrasting edge of the ossification area. Bill shape may differ between adults and juveniles with juveniles sometimes having a less pronounced hook on the upper mandible (Fig. 5.5).

Birds can often be sexed by plumage differences with the sex that incubates most (usually the female) being the most inconspicuous. During the breeding season they can also be sexed by cloacal protuberances, with the cloaca being more steep sided in the male while tapering in the female towards the vent (Svensson 1992), and by the presence of a wrinkly brood patch (hold the bird upside-down with the head away from you and blow through the legs against the feathers on the belly); if both brood the eggs then both sexes may have a bare belly, but only the female will be wrinkly. In many species the sexes differ in size and can then be sexed by measuring bill length or wing length. The males tend to be larger but the reverse is true in some groups such as waders and raptors.

Birds (except ratites, such as ostriches) may be sexed by a chromo-helicase-DNA binding gene (CHD) linked to the W chromosome (females possess a W sex chromosome while males are ZZ) (Griffiths *et al.* 1996) but this is by far the most expensive and elaborate technique (Ellegren & Sheldon 1997). This gene can be amplified by polymerase chain reaction primers (Section 5.15.1), enabling small samples, such as a feather or faeces, to be used. Even embryos can be sexed with this

Juvenile

Adult

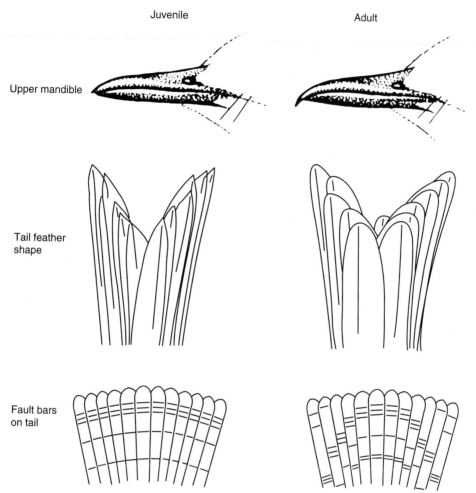

Upper mandible

Tail feather
shape

Fault bars
on tail

Fig. 5.5 Means of ageing birds. Juvenile left and adult right. Possible differences include whether there is a hook on the upper mandible, whether tail feathers are pointed or round and whether the growth bars on the tail feathers are synchronous (from Svensson 1992).

technique. This technique was used to show that the last wild Spix's Macaw *Cyanopsitta spixii* was a male using DNA from a moulted feather (Griffiths & Tiwari 1995). A female from captivity was subsequently released. It was also used to indicate that the critically endangered Taita Thrush *Turdus helleri* has a highly skewed male-biased sex ratio in the most disturbed site (Lens *et al.* 1998). Kakapo *Strigops habroptilus* have been sexed from DNA extracted from faeces and then amplified (Robertson *et al.* 1999).

Corpses can be sexed by cutting open the body cavity and displacing the intestines. The gonads are above the kidneys. Males have two oval testes while females have a single irregularly lobed ovary on the left side. The adrenal glands are in a similar location but are yellowish.

5.8.7 Ageing and sexing mammals

Many mammals can be aged in seasonal areas by counting the number of growth rings 'annuli' in the teeth (Pletscher 1995). Large teeth can be ground longitudinally, polished and the annuli counted. For smaller teeth remove the tooth and root from the socket, decalcify in a weak acid solution, rinse to remove acid, cut off the root at the gum line and cool or freeze before cutting a 10–30 micron longitudinal section with a microtone. After staining, e.g. with a metachromatic stain (Thomas 1977), count the annuli. This is often clearest at the root tip. For ruminants use an incisor, while for carnivores use a canine or premolar. Wet eye lens weight can provide a good measure of age but the relationship between lens weight and age of known individuals needs to be determined (Mallory *et al.* 1981).

Sexing is straightforward in those species in which the penis, testes or nipples are obvious, although males of most species have vestigial nipples and male fruit bats may even produce some milk (Francis *et al.* 1994). In marsupials only the females have nipples. In very many species the testes are often not visible externally, or only for the breeding season. In many taxa the distance between the genital opening and anus is greater in males than females. Cetaceans may be sexed by the number of ventral slits in the inguinal surface: females have three (two containing mammary glands and one genital vent) while males have one (Kunz *et al.* 1996). Internally, male mammals have a pair of testes near the penis (often held within a scrotum) and connected by the vas deferens while females have a pair of usually much smaller ovaries next to the kidneys and connected to the vagina by a uterus.

Mammals are usually polygynous, with fights between males. As a result, males tend to have features such as a broader head, stronger head musculature, and larger canines, incisors, horns or antlers. Scent glands, such as the anal glands or antorbital glands, are often conspicuously larger in males.

5.9 Pollination biology

5.9.1 Determining the breeding system

The first stage in studying the breeding system of a plant is to determine whether stamens and stigmas are present in each flower or whether there is another combination. The female part, or stigma, is typically in the flower centre and results in seeds or fruit. The male part, or stamen, consists of a pollen-producing anther on a stalk (filament). Most species are bisexual in bearing both sexes in one flower. Monoecious species have separate male and female flowers on the same individual. Dioecious species have male and female individuals. Andromonoecious species have male and bisexual flowers on the same individual, although in a number of species these are actually dioecious as the hemaphrodites have non-functional pollen (Anderson & Symon 1989, Schlessman *et al.* 1990, Mayer & Charlesworth 1991). Gynodioecious species have female and bisexual individuals. Kearns & Inouye (1993) give a comprehensive review of techniques for all aspects of pollination biology.

The usual next stage is to consider the extent of self-versus cross-fertilisation. There are four main possibilities.

Obligate outcross. Self-fertilisation cannot occur. This obviously includes all species that are genetically self-incompatible and those that are dioecious (separate male and female individuals).

Facultative outcross. Both self- and cross-pollination may occur.

Obligate selfcross. Cross-pollination cannot occur. Cleistogamous flowers are those that do not open and are always self-pollinated.

Apomictics. Seed production takes place without sexual reproduction. Apomictic species typically possess nectar, pollen and colourful petals such that they cannot be distinguished from sexually reproducing relatives.

Some indication of the breeding system can be obtained from examining the flower structure (Table 5.1). As shown in Table 5.2, these breeding systems can be distinguished by studying the seed production of plants that have been bagged, emasculated and artificially pollinated in various combinations.

The importance of cross-pollination is usually studied by comparing typical levels of seed set with plants from which pollinators are excluded. Visitation by insects can be prevented by

Table 5.1 Characteristics that suggest whether a plant species is predominantly selfing or outcrossing.

Predominantly selfing	Predominately outcrossing
Small inconspicuous flowers	Large showy flowers
Anthers and stigma in close proximity	Anthers and stigma spatially separated
Anthers shed pollen after the stigma becomes receptive	Anthers shed pollen before the stigma becomes receptive
No heterostyly	Heterostyly (two or more flower types differing in the location of anthers and styles)

Table 5.2 Tests to determine the breeding system of plants. Ticks indicate seed production.

Treatment	Breeding system			
	Obligate outcross	Facultative outcross	Obligate selfcross	Apomictic
Cross pollination allowed	✓	✓	✓	✓
Cross pollination allowed. Emasculated	✓	✓	✗	✓
Cross pollination prevented	✗	✓	✓	✓
Cross pollination prevented. Emasculated	✗	✗	✗	✓
Cross pollination prevented. Self-pollinated	✗	✓	✓	✓

surrounding with mesh, such as mosquito netting or bridal veil. Synthetic fabrics retain less water and thus are not as heavy after rain. Chicken wire excludes birds and bats but not insects. Plastic or paper prevents wind pollination but may also affect the microclimate. Whether exclosures result in microclimate changes which affect seed production can be tested by hand pollinating some flowers and measuring seed set.

Emasculation is carried out by removing the anthers before they can produce pollen. Emasculating all the flowers on an individual (or bagging or removing flowers that are not emasculated) prevents self-pollination.

Artificial pollination has been carried out by a wide range of techniques (Kearns & Inouye 1993). Three basic methods are:
• Collect pollen (or a pollinia of orchids and milk-weeds) on a toothpick, needle, paintbrush or on tissue around forceps, and transfer to a receptive stigma. For transporting, suspend the implement through a lid of a pot.
• Collect pollen by tapping anthers over a Petri dish then apply pollen to stigma.

• Remove entire anther or flower and rub anthers over receptive stigmas. This is easier with locking forceps.

The role of pollinators in limiting the seed set can be assessed by artificially cross-pollinating some randomly selected individuals while still allowing natural pollination and comparing their seed production with other individuals that had just natural pollination. If the addition of artificial pollination increases seed production and natural seed production is low this indicates that consideration should be given to enhancing the pollinator population. Thus studies of a threatened species of mistletoe *P. tetrapetala* showed that seed set was low, was improved by hand pollination and that natural visitation rates by bird pollinators were low, probably due to the recent decline in native birds (Robertson *et al*. 1999). For critically threatened populations artificial pollination might be justified until the pollinator population is higher.

There are two types of self-incompatibility. Gametophytic is the commonest system, in which the growing pollen grain is inhibited if it possesses the same allele as the plant on which it has landed

Table 5.3 Differences between gametophytic and sporophytic self-incompatibility (based on information in Richards 1999).

Trait	Gametophytic	Sporophytic
Self-incompatibility determined by genetic composition of:	Pollen	Parent that produced pollen
Stigma	Usually wet papillae	Usually dry and covered by cuticle
Usual number of nucleus in pollen grain	Two	Three
Pollen viability	Usually long	Usually short
Outcome of cross between two parents	Partly compatible	Always either fully compatible or incompatible

(all grasses are gametophytic). As plants must be heterozygous for incompatibility, some pollen may be incompatible while other pollen from the same plant is compatible. In sporophytic incompatibility the response is to the proteins on the pollen coat derived from the anther and thus is identical for all pollen produced by one individual (Richards 1999). Sporophytic incompatibility occurs in 13 families including probably all Asteraceae and Brassicaceae (Richards 1999). Table 5.3 describes how to distinguish these two types.

Incompatibility is often weak before the flower opens and horticulturists frequently pollinate immature stigma in the bud to overcome incompatibility (Lawrence 1968). Treating the pollen with acetone to remove the surface proteins can overcome sporophytic incompatibility. Incompatibility has also been overcome by increasing humidity (Carter & McNeilly 1975), heat (Leduc *et al.* 1990) or carbon dioxide (O'Neill *et al.* 1988).

5.9.2 Identifying the pollinators

The main type of pollinator can often be determined by examining the flower (Proctor *et al.* 1996).

Moth pollinated. Usually spurred, white and with strong sweet scent at night.

Bird pollinated. Often sturdy, brightly coloured (often orange or red) with abundant dilute nectar and no scent.

Bat pollinated. Exposed, strong, dull-coloured flowers with abundant easily accessible nectar and pollen and an unpleasant musty smell. They are usually trees or woody climbers.

Fly pollinated. Usually flat or bowl-shaped with readily accessible nectar.

Butterfly pollinated. Usually brightly coloured (red, blue or yellow), sweet scented and spurred.

Beetle pollinated. Usually dull-coloured, flat, often with fruity smell.

Bee pollinated. Difficult to distinguish as they vary from being small and inconspicuous to large and showy.

Wind pollinated. Usually have large, exposed stamens which release pollen if tapped and the stigma is often large, exposed and finely divided to capture passing pollen. They are usually small, inconspicuous and odourless flowers. Pollen grains of wind-pollinated species tend to be smooth and dry while animal-pollinated ones are sticky and often sculptured. All conifers, grasses, sedges and rushes are wind pollinated.

Determining the pollinator, or pollinators, usually involves watching a group of flowers and recording the visitors. If the observations are carried out on a known number of open flowers for a measured period then this can be converted to visitation rate. Species may visit but not pollinate. It is necessary to determine that the pollen is transferred to the visitor. Exclusion (see Section 5.9.1) can also be used to determine the type of pollinators, such as using different meshes to see if the pollinator is large or small or applying exclosures at different times to see if the pollinator is diurnal or nocturnal.

Another technique for identifying the pollinators is to capture potential pollinators and look for pollen on their bodies. Pollen from different species often varies markedly in size, shape and the location and number of pores. Pollen can often be identified to species or genus under a light or electron microscope. It is thus necessary to compare the pollen collected from the animal with a reference collection.

For both the field sample of pollen and the reference collection, make a jelly by heating 50 g gelatine in 150 cm^3 water until dissolved (Kearns & Inouye 1993). Add 150 cm^3 glycerine while still warm. Add crystalline basic fushsin stain (also known as para-soniline) to produce a deep red solution and leave to solidify. It is often useful to make a range of intensities. Traditionally 5 g of crystalline phenol has been added as a preservative but it is highly poisonous. Cut about 2 mm^3 of the jelly and dab it onto the pollinator's body or anthers to collect the pollen. Place on a glass slide and warm gently so that the jelly melts but does not boil. Lower a coverslip on top. The sample can be sealed permanently by melting paraffin wax adjacent to the coverslip.

For larger species such as birds and mammals, pollen can sometimes be removed directly or by pressing clear sticky tape against the pollen.

5.10 Marking individuals

The main approaches for marking are: add numbered, lettered or coloured tags; paint or dye; clip parts; identify without capture by using natural variation in markings or structure.

One technique for marking individuals is to place coloured dots on up to ten unambiguous locations (for example five locations on each underside of a butterfly wing), allocating the colour according to a particular code, say red, for 100s, blue for 10s and yellow for 1s. Thus a red dot in position 4, blue in 9 and yellow in 1 is individual 491.

Before starting a scheme consider the ease of distinguishing individuals, bearing in mind that they may be observed from a distance or the marks may alter. It can be useful to test beforehand how marks disappear or change. In studies of dispersal, it is usually important to be confident that the occasional observation of an apparent long-distance disperser reflects a real movement rather than some ambiguity in the marks. To be confident of this it is often sensible to use different colours in different study sites.

An increasingly popular approach is to use Passive Integrated Transponder (PIT) tags, which are 2×12 mm electronic devices encapsulated in glass. The tags can be supplied in sterile needles ready for insertion by syringe either into the peritoneum or subcutaneously. The unique code of each tag is read with a hand-held scanner which can store information on date and time to download onto a computer. This technique requires a licence in many countries. This system is widely used for marking pets, and farm, zoo and laboratory animals and is becoming more widely used for ecological studies but the long-term effects on small species are poorly known.

The ethics of marking should be considered. Marking can cause suffering, mortality or otherwise reduce the fitness of the animal, for example by altering the attractiveness to the opposite sex, and it may also need licensing in some countries. Training is important to ensure the safety of the individuals being marked and is obligatory in many countries for marking groups such as birds. It is necessary to consider whether there are negative consequences of marking and if so whether the information to be gained will be sufficiently important to justify it. It can be useful to mark captive individuals to see if there are any obvious problems with the technique. If using an unconventional means of marking or if unsure that any marking is harmless then it is sensible to try to study its consequences in the field.

5.10.1 Marking plants

Plants can be marked by tying labels to a tree or shrub (ensuring it cannot constrict growth), placing a wire hoop round the base or by placing markers in the adjacent soil (if at sufficiently low density for this to be unambiguous). Coloured wire or small numbered tags (such as those used to mark small fish or those used by telephone engineers to mark wires) can be used for small plants. Polyethylene fades more rapidly than vinyl but survives for longer at cold temperatures (Kearns & Inouye 1993).

5.10.2 Marking invertebrates

Invertebrates with hard surfaces are usually marked by painting or writing dots, numbers or letters. Artists' oil paint is widely used and applied with a single bristle, a sharpened match stick or an entomological pin but felt tip pens can also be used. Numbered tags glued on have been used for a range of taxa. Insects can be marked with a lightweight diode that can be detected up to 20 m in the air and 10 m underwater (John Gee personal communication). With this method it is not possible to separate individuals but it may be used to locate individuals that also have unique marks. Any marks are obviously lost if moulting occurs. Sedentary aquatic invertebrates are often identified by mapping. Butterflies and moths can be marked by scraping off scales.

5.10.3 Marking fish

Use of dye marks, such as Alcian Blue, is an important technique. Fish can be marked using a Panjet needle-less injector (Hart & Pitcher 1969) in which the jet of dye penetrates the skin. There is a risk of damaging the organs of fish less than 15–20 cm long. Dye may also be injected beneath the skin with a syringe after anaesthetising. An alternative is to inject within fin rays. Batches of fish can be dyed by dipping a net full of fish in dye for a few seconds; this marking usually lasts for 2–3 weeks.

Individuals in small populations can often be recognised by differences due to scale loss, e.g. due to parasite injuries. This technique is widely used in studies of coral reef fish. Fin clipping and spine clipping are also sometimes used.

There are a range of tags such as the Petersen tag which goes through the forsal musculature. PIT tags (see Section 5.10) are becoming more widely used. Short (1–2 mm) lengths of insulation from electrical wires can be inserted onto fins and glued into position.

5.10.4 Marking amphibians

A range of techniques have been used but most have been tested on just one or a few species (Halliday 1996). Light-skinned species can be tattooed by injecting dye beneath the skin using an electric tattooer. This is cheap and can be applied to many individuals. The tattoos are visible for at least 3 years (Joly & Miaud 1989). Coloured fluorescent dust has been used to mark skin by applying with compressed air. The marks last for 1 or 2 years (Nishikawa & Service 1988).

Toes have been removed in unique combinations. The toes can also be used for ageing and DNA analysis. This is clearly unacceptable for arboreal species and is increasingly considered unacceptable for all species.

For studies lasting a few days or weeks, colour-coded or numbered elastic waistbands can be used so individuals can be identified from a distance (Emlen 1968). Stretchable thread can be used to tie a small tag to the knee (Elmberg 1989) of frogs and toads. PIT tags (see Section 5.10) can be used for marking a range of amphibian species (Camper & Dixon 1988, Fasola et al. 1993).

Some species have considerable natural variation in skin patterns and a catalogue can be built up of photographs (Hagström 1973). Salamanders and newts may be best placed in a glass tube or in a foam ridge within a transparent plastic box before photographing.

5.10.5 Marking reptiles

Reptiles are usually marked with paint or nail polish but there is evidence for high mortality from xylene-based paints (Boone & Laurie 1999). They can also be tattooed using a small tattoo gun. PIT tags (see Section 5.10) which are inserted under the skin on the side anterior to the cloaca or, for small individuals, in the abdominal cavity, are likely to be used increasingly (Camper & Dixon 1988, Reading & Davies 1996).

5.10.6 Marking birds

Birds are usually marked with either individually numbered metal leg rings (also known as bands) giving an address so the finder can report the details and usually produced as part of a national scheme, or with plastic coloured leg rings so that the individual can be recognised at a distance. Colours should be selected which can be easily distinguished. Some colours fade and then resemble others. Inexperienced observers often confuse right and left legs so it is preferable not to have symmetrically identical combinations. The frequency with which non-existent combinations are reported is useful for estimating the frequency and sources of errors and so should be documented.

Feathers may be dyed, for example with picric acid, to make individuals more obvious. Captive bred Lammergeirs *Gypaetus barbatus* have had primaries bleached to allow individual recognition once released into the wild.

Some species have sufficient individual variation that it is possible to recognise individuals. Thus the fifty or so Lesser White-fronted Geese *Anser erythropus* in Fennoscandinavia are monitored by the pattern of black markings on the belly.

5.10.7 Marking mammals

Small mammals have traditionally been identified by toe clipping but this is rapidly becoming unacceptable. Clipping patches of fur has the advantage that it regrows. Coloured or numbered ear tags are often used and can be read at a distance. Some large species are best dyed or, if lacking a hairy coat, painted with non-toxic paint.

Microchiropteran bats are usually marked with special lip-edged rings/bands in which the rings/bands are folded back so that they do not cut into the skin. They are placed over the shaft of the forearm leaving a small gap between the ring and membrane (Kunz 1996). This is unsuitable for species with a large propatagium (membrane in front of the forearm). Fully grown megachiropterans can be marked with monel define or stainless steel butt-end bands over the thumb or using stainless steel ball chains

acting as a necklace which include a numbered ring/band, fitting the chains tightly so they do not abrade the skin (Kunz 1996). Necklaces can also be used on microchiropterans.

Transponders (see Section 5.6) can be used for mammals which use regular sites such as bat boxes or small mammal boxes. The location of an individual can be determined without disturbing it.

If likely to have contact with bats or carnivores, it is recommended to be immunised against rabies.

5.11 Studying the fate of individuals

Studying the fate of individuals can be used to measure mortality, age of first reproduction and breeding output. The commonest approach is to have one or more study areas in which as many individuals as possible are studied.

5.11.1 Measuring breeding output

Nesting success used to be measured as the proportion of nests found that are eventually successful, but there are serious biases in this method. One fault is that, unless all nests are found from the beginning, this exaggerates success as nests which fail early are likely to be excluded. Conversely, if some nests are only followed for a period in the middle, then if they fail within this period they will be included but if successful they will be excluded as the final outcome is unknown. Percentage success is thus only suitable if every attempt and its outcome is known.

A good approach for incomplete data is to estimate daily survival rates using the Mayfield method (Mayfield 1961, 1975).

$$\text{Daily survival rates}, S = 1 - \frac{\text{number of failures}}{\text{total nest days}}$$

Thus if eight nests are watched for periods of 8, 11, 14, 3, 10, 11, 9 and 14 days, a total of 80 nest days, and four fail then

$$\text{daily survival rate}, S = 1 - \frac{4}{80} = 0.95$$

Survival rate over a period of J days $= S^J$. Thus if the period between first egg and the young leaving is

20 days then the fraction of successful nests $= S^J = 0.95^{20} = 0.36$. The estimate is that 36% of nests were successful. For birds the survival rate is often calculated separately for eggs and young as the rates may differ. These are then multiplied to give total breeding survival.

Only those nests visited at least twice are included in the Mayfield method. The total nest days is the total period that all nests are extant and documented. If a nest fails between observations, then it is standard to assume it fails midway in the calculation of nest days. Other methods (e.g. Johnson 1979) are more sophisticated but in practice it is accepted that the error from this assumption will usually be small (Johnson 1979, Beintema 1992). The other major assumption of the Mayfield method is that failure rate is constant throughout the period. Thus, if nests are particularly likely to fail in the first few days but nests tend to be found later, then the calculated success will be an overestimate.

Daily survival rates from two sites, seasons or years, S_1 and S_2 can be compared by calculating

$$Z = \frac{(S_1 - S_2)}{\sqrt{\dfrac{S_1(1 - S_1)}{\text{nest days}_1} + \dfrac{S_2(1 - S_2)}{\text{nest days}_2}}}$$

where the significance of the value of Z can be determined from standard statistical tables (Johnson 1979). Important values of Z are 1.645 ($P < 0.05$), 2.33 ($P < 0.01$) and 3.08 ($P < 0.001$). Over a thousand nest days per sample are likely to be necessary to detect a difference in daily survival of 0.01 (Beintema 1992). For detecting differences, increasing the number of nests found is likely to be better than increasing the frequency with which nests are visited. The statistical tests suggested in Mayfield's original papers should not be used (Johnson 1979). For studies that include individuals of known age, either because they can be aged or were marked when young, it is useful to produce a fertility table in which the mean breeding output of each age class is tabulated.

There is a range of techniques for determining why eggs and dependent young fail. In particular, it is often possible to recognise if, and which, predators are responsible from remains of eggs or young,

footprints and faeces. The pattern of damage to the shells of birds and some reptiles can provide clues as to the cause of death, since different predators often open the shell in characteristic ways. Trampling by humans and other animals can be identified. A video camera (Fig. 5.6) or a camera with a trip switch can be placed next to the nest. Predation on birds eggs can be documented by creating artificial eggs out of children's Plasticene (Møller 1987) and then painting them (see Fig. 5.7). The tooth or bill marks can be matched to skulls or compared with Plasticene eggs given to captive animals. Plasticene eggs cannot be used for assessing actual predator rates or even the importance of different predator species as there will be various biases in the extent to which different predators are fooled, but it is still a very useful technique for indicating likely predators or the likely distribution of predation risk.

If the objective is to measure recruitment, then using breeding success to determine recruitment has the disadvantage that it is necessary to incorporate three measures that are often difficult to estimate accurately: the proportion that do not breed, the number of breeding attempts in a season and the (usually considerable) mortality of juveniles before recruitment into the adult portion of the

Fig. 5.6 Deer mouse *Peromyscus maniculatus* eating nestlings of clay-coloured sparrows *Spizella pallida* as recorded by an adjacent video camera. Photograph taken 1.5 hours before sunrise using infra-red light-emitting diodes (photo: Pamela Pietz).

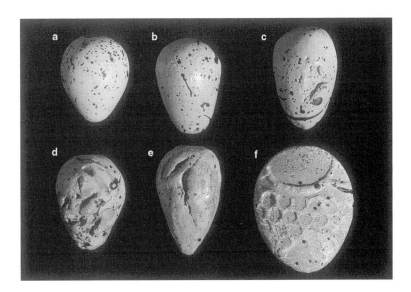

Fig. 5.7 Use of Plasticine eggs to study causes of mortality as part of a study by Durwyn Liley; (a) Ringed Plover *Charadrius hiaticula* egg; (b) imitation plasticine egg; (c) egg chewed by mouse (tooth marks compared with Plasticine chewed by a tame rat and compared with mouse skulls); (d) egg chewed by Stoat *Mustela erminea* (identified as a mustelid by deep canine indentations; the measured distance between canines is the same as measured from stoat skulls); (e) egg chewed by hedgehog (long claw marks on egg); (f) egg trampled by human.

Once one egg of the right size had been made it was weighed so others could be made to the same size. They were painted with car paint of the same colour (champagne beige), splattered with black paint and once dry left in a bird bag used for holding adult Ringed Plovers to transfer the scent. These eggs were sufficiently realistic for the fake clutches to be documented by ornithologists and for Durwyn to be reported for egg collecting (photo: Sheila Davies).

population. Often the best way to measure recruitment is to estimate the proportion of juveniles after one breeding season but before the next, ideally as late in this period as possible. The estimate of the proportion of young will be biased if, as is often the case, the juveniles are easier to see or catch. For comparing measures of recruitment, such as between years or sites, such biased estimates may be acceptable if the bias is likely to be equal for all estimates. If an unbiased measure of recruitment is needed it is necessary to either quantify the biases or remove the biases by devising a sampling routine that ensures random sampling.

5.11.2 Measuring mortality

In practice it is common to calculate mortality rates for different time periods and it is then useful to

convert them to a finite rate for a constant period (Krebs 1999). This is best done by first converting to an instantaneous mortality rate.

Adjusted instantaneous mortality rate

$$= \log_e \frac{\dfrac{(\text{number surviving})}{\text{number marked}}}{\dfrac{(\text{standardised time interval})}{\text{observed time interval}}}.$$

The finite survival rate for the standardised time interval can then be calculated

$$\text{finite survival rate} = 1.0 - e^{-\text{instantaneous survival rate}}$$

This is useful in standardising mortalities to say, 30 days, when the observation time intervals were 23 and 37 days in different areas. It is misleading to extend the periods considerably, say from 23 days to a year.

There are five main approaches to estimating mortality rates.

1 Estimating mortality from recorded deaths or loss of individuals. If of n known or marked individuals, it is known that d have died and a are alive then survival rates over the period can be calculated as a/n. This gives the annual mortality rate if calculated over a year. If individuals are of known age or sex then the mortality of that age or sex class can be calculated. The standard error of the survival rate estimate is

$$\text{s.e.} = \sqrt{\frac{ad}{n}}$$

For shorter periods of time, the Mayfield method (see Section 5.11.1) can be used. Thus the Mayfield method can be used for the analysis of radio-marked individuals, where different individuals are tracked for different periods and the time of death is known.

A better approach for estimating survival when individuals are recorded for different periods, such as during radio tracking, is the Kaplan–Meier method (Pollock *et al.* 1990). The population is checked on n occasions and at each occasion i, the number d_i that died during that time period and the number r_i known to be alive are recorded. If an individual cannot be detected, for example as a result of radio failure or the radio falling off, then it is excluded from the analysis for that time period. The survival S can be calculated as:

$$S = \prod_{i=1}^{n} \left[1 - \left(\frac{d_i}{r_i} \right) \right]$$

and the standard error, a measure of the accuracy of the value, of the survival rate is

$$\text{s.e.} = \sqrt{S^2 \left[\sum_{i=1}^{n} \left(\frac{d_i}{r_i(r_i - d_i)} \right) \right]}$$

This method is particularly useful for analysing radio-tracking data in which individuals are regularly added while others are lost as radios fail or fall off. If dead animals are not located and are thus omitted under conditions when live animals would be recorded then this will overestimate survival.

2 Estimating mortality from the change in population size. As in the previous section, if of n individuals present in the last time period, a are alive and d dead then the survival rate over that period obviously equals a/n. The standard error of the survival rate is again

$$\text{s.e.} = \sqrt{\frac{ad}{n}}$$

assuming survival probabilities follow a binomial distribution. If there is net movement in or out of the area then this will also bias the estimate. Thus 18 Whooping Cranes *Grus americana* (14 adults and 4 young) were alive in the first census year of 1938 and 15 adults were present the following year, giving a survival rate of 0.83 (15/18) and standard error of 0.09. The same analysis has been carried out for every subsequent year (Johnson 1996). Used this way, the method gives a combined mortality rate for adult and juvenile cranes, although in reality juveniles often have higher mortality.

3 Estimating mortality through mark–release–recapture. Loss of individuals through mortality and emigration can also be calculated from the mark–release–recapture analysis (see Section 4.9) if there are more than two capture periods (Pollock *et al.* 1990, Lebreton *et al.* 1992) using the Cormack–Jolly–Seber analysis. This methodology is undergoing rapid development and is now an extremely powerful technique giving good survival estimates and the capacity to examine how survival varies with other factors. This method is likely to be used increasingly by conservation biologists. It requires one of the relevant computer packages (e.g. MARK or POPAN). Recaptures can be from recoveries of dead individuals, recapture, resighting or relocation by radio tracking or even a combination of these.

4 Estimating mortality from the age structure. If there is 50% annual mortality, then the number of each successive age will halve. Hence for a population that can be aged accurately, for example using annual growth rings, the age structure can be used to estimate mortality. A crucial assumption is that there has been no increase or decrease over time in the average birth or death rates at the level of the population as a whole so that the age structure is stable. This requirement often makes this method unsuitable for those conservation studies in which the subject is being studied because of a population decline. It can be useful to plot the age structure. If individuals are predominantly young it points to high mortality or a recent increase in either recruitment or survival of young individuals. If the age structure is predominantly old it points to low recent recruitment or a recent increase in mortality of young individuals. A further assumption is that there are not serious biases towards being more likely to record either old or young individuals.

The numbers of each age class can then be converted to natural logarithms (Richer 1975) and replotted (Fig. 5.8). If mortality does not change with age the relationship will be linear and the slope, i, can be calculated. With an increase in mortality with age the points will bend downwards. The

estimated annual survival S can be calculated from $S = e^{-i}$. At least 150 individuals of known age should be used for this method (Caughley 1977). For female Petrale Sole *Eopsetta jordani*, as shown in Fig. 5.8, the slope is -0.36056. Hence $S = e^{-0.36056} = 0.70$ per year (Krebs 1999). Incidentally, one afternoon when feeling trivial I used this method to study the mortality of internal mail envelopes, which have 12 address boxes which showed that the mortality per trip was 0.71 (i.e. 29% were thrown away) and that the mortality per trip was independent of the number of times reused (Sutherland 1990).

5 Estimating mortality from the recovery of marked individuals. Individuals, usually birds, are marked and reported by the public when found dead. The mortality rate can then be estimated from the decline in numbers found with time (Brownie *et al.* 1985). One key assumption is that the marks do not disappear or become illegible with time or mortality will be overestimated. Simple marking of a number of young in one year and determining the age-specific mortality from the number found dead each subsequent year is unsatisfactory as it combines age differences and time differences, such as weather effects and changes in

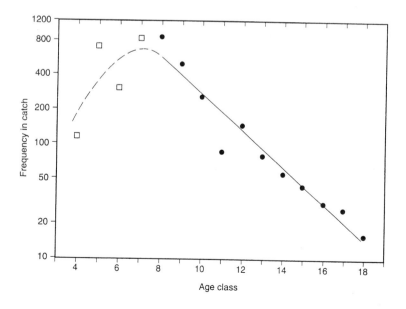

Fig. 5.8 Estimating mortality from age distributions. This shows the number ($\log_e$) of different age Petrale Sole caught in one year. Young fish were unlikely to be caught in the nets. For fish over 8 years old the slope, i, of the line is -0.36. (from Krebs 1999).

reporting rates. This method is most accurate if both adults and juveniles are caught in a number of years (Anderson *et al.* 1985). It is most appropriate for extensive marking programmes in which considerable numbers of individuals are recovered and is thus not widely used by conservation biologists.

5.12 Determining the cause of illness or death

Occasionally epidemics or poisoning events are spectacular and unmissable. Large numbers of animals die in full view and over a short period of time. Examples of such cases have included the phocine distemper outbreak in Common Seals *Phoca vitulina* in the North Sea in 1988, the outbreak of canine distemper in Lions *Panthera leo* in the Serengetti in 1995, die-offs in waterfowl due to botulism or pasteurellosis, and acute poisoning incidents. However, these dramatic, often high-profile events give a misleading impression of the nature of most wildlife diseases. Most diseases, including those that are likely to have the greatest impact on population viability (McCallum & Dobson 1995), are much more insidious in nature. They do not cause sudden widespread mortality but may erode the health of affected individuals, leading to sporadic deaths or reductions in reproductive output that without detailed post-mortem examinations may easily be attributed to other causes (e.g. Kirkwood *et al.* 1997). Some diseases may cause easily observed signs, such as intermittent fits or profuse nasal discharges but others may cause only a vague malaise, the signs of which are not likely to be apparent until a late stage or may be missed completely.

Many diseases of mammals and some of other groups can be transmitted to humans so carcasses should be treated as possible health hazards. Use inverted plastic bags for covering hands when picking up specimens. If possible wear a mask, rubber boots, overalls and gloves. Wash hands well after touching live or dead animals or anything they have been in contact with.

Effective control of disease depends on accurate diagnosis of its cause. The key to detecting a problem is a good knowledge of the normal behaviour and biology of the species and close and attentive monitoring at individual and population levels to detect changes ranging from subtle behavioural alterations to unexplained reductions in survival or breeding output. Diagnosis is likely to depend on detailed examinations of samples collected from affected animals or from carcasses. However, good information about the history of the incident and the signs observed in affected animals help greatly by indicating priority avenues of investigation. So as soon as some potentially serious abnormality is detected, it is important to keep notes of all salient information. This should include the following:

1 Date first observed or suspected and time course of significant events. If dealing with more than one sick individual, mark each (or keep their containers marked) and keep separate records so that the change in each and response to treatment can be studied. Another advantage is that if released animals are marked then it is possible to relate survival to the severity and nature of the problem and to the treatment given. This information can be used for comparing the success of different treatments or to determine whether animals with certain symptoms are not worth treating.

2 Where the incident occurred and any unusual occurrences there, or nearby, prior to the incident. Look for other clues to the problem such as evidence for shooting, poisoning or food shortage. Try to determine the feeding location and diet. If it is an aquatic species, record colour, odour and temperature of water and if possible salinity, pH, dissolved oxygen, conductivity and nitrates. Photographs are often useful, especially as they can be re-examined later to answer specific questions.

If the animal has died through predation it is usually possible to see the physical damage. The predator involved may be identified by looking for footprints, hair or feathers, bite marks or faeces. Shooting is not always straightforward to detect.

3 The species, sex, age, other characteristics and whether ill or dead, for each individual affected.

4 The exact signs of illness described in detail, including lack of symptoms. Are animals thin or close to normal weight range for individuals of the age and sex and at that time of year? For mammals see if the rib cage looks or feels prominent; for birds feel whether the breast bone feels sharp or surrounded by fat and muscle, for fish see whether the belly seems hollow or rounded. Photographs or video recordings are often useful.

If animals are brought indoors then allow them to acclimatise for 5–10 minutes in an environment without sudden noises or movements to exclude change in behaviour due to stress. Examine species that are readily stressed in dim conditions. Compare each with a mental picture of a healthy individual. How does its behaviour, posture, skin texture, eyes and faeces compare? Is there any sign of physical damage? Carry out a full physical examination, searching for differences from a typical individual by feeling for any abnormalities and examining inside the mouth.

After initial information about the history and pattern of a disease incident has been assembled, this is a good time to contact an appropriate wildlife disease investigation agency (e.g. through the IUCN Veterinary Specialist Group) to discuss approaches to further investigations. From the description and history of the incident, it may be possible, in discussion by phone, email or fax with a wildlife vet, to focus diagnostic investigations, refine techniques (of the kind that are outlined in Section 5.12.2) and to be alerted to possible risks to human or other animal health and measures to protect against these.

5.12.1 Collecting material for examination

The priority is to collect material as soon as possible. Blood tests such as serology, haematology and serum analysis can only be carried out if the animal is alive or very freshly dead. Recently dead specimens kept in the refrigerator are suitable for necropsy, histology or parasitology. Frozen specimens are less suitable for necropsy, histology or identification of some bacteria but most infection agents can be identified from them and they are suitable for identification of parasites and measuring of toxins. Storing the entire animal in 70% ethanol allows identification of parasites. For larger species organs may be collected and either frozen or stored in formalin (but see Section 2.4.2). See Friend (1987) or Roffe *et al.* (1996) for more details.

To summarise, there may be advantages in transporting the live animal to a laboratory for diagnostic investigations, but animal welfare or logistical problems may preclude this. If so, refrigerate carcasses on wet ice and transport to the lab within 4–5 days, or, if this cannot be done, freeze some carcasses and place others in 70% ethanol. The smaller the animals, the more specimens are needed for testing.

Plants can be photographed to show symptoms. Plants with fungal infections can be dried and pressed (Section 2.4.3) for identification.

Sending material abroad often requires licences from the accepting country and may require one from the country of origin. Movement of biological samples is regulated by animal and plant health legislation and licences are likely to be required from the state veterinary or agricultural services. If the species is listed in CITES (see Section 10.8.2) then a licence from them is required.

5.12.2 Autopsies

Local vets are likely to be useful in carrying out post-mortem examinations. They may not, however, have specialist knowledge of wildlife diseases or the resources to identify the causes of wildlife incidents. The best combination is to contact an expert on wildlife diseases and place the local vet in touch.

Autopsies carry serious health hazards. Carrying out an autopsy without training is unwise. Carry out post mortems on plastic sheets. Double wrap all samples in two plastic bags, one sealed within the other and disinfect with chlorine bleach or commercial disinfectant. Dampening specimens with disinfectant or soapy water reduces risk of inhaling pathogens. Use sterile instruments to collect samples for microbiological investigations. Instruments can be sterilised chemically (e.g. 10% household bleach). Rinse in sterile water after disinfection or the disinfectant may kill agents in the samples. These instruments should just be used for autopsies. Dispose of carcasses by incineration or bury at a depth that will not be excavated by scavengers. Wear disposable gloves, mask and overalls.

Describe and record the autopsy, giving details such as colour, size, texture, shape and consistency and absence of symptoms. This is easiest if a second person records the observations. Dictation into a tape recorder can also be used but this runs the risk of the data being lost if the tape recorder malfunctions. A dissecting microscope or magnifying lens is useful for small species. A camera with flash, tripod and close-up facilities is useful for documenting.

Table 5.4 lists the samples that can be taken and the means of preservation. Tissue samples should be 4–6 mm thick, made with a sharp blade, should include both the abnormality and the neighbouring tissue and be placed in ten times their volume of 10% buffered formalin (but see Section 2.4.2). All samples

Table 5.4 The suggested means of preserving samples when the entire carcass cannot be submitted (modified from Roffe *et al.* 1996).

Sample	Projected tests	Method of preservation	Comments
When microbial infections are suspected			
Observed lesions (abnormal-appearing tissue)	Microbiology	Chilled/frozen[a]	A portion of each lesion should be saved frozen and fixed.
Heart	Bacteriology	Chilled/frozen[a]	Entire heart from birds and small mammals; selected portions from larger mammals.
Liver	Bacteriology	Chilled/frozen[a]	Entire lobe from birds and small mammals; several pieces up to 2 cm^2 or larger in larger mammals.
Blood/serum	Bacteriology/virology	Chilled/frozen[a]	Serum also useful for serology.
Spleen	Bacteriology/virology	Chilled/frozen[a]	Entire spleen from birds and small mammals; selected portions from larger mammals. Fix remainder.
Intestine	Bacteriology/virology	Chilled/frozen[a]	Segments from middle or distal (ileum) of the small intestine.
Brain	Bacteriology/virology	Chilled/frozen[a]	If animal exhibited abnormal behaviour, save entire head; send intact head for removal of brain by laboratory.
When toxicants are suspected			
Lesions	As appropriate	Chilled/frozen[a]	Lesions (abnormal-appearing tissue): a portion of each lesion should be saved frozen. Fixed tissue important.
Liver	Heavy metals (Pb, Tl)	Chilled/frozen[a]	Entire liver from birds and small mammals; selected portions from larger mammals. Fixed tissue important.
Kidney	Heavy metals (Pb, Hg, Tl, Fe, Cd, Cr)	Chilled/frozen[a]	Entire kidneys from birds and small mammals; selected portions from larger mammals. Fixed tissue important.
Crop/stomach contents	Organophosphates, carbamates, plant poisons, strychnine, cyanide, mycotoxins	Chilled/frozen[a]	Save entire contents. Samples to be checked for cyanide or H$_2$S must be placed in air-tight container to prevent loss of these toxic gasses into the air.

(*Continued*)

Table 5.4 (*Continued*)

Sample	Projected tests	Method of preservation	Comments
Brain	Brain cholinesterase, organochloride residues, organomercuric compounds	Chilled/frozen[a]	If brain is removed for chemical analysis, the brain must be wrapped in clean aluminium foil then placed inside a chemically clean glass bottle. Fixed tissue important
Blood	Lead, cyanide, H_2S, nitrites	Chilled/frozen[a]	Samples to be checked for cyanide or H_2S must be placed in air-tight container to prevent loss of these toxic gasses into the air
For microscopic study			
Lesions	Specimen is fixed, sectioned, and stained for microscopic study	10% formalin	Lesions (abnormal-appearing tissue): a portion of each lesion should be saved frozen
Liver	As above	10% formalin	Specimen portions should not exceed 6 mm in thickness
Kidney	As above	10% formalin	As above
Gonads	As above	10% formalin or Bouin's[b]	As above
Heart, lung, skeletal muscle, lymph nodes, spleen, thymus	As above	10% formalin	As above
Intestinal tract	As above	10% formalin or Bouin's[b]	Snippet of stomach at the ileocecal junction, piece of duodenum (near the pancreas), and colon
Brain, nervous tissues, eyes	Formalin-fixed material will be sectioned and stained.	10% formalin	Divide brain in half (sagittal); place one half in formalin and save the other half frozen
Impression smear	Can be made by touching glass slide to cut surface of any organ.	Air-dry	Air-dried slide can be used for many laboratory tests

[a]Chilled specimens are preferred, however the specimen should be frozen if it cannot be delivered to laboratory within 48 hours.
[b]Bouin's fixation: picric acid, saturated aqueous, 75.0 ml; concentrated formalin, 25.0 ml; glacial acetic acid, 5.0 ml.

from the same individual are conventionally given the same number.

Blood smears are made by placing a small drop of blood on a microscope slide. Hold another slide end on so the slide forms an angle about 45° above the drop. Pull the upper slide towards the drop until it touches and the drop runs along the end of the slide with surface tension. Push the top slide away from the drop creating a uniform smear. Dry in the air and keep away from dust.

The post-mortem examination should be carried out methodically and systematically (Ginsberg *et al.* 1997):

1 Look for any obvious external signs. Look for wounds, damage or soiling. Examine the orifices and describe all discharges. Record body weight

if practical and whether thin or fat. Feel for any abnormalities such as fractures or joint problems.

2 Lay the animal on its back and cut down the midline, pulling back the skin if appropriate.

3 Open the abdominal cavity along the lowest rib/ sternum. View abdominal organs.

4 Cut through diaphragm and remove right side of rib cage (this may require a hacksaw for large species) or for small mammals and birds cut through ribs on both sides and lift off with sternum to expose thoracic (chest) cavity. Examine organs and note appearance.

5 Remove heart and lungs.

(a) Examine tongue and oral cavity.

(b) Dissect out thyroid and parathyroid. Take tissue sample.

(c) Examine oesophagus.

(d) Examine thymus. Take tissue sample.

(e) Find lymph glands near lungs and take sample.

(f) Feel lungs. Note colour and texture. Take sample.

(g) Open tracheae and examine contents. Extend incision into the lungs and through bronchi.

(h) Open pericardium (fibrous sac enclosing heart) and look for abnormalities in fluid. Take swabs if appropriate.

(i) Examine heart. Examine surfaces for haemorrhages and all cut surfaces for pale patches. Determine if heart shape is normal.

6 Examine abdominal organs, leaving the intestine until last in case it contaminates the other samples.

(a) Remove and examine spleen. Cut through parenchyma and take tissue samples.

(b) Remove and examine liver. Cut slices as if slicing tomatoes but without cutting right through and take samples. Open gall bladder last as contents will damage tissues. Sample gall bladder if it looks thickened.

(c) Remove kidneys and adrenal glands together. Examine surfaces of kidneys. Freeze kidneys if planning chemical analysis such as for heavy metals. Cut adrenal glands in half and examine.

(d) Examine outside of bladder. Collect urine if it appears abnormal.

(e) Remove stomach and intestines. Open stomach and continue along entire length to rectum. Take tissue samples.

(f) Examine and measure reproductive tracts. In breeding season cut through testes and through female reproductive tract to look for eggs or young.

7 Examine articulating surfaces of some joints.

8 Sample bone marrow by cracking open femur.

9 Sample brain by removing skin from back of head and removing neck muscles.

5.12.3 Identifying plant pathogens

It is usually not easy to tell if a symptom is a result of a transmissible disease or poor growing conditions, such as a mineral deficiency. The pattern of occurrence may give some clues, for example, does there appear to be a centre of infection or are the symptoms related to growing conditions? Lesions caused by fungi often tend to have a fuzzy outline, bacteria tend to have sharp edges to the lesion while viruses sometimes resemble mineral deficiencies causing yellowing, spotting, distortions or stunting. Furthermore even if a fungus, virus or bacteria is present, it may not be the causal agent of the particular symptoms.

The standard method to relate symptoms to a pathogen is to perform the following four tests, known as Koch's proofs of pathogenicity.

1 *Confirm that the organism is consistently associated with the symptoms.* Examine sections of the infected areas with a microscope. For fungi it should be possible to see hyphae and perhaps spores. Bacteria may be visible on the tissue but can usually be isolated from the plant tissue by placing infected material in a solution of peptone (or any nutrient solution) before examining. Viruses can only be seen under an electron microscope.

2 *Isolate a pure culture of the pathogen (also called an axenic culture) and identify.* Ideally cut a section from the growing front where there is less chance of a secondary contamination infection by another pathogen. Place in agar and incubate. Bacterial contamination of a fungal culture can usually be minimised by using acidic agar (pH 4–5) or adding an antibiotic, such as streptomycin. Aseptic techniques are obviously essential to prevent contamination of cultures.

3 *Reinoculate with the isolated axenic culture and show that it develops the same pathogenic symptoms as the diseased plant.* The means of inoculation depends upon the pathogen. Techniques include: (a) add fungal spores to the leaf surface; (b) press agar sections against the plant; (c) sterilise plant surface with alcohol, sterilise scalpel blade in flame and cut flap in plant, insert a section of the isolated infected material then tape to seal. Pathogens usually grow faster at high temperatures, e.g. 25°C for fungi and 30°C for bacterial infections. Controls are necessary to check that the symptoms do not appear in the absence of reinoculation.

4 *Reisolate pathogen from reinoculated plant and identify again.* This is a repeat of stage 2 to confirm that the pathogen isolated in stage 2 is responsible and that there has not been contamination during reinoculation in stage 3. In practice this stage is often omitted.

Some pathogens, such as obligate parasites, do not grow in axenic culture agar so only stages 1 and 3 are possible, with inoculation directly from a diseased plant but this is much less rigorous.

5.12.4 Determining why eggs fail

If there is low reproductive success it can be useful to try and understand why eggs have failed. By drawing or photographing the developmental stage of dissected eggs of known age, or those of a similar species, it is possible to estimate the age of death. Bird and reptile eggs can be held up to a strong light, a process known as candling, to roughly estimate the developmental stage. Eggs that have not developed can be dissected for a more accurate estimate of the age at which they failed.

Eggs that show no sign of embryo development may either be infertile or the young embryo may have died, but distinguishing these is difficult (Birkhead *et al.* 1995). Fresh eggs can be dissected and the presence of a developing embryo is determined by the appearance of the germinal disc, also known as the blastoderm (Koslin 1944). The yolk can be separated from as much of the albumin as possible and then placed in formalin for fixation (but see Section 2.4.2). After 24 hours the vitelline

(surface) membrane is removed and the germinal disc is stained (e.g. with fluorescent dye). Cell division starts soon after fertilisation so by the time of laying the germinal disc consists of several thousand cells and fertile eggs can thus be identified by scattered clumps of cells. In fish and amphibia, the first division of the embryo often occurs within a day (but later in colder conditions) and is easy to see, sometimes with the naked eye. In many taxa the egg surface may change within minutes of fertilisation (Balinsky 1970). In frogs the privitelline space, the space between the egg cytoplasm and membrane may increase markedly after fertilisation, lifting the membrane from the egg surface, while in fish the membrane may darken and thicken.

A lack of fertilisation may be due to either no or insufficient sperm reaching the egg. To count the density of sperm, examine the perivitelline layer of an unincubated egg by cutting out a section around the germinal disc, wash for example in calcium and magnesium free Dulbecco's phosphate-buffered saline and stain with a fluorescent dye (e.g. 3,4-diamidino-2-phenylindole) to establish the presence or absence of sperm (Wishart 1987). Sperm nuclei have a distinct rod-shaped shape while other nuclei are round. There is a good relationship between the density of spermatozoa trapped in the vitelline layers of laid domestic fowl eggs and the probability that the egg has been fertilised (Wishart 1987); the probability of fertilisation was 50% at about 0.4 sperm per 5.5 mm^2 of membrane but over 95% above 1.6 sperm per 5.5 mm^2.

During the breeding season it is possible to confirm that male birds are producing sperm by inspecting the fresh droppings under a microscope. Simply dab a glass slide onto a fresh dropping and examine with a microscope at ×400. Sperm are easiest to detect with a fluorescent dye examined using fluorescent microscopy.

5.13 Modelling population changes

Population models are a very powerful technique in conservation. The major use of population models is to predict the consequences of damaging processes such as harvesting (see Chapter 13), habitat loss or habitat deterioration or the effectiveness of various conservation measures or the likelihood of extinction under existing conditions. For example, much of the debate about logging in the north-west United States was based on population models of the Northern Spotted Owl *Strix occidentalis caurina* (Lamberson *et al.* 1992, Harrison *et al.* 1993).

In reality the processes of population modelling (including population viability analysis), diagnosing declines (see Chapter 6) and writing species action plans (see Section 7.3) are each different ways of collating the evidence and deciding what actions are of higher priority. Where the data are available, population modelling is often an integral part of diagnosis or planning. In general, it is probably best to start with a process of planning and diagnosis and then create models to answer specific questions.

5.13.1 Principles of population ecology

Population ecology appears deceptively easy (see Fig. 5.9). The critical component is density dependence, the common-sense phenomena that, as the density increases, survival probability or mean breeding output will decrease, for example, as a consequence of competition for food or space. Thus innumerable studies (reviewed by Sinclair 1989) of a wide range of animals have shown that density dependence can act through diverse mechanisms; depending upon the species, at higher densities juveniles may be less likely to survive, adults may be less likely to obtain a territory, adults may wait for more years before breeding, adults may be less likely to obtain a breeding site or each breeding female may produce fewer offspring. Similarly, plants grown at high densities are more likely to die and the average seed production is reduced (Watkinson 1985).

The fundamental concept of population ecology is that population size is dictated by the interaction between density-dependent birth and death rates. As shown in Fig. 5.9(a), if the population is below the equilibrium, the birth rate will exceed the death rate so the population will increase; if the population is

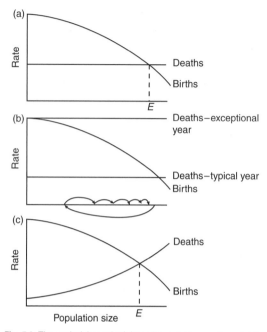

Fig. 5.9 The underlying principles of population ecology. (a) In this example the *per capita* birth rate (mean number of young produced per adult) is density dependent, it decreases with population size, perhaps because of shortage of breeding sites or competition for food. The *per capita* death rate is density independent. The equilibrium population size, *E*, occurs where these two rates equalise. (b) As in (a) but with variation between years. In this case most years are the same but in one year there is a much higher mortality, for example as a result of a cold winter, drought, food shortage or disease. The population declines and then gradually recovers. With constant conditions it will equalise to an equilibrium but with varying conditions it will not and then at any one year the actual population is unlikely to be at the point where the curves intersect. (c) Other possibilities are that the death rate is density dependent while the birth rate is density independent or, as shown here, that both birth and death rates are density dependent.

above the equilibrium, the death rate will exceed the birth rate so the population will decline. At the equilibrium the birth rate equals the death rate.

The necessary existence of density dependence can be understood by imagining a case with no density dependence but with births happening to balance deaths. Any change which minutely increases the death rate results in inevitable extinction, conversely any reduction in death rate leads to an infinite population. Without density dependence it is extremely unlikely that births and deaths will happen to balance exactly.

The idea of an equilibrium population is based on the assumption that conditions are constant. However, in reality, in some years there will be frosts, droughts, disease or food shortages, all of varying severity and frequency. These can be incorporated within the same framework. Figure 5.9(b) shows the consequences of a single year of higher mortalities, say from atypical weather in an otherwise constant system. The population is likely to take some time to recover to the old equilibrium. Although the concept of a stable equilibrium can be useful conceptually, in reality both the density-dependent birth rate and density-dependent death rate will often vary between years with the population continually moving towards, but rarely reaching, a shifting equilibrium.

The rate at which a population recovers from a disaster depends upon the finite rate of increase. At a stable equilibrium, such as in Fig. 5.9(a), the birth rate exactly balances the death rate. If there is a disaster one year, such as in Fig. 5.9(b), the population temporarily drops but once the disaster is over and it returns to the old equilibrium, the birth rate exceeds the death rate. The maximum difference between the birth and death rates determines the maximum rate at which the population can increase, the finite rate of increase, λ (sometimes also called R). The finite rate of increase can be viewed as the expected proportional change from one year to the next in the absence of competition: thus if $\lambda = 1.04$, an annual 4% increase is expected. Even this apparently straightforward measure is complicated by the fact that it only applies to the average population in a stable environment and as variability increases it becomes more likely that the typical population will have a growth rate below that expected from λ (Tuljapurkar 1990).

To summarise, a basic understanding of population ecology requires an understanding of the density dependence, the annual variability in the density dependent response and the finite rate of population increase. Reality is more complex than these simple models, for example, it is often

necessary to consider how fecundity and mortality rates change with age or to incorporate dispersal between areas.

This basic framework can then be used to think about conservation problems (see Fig. 5.10). Exploitation can be thought of as an increase in the death rate, resulting in a reduced population (see Chapter 13). Habitat loss can be thought of as compressing a given population into a reduced habitat. Thus if half the habitat is destroyed a given population size will now be at twice the density in the remaining habitat and will experience greater density dependence than before (Sutherland 1996b) (Fig. 5.10). Habitat deterioration, such as draining or deforesting an area can be thought of as a version of habitat loss. If the deterioration is so severe that the habitat is no longer used at all, then the consequences of habitat deterioration are the same as habitat loss and the curve in Fig. 5.10(b) again applies. If the drained or deforested area are still used but proportionately less, then this results in a curve intermediate between the two cases shown in Fig. 5.10(b) (Sutherland 1998b). It is thus usually much harder to predict the consequences of habitat loss or deterioration than exploitation.

The Achilles' heel of population ecology is that density dependence is of critical importance, yet almost impossible to measure. For example, the most widely used technique for studying density dependence in wild populations is to estimate population size over a series of years and then calculate the rate at which the population has increased or decreased since the previous year and plot this against population size. With density dependence the population will decrease at high populations and increase at low. However, this method has such severe analytical problems resulting from any errors in population estimates that it can rarely be used (Freckleton, Watkinson & Sutherland unpublished). Easily the best way of detecting density dependence is through experiments, in which individuals are added or removed. Not only is this difficult but it is rarely acceptable for species of conservation interest, although useful information can be gained during manipulations. For example, the conservation programme to remove some Seychelles Warblers

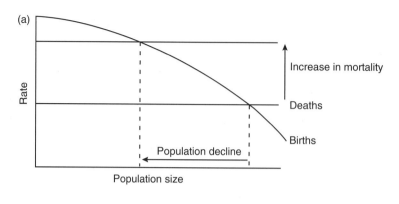

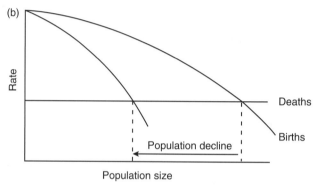

Fig. 5.10 How the framework shown in Fig. 5.9 can be used for understanding conservation problems. (a) As a result of increased mortality, say from harvesting or pollution, the per capita death rate has increased resulting in a reduced equilibrium population size. (b) As a result of loss of half the habitat the density-dependent birth rate function has shifted to the left. This for example could occur due to there being fewer breeding sites or reduced food abundance. This results in a new population size.

Acrocephalus sechellensis (Komdeur 1992, Komdeur *et al*. 1995) and Seychelles Magpie Robins *Copsychus sechellaum* (Komdeur 1996) from certain islands to establish populations on other islands, provided some excellent information on the processes determining density dependence. In both species, at high densities many juveniles stay as non-breeding individuals on territories, while at low densities they move to vacant territories and breed.

The finite rate of increase is also difficult to estimate. One problem is that this measure is also dependent upon measuring the density dependence. Without being able to measure density dependence it also becomes difficult to estimate the finite rate of increase. As shown in Fig. 5.9(a), for a population in equilibrium the birth rate must equal the death rate

as the population is not growing. To predict how fast the population is capable of growing, the finite rate of increase, it is necessary to understand how births and death rates change with density.

A separate problem in estimating the finite rate of increase arises from the estimation of births and death rates, for example, by using the methods of Section 5.11. The errors in such estimates are such that even for stable populations there is usually a large discrepancy between estimated birth and death rates, resulting in the prediction of dramatic changes in population size.

Many models incorporate some variation, such as the occurrence of disasters, yet in reality we have little understanding of the frequency, severity and especially the demographic consequences of disasters. Thus each of the three essential ingredients of the models, finite rate of increase, density dependence and variability, are hard to quantify.

5.13.2 Creating population models

Population models can be produced by using one of the commercially available packages such as VORTEX (Lacy 1987), GAPPS (Harris *et al.* 1986) or RAMAS (Akçakaya *et al.* 1999), by writing a model from scratch using a computer language or using a computer spreadsheet. Creating a model from scratch or by using a spreadsheet has the considerable advantages that the assumptions are explicit and the relationship between the assumptions and conclusions are usually clearer to the programmer. Spreadsheets are perfectly good for analysing deterministic (non-random) problems.

Random (stochastic) variation cannot be included within a standard spreadsheet analysis, so if stochasticity is considered important is better to use one of the available packages or write your own program. For large populations, stochastic processes are unimportant and deterministic models are reasonable. Stochastic models usually use integers (whole numbers) with individuals having given probabilities of surviving and breeding while deterministic models use real numbers (allowing parts of an individual) with rates of survival or breeding. Thus if there are three individuals and the survival rate is 0.5, for a stochastic model there is a 0.125 chance that all die, a 0.375 probability of one survivor, a 0.375 probability of two survivors and a 0.125 probability that all survive. For a deterministic model the number of survivors is 1.5.

The principles of modelling are similar whether using a spreadsheet, writing a model from scratch or using an available package. A conventional matrix model (Caswell 1989) is really a version of the same technique. Models differ depending upon the ecology of the species and the questions being asked, but the structure of most population models are reasonably similar and the following generic outline can provide the base for most models:

1 Create a table to record the numbers of each age. This will be a matrix in a computer model and a column in a spreadsheet. If there are sex differences in demography, use a separate column for males and females. A common approach is just to create a model of the females if it is reasonable to assume that the likely variation in the number of males will not influence female survival or breeding output.

2 Add further columns giving the survival rate of each age class of females and males (if included) and the reproductive output of each age of female. The stages may be subdivided, for example by specifying winter and summer survival or the different stages of recruitment.

3 If there is any evidence that survivorship or breeding output are density dependent then the equation that best fits the data should be included.

4 For each year of the simulation the number of each age is determined by multiplying the number of the previous year by the survival rate for that age. Mortality can be made stochastic by allocating a random number between 0 and 1 to each individual each year. An individual dies if the random number exceeds the value of the survival probability for its age. If there are 20 individuals then this process is repeated 20 times to determine whether each survived.

5 The number of young produced by each age of female is derived from the number of females of each age and their reproductive output. These are combined to give the total number of young produced by all age classes. Breeding output can be made stochastic by incorporating probabilities for each number of young produced.

6 The initial numbers of each age are either obtained from the actual numbers or by starting with, say 1000 age 0 of each sex, and multiplying by the survival rate of each age class in turn to give a stable age distribution and then running the model.

7 The mortality when young is either incorporated into the reproductive output or the first year survival.

8 If there is no density dependence then the finite rate of increase λ can be roughly estimated by running the model and calculating the rate at which the population increases or decreases. If density dependence is included then the equilibrium population size can be assessed. If stochasticity is included, the probability of extinction can be calculated.

9 This basic model can be extended in innumerable ways, for example by incorporating inbreeding depression or by running a series of subpopulations simultaneously with dispersal between them.

One major application of such models is to predict the response to changes. For example, what would be the consequences of changing the mortality rate by increasing or reducing the exploitation? This requires firstly developing the model incorporating sufficient details to produce a realistic population size and then changing the model to the changed conditions (for example by increasing mortality due to exploitation) and determining the change in population size.

In reality, as described in the previous section, survival and breeding output are difficult to measure, particularly as it is necessary to determine the number of breeding attempts and the number of adults that do not breed at all. Incorporating the best estimates usually either results in a dramatic increase or dramatic decline. The range of estimates within the confidence limits will often result in a wide range of predictions from increases to declines (e.g. Green *et al*. 1996, Moss *et al*. 2000).

Many models estimate the sensitivity or elasticity which has two main uses. Firstly sensitivity models are very useful for assessing the consequences of errors in model parameters (for example, what if the survival rate is actually 10% higher or lower than estimated?) which is invaluable for evaluating whether the model results can be treated with confidence. Secondly sensitivity or elasticity can be used for deciding where to allocate conservation effort to greatest effect. Sensitivity is the absolute change in population growth rate given a proportional absolute change in a parameter (Caswell 1978). Elasticity analyses are better for comparing different components which differ in absolute values to detect which components have greatest influence on population growth rate. They involve changing components of the model such as juvenile survival by say 2% and calculating the proportional change in population growth rates (de Kroon *et al*. 1986, Crouse *et al*. 1987).

The elasticity is calculated as percentage change in population growth divided by the percentage change in the parameter. The elasticities will sum to one. Elasticity analyses have been useful for determining which components of life history should be the focus of attention by conservationists (Benton & Grant 1999). They do, however, need to consider the extent to which conservationists are likely to be able to change each component (Mills *et al*. 1999).

Another major application of population models is to see whether the anecdotal explanations are plausible. For example, suppose a population is declining at 5% a year, while studies show an introduced predator has reduced the breeding success by 20%. Is this a sufficient explanation for the decline? The model could be run with the two values of breeding success and the population change calculated in each case. For example, the population of Wandering Albatross *Diomedea exulans* on Bird Island, South Georgia, declined by 22% between 1961 and 1990. Many individuals, especially immatures, were caught by long-line fishing, but was this why the population was declining? Mortality of immatures was shown to be higher after long-line fishing was introduced in the feeding grounds of this population and a model showed that this measured increase was sufficient to explain the population decline (Croxall & Rothery 1991). Convincingly showing the cause of population decline is powerful ammunition when trying to bring about policy changes.

5.14 Risk of extinction

As well as considering whether populations are likely to increase or decrease, models are often used to estimate the probability of a population extinction. The approach thus extends that of the previous section but is made even harder by the fact that some further processes are important in small populations.

5.14.1 Processes in small populations

Various studies have shown that small populations are more likely to go extinct than large ones (e.g. Berger 1990). Although this seems obvious, as large populations must become small before going extinct, there are however particular problems that small populations face which make extinction especially likely. Small populations are more vulnerable to chance events, such as all individuals happening to be one sex or all failing to reproduce or survive (demographic stochasticity). Small populations are also less likely to be viable after an environmental disaster such as a storm or flood (environmental stochasticity) but it is unlikely that a large population will be reduced so low that it goes extinct. Catastrophes can be considered an extreme form of environmental stochasticity and are a regular source of extinctions of small populations. For example, Hurricane Iniki eliminated five bird species and subspecies after striking the Hawaiian island of Kauai in 1992 (Simberloff 1998).

Small populations may also suffer from the Allee effect (Allee 1931), the decline in survival rate or mean reproductive output at small populations due to a range of processes such as increased predation, reduced ability to find mates, reduced hunting ability or reduced breeding success in small groups (Courchamp *et al.* 1999, Stephens & Sutherland 1999, Stephens *et al.* 1999).

There are a range of genetic issues linked to small populations. Here is a one-paragraph refresher course in genetics terminology. The gene determining the variation in basic human eye colour comes in two forms (alleles): brown A and blue a. Each individual has two copies of each gene, one inherited from the mother and one from the father. In homozygotes both copies are identical, thus AA individuals have brown eyes and aa individuals have blue. Heterozygotes have two different alleles Aa and their appearance is determined by which allele is dominant. In this case brown is dominant (hence the capital A) and blue is recessive (with a lower case a) so heterozygotes have brown eyes.

The main genetic concerns with a population becoming small is a loss in genetic variation (Otto & Whitlock 1997) and an increase in the likelihood that a fertilisation will be with a relative (Fenster & Dudash 1994, Frankham 1998). Both result in an increase in homozygosity. Deleterious alleles resulting from mutation that reduce survival

Inbreeding can undoubtedly have disastrous consequences for small populations but some argue that this process has been overemphasised. Poor breeding success in the Cheetah was attributed to inbreeding following an evolutionary bottleneck. The low success in the wild is now considered largely due to predation while the poor success for captive animals was due to poor husbandry but has improved (Caro 2000) (photo: William J. Sutherland).

or fecundity are, unsurprisingly, rare; they are also usually recessive. Although each deleterious allele is rare, there is such a large range of different possible mutations that most individuals are likely to contain some deleterious alleles. Thus on average each European Bison *Bison bonasus bonasus* contains six lethal recessive genes (Slatis 1960) and each Speke's Gazelle *Gazella spekei* contains five (Templeton & Read 1983, 1984). If there is a large pool of prospective parents it is statistically highly unlikely that both parents possess the same deleterious allele and as a result homozygote recessives of rare alleles are extremely uncommon. In small populations the parents are quite likely to be relatives and contain the same deleterious alleles, so that homozygous recessives of deleterious alleles become much more likely and so survival and fecundity is reduced. There is evidence of fitness reduction from inbreeding from a range of groups including plants (Newman & Pilson 1997), insects (Latter & Mulley 1995), snails

(Chen 1993), fish (Kincaid 1976), reptiles (Madsen *et al*. 1996), birds (Keller *et al*. 1994) and mammals (Jiménez *et al*. 1994). For example, inbreeding reduced larval survival, adult survival and egg hatching rate of the Glanville Fritillary butterfly *Melitaea cinxia* and thus led to local extinction (Saccheri *et al*. 1998). Inbreeding depression undoubtedly occurs but some consider its importance to have been overstated (e.g. Britten 1996).

Although it is commonly believed that extinction is inevitable if the population becomes small, this is not the expectation from either the theory (Mills & Smouse 1994, Frankham 1995) or the empirical evidence. Many species have persisted for long periods at low population sizes (Brown 1995), have recovered from small populations (e.g. Haig *et al*. 1993) or have flourished from small founder populations (e.g. van Aarde 1979). The greatest problems are likely to arise when a large population is suddenly reduced. Inbreeding is a particular problem for captive breeding, with the current policy being to minimise mean relatedness (Ballou & Lacy 1995). Reduced populations lead to an increased likelihood of lower fitness rather than guaranteeing it (Hedrick & Miller 1992, Frankham 1995). Much depends upon luck as to the genetic composition of the remaining population, who mates with who and which alleles are passed on by heterozygote parents. It is possible to walk to a cliff edge, blindfold yourself, run around at random and get back safely, but it is not recommended.

The role of genetics in conservation has been vigorously debated with some arguing that it is relatively unimportant compared with demographic events (Lande 1988) or that it is more important to consider why species have declined than the reasons, such as inbreeding, for the final extinction (Caughley 1994). Others have argued that it is still of considerable importance on both theoretical (Mills & Smouse 1994) and empirical grounds (Frankham 1995). However this debate is resolved, it is certainly true that it is unwise to allow populations to drop to low levels. It is also likely that genetic and demographic processes interact so that as populations decline it becomes increasingly harder to bring about recovery (Gilpin & Soulé 1986; Lynch *et al*. 1995).

Many plants are incapable of fertilising themselves even if their own pollen reaches the style (Section 5.9). The usual mechanism is that pollen cannot grow within a style which shares the same self-incompatibility allele ('S allele'). There is usually only a single gene involved (Richards 1999) but there are typically a large number of different alleles, as many as 45 for species with gametophytic self-incompatibility but generally fewer for sporophytic self-incompatibility (Weller 1995) so that cross-fertilisation is possible between most individuals in a large population. This diversity is maintained by negative feedback: if one allele becomes more frequent, then a reduced proportion of its crosses will be viable and it will thus decline in abundance. However, in small populations, this diversity may be lost so that many or even all crosses are self-incompatible and this may lead to extinction.

Self-incompatibility is a particular problem for species with clonal growth, as a population may then consist of only a few genetic individuals. As well as the usual concerns of a lack of genetic diversity, an absence of seeds restricts dispersal and excludes the insurance of a seed bank should the adults die. All the Illinois populations of the Lakeside Daisy *Hymenoxys acualis* var *glabra* went extinct largely due to habitat loss, but plants were collected beforehand from three patches which have been kept in gardens. Experimental crossings showed that these were all of the same mating type and self-incompatible (DeMauro 1995). Thus if these were used for re-establishment then it would result in a self-incompatible population. Crossing with populations from Ohio, where there are many more mating types, is one solution.

5.14.2 Population viability models

Population viability analysis (Gilpin & Soulé 1986, Schaffer 1981, 1987) is widely used to provide estimates of the likelihood of extinction, usually by either estimating the time to extinction or the probability of extinction within a given period, such as 100 years. These models can also be useful in convincing policy makers that extinction is a real possibility (Lindenmayer *et al*. 1993). The process of

CASE STUDY

Estimating the population viability of a re-established White-tailed Eagle population

In the early twentieth century the White-tailed Eagle *Haliaeetus albicilla* became extinct in Britain as a consequence of human persecution. Eighty-two birds, mainly juveniles, were imported from Norway between 1975 and 1985 in order to re-establish the population (Green *et al.* 1996). By 1992 there were about 19 individuals, including eight pairs, occupying fixed home ranges and an unknown number of immatures. The question was raised as to whether this population was viable and whether releasing further birds was sensible.

Annual survival rate was estimated at 73% for young birds prior to settlement and 94% after that, but these survival estimates were imprecise due to small numbers of individuals and some individuals losing their marks. Older individuals raised more young up to the age of six, after which the average was 0.465 fledged young per pair per year. There was considerable variation in breeding success between pairs and territories. As a result of differences in numbers of the two sexes arising by chance, 13% of bird-years on occupied territories consisted of unmated birds.

A model was created (see Fig. 5.11) to consider whether this population was likely to persist (Green *et al.* 1996). Mortality was made stochastic (random) and the number of young fledged per territory was assumed to follow a Poisson distribution. The data was insufficient to confirm that the population was sustainable as the 95% confidence limits for estimated growth included both an annual 11% increase and a 11% decline. Releasing further birds would increase the pre-cision of the population estimates and thus show the likelihood of survival. Furthermore, at low population sizes stochastic events were likely to result in extinction. They calculated a 0.6 probability that the population would become extinct within 100 years. Introducing a further 60 juveniles would greatly reduce this risk of extinction and as a result of the results of this model these translocations were carried out.

Release of juvenile White-tailed Eagles (with wing tag to document fate) in Scotland to augment the population carried out after a model showed that the existing established population had a 60% chance of extinction within a century (photo: L. Campbell).

(*Continued*)

(*Continued*)

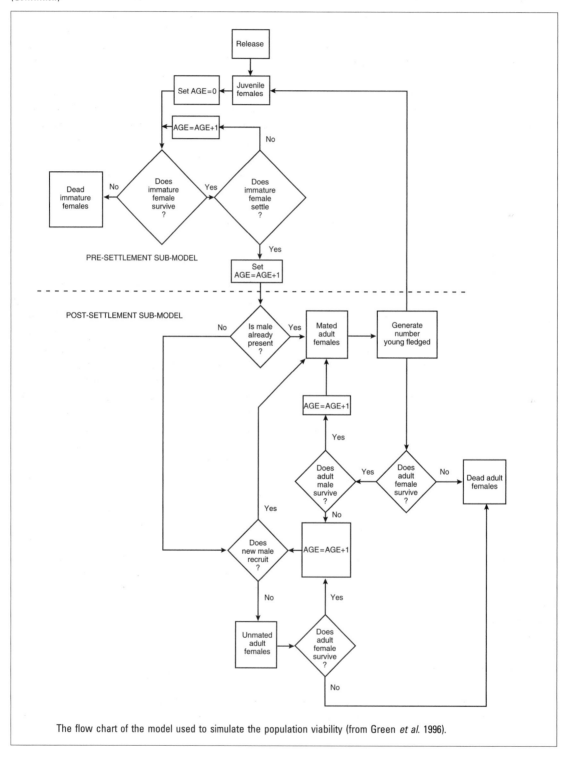

The flow chart of the model used to simulate the population viability (from Green *et al.* 1996).

carrying out a population viability analysis is useful for forcing individuals to think about the demographic parameters, analyse data, identify gaps in the knowledge, consider how further data can be collected, understand the consequences of components and understand the consequences of changes in parameter values (Akçakaya & Burgman 1995). As shown in the case study, population viability models can be very useful in determining conservation actions.

Population viability analysis incorporates two processes that cause extinction. Deterministic extinction: births are insufficient to balance deaths, so the finite rate of increase $\lambda < 1$, and extinction is thus inevitable. Stochastic extinction: births are on average sufficient to balance deaths but sometimes by chance too many individuals die or fail to breed and extinction occurs. Stochastic extinction is only likely in small populations.

There are a range of approaches to population viability analysis modelling. One very straightforward and easy approach is to use one of the commercially available analysis packages such as VORTEX (Lacy 1987), GAPPS (Harris *et al.* 1986) or RAMAS (Akçakaya *et al.* 1999).

There is considerable disagreement over the usefulness of population viability analysis (Boyce 1992, Ludwig 1999). Population viability analysis depends critically upon the interaction of population growth rates, density dependence and the variability in these, yet all of these are difficult to measure (Beissinger & Westphal 1998). Allee effects, the decline in survival or breeding output at low populations (Courchamp *et al.* 1999, Stephens & Sutherland 1999), are also likely to be important yet these are often not included. Brook & Kikkawa (1998) carried out a population viability analysis on the Capricorn Silvereye *Zosterops lateralis chlorocephala* on Heron island, Australia and showed that using data for a five-year period gave markedly different extinction probabilities compared with using the full 26-year data set. As they point out, most population viability analyses are based on short-term or small data sets. Even when using the same demographic data, different population models may give considerably different predictions (Lindenmayer *et al.* 1995, Mills *et al.* 1996) as a result of differing assumptions and constructions.

Population viability analysis can undoubtedly be useful but can also be over-enthusiastically applied to very limited data sets with little insight into the conservation problems or solutions. It has been argued that extinctions are usually the inevitable consequences of continual patterns of decline and that it is more useful to consider the reasons for these declines than the chance processes that result in the final extinction (Caughley 1994, Harcourt 1995). The emphasis is now on using diagnosis, as described in Chapter 6.

5.15 Molecular techniques

Molecular techniques can be used in conservation to identify units such as species, subspecies, races, populations, families or individuals. Which molecular technique should be used depends upon the question. An important issue determining choice is the rate at which the DNA changes. Some sections of DNA change rapidly, especially minisatellites or microsatellites, and thus are highly variable so that even related individuals differ, but this variability confuses comparisons between populations or species. Other sections change more slowly, so are poor for distinguishing individuals but good for examining relatedness between populations and species. My objective in this section is to outline the general methods and possibilities. Collaboration with a molecular lab is obviously necessary to carry out this work. Avise (1994) is a good summary of the molecular approach, Avise & Hamrick (1996) and Smith & Wayne (1996) review the relevance for conservation and Karp *et al.* (1998) provide a detailed description of the range of methods.

5.15.1 Identifying individuals and relatives

Minisatellites are short sequences of DNA, 10 to 100 bases long, which occur in a repeated sequence up to a few hundred times long. A restriction enzyme is used to cut the DNA wherever a given sequence of bases (say GACCAT) occurs. The erratically cut pieces of DNA are placed in slits in a gel, usually of starch or acrylamide, to which an electric current is applied. Being negatively charged, the sections of DNA are attracted towards the anode, with shorter lengths moving more quickly through the gel and thus travelling further. Probes are added which attach (hybridise) to the repeated minisatellite sequence and make a visible band. The chromosomes are paired, one set from the mother's egg and one set from the father's gametes. Thus each DNA length occurs in two versions (the sex chromosomes are different) and each of the cut lengths of the DNA from the mother that contains the minisatellite produces a single band and the DNA from the father produces a second band. If, say, the mother's chromosome has fewer repeats of the minisatellite sequence at that location then that length of DNA, being shorter, will travel further and the gel will show two bands. The same minisatellite sequence often occurs in a number of locations in the genome and thus a number of the cut lengths of DNA will produce visible bands. The ensuing barcode-like pattern, or DNA fingerprint, is extremely unlikely to be shared by any two individuals apart from identical twins (Jefferies *et al*. 1985a,b) so that it is possible to identify individuals.

As each band in the fingerprint must be derived from either the mother or father (except for a rare mutation), minisatellites can be used for identifying relatives. This technique has been used in population studies, in seeing whether individuals added to the population succeed in breeding, in determining relatedness before selecting breeding pairs and for unravelling the social structure and dispersal. Fingerprints can be analysed to determine how inbred a population is (Fig. 5.11).

DNA techniques have been used to show that birds which individuals claim have been bred in captivity are not of the alleged parents and thus presumably have been taken from the wild (Wetton & Parkin 1997). The first such prosecution in the world was a man from Liverpool, England who claimed to have bred four Goshawks *Accipiter*

gentilis. DNA fingerprinting showed that the birds were siblings but not related to the alleged mother. When presented with this evidence the man pleaded guilty, admitted they were taken from the wild and was convicted, but he did not relinquish the birds as he claimed the young Goshawks were stolen between being tested and the trial. The following year another falconer was discovered with a Goshawk which was said to be more than 10 years old despite being in juvenile plumage! Under United Kingdom law, captive bred birds must be fitted with rings and this bird wore a ring which did indeed suggest it was an old bird. DNA analysis showed that this was one of the 'stolen' birds. The ring had been transferred from a captive bird that had died. A parentage analysis of 300 captive bred Peregrine Falcons *Falco peregrinus* in the United Kingdom showed that 39 (13%) were not related to their declared parents and over a dozen more were likely to be wild birds but the declared parents were 'stolen' or mysteriously disappeared prior to testing (Shorrocks 1997). There is evidence that this technique is greatly reducing the number of birds being taken from the wild and then claimed to be captive bred.

DNA fingerprints are difficult to analyse, and require a lot of good-quality DNA, so an alternative is to concentrate on a single region of repeated DNA sequences using 'single-locus probes'. These give just a pair of bands, one from each chromosome, but this is still sufficient to identify individuals and relatives with reasonable confidence. One problem is that the single-locus probes often have to be developed for a particular species while the fingerprinting probe initially used by Jefferies *et al*. (1985a) can be widely used.

Microsatellite repeats are shorter (less than ten bases and usually less than five) than minisatellites, but the principles and applications are the same and are likely to become the preferred technique. In practice, the usual important distinction is that microsatellites are analysed by polymerase chain reaction (PCR) in which a specific sequence of a single strand of DNA can be copied ('amplified') into innumerable copies. For PCR to work it is first necessary to determine and create the sequence of bases ('primers') on each side of the repeat region. Creating these primers can be very time consuming and a short cut is to try primers created for related species. Primers often work across different fish and mammal species and sometimes across bird species (Primmer *et al.* 1996). The DNA is heated, so that the double strands separate, and are then cooled in the presence of primers, bases and the *taq* enzyme. Primers attach to the single strand of DNA while the *taq* enzyme causes the extension of existing strands out from the primer to form a double DNA strand. The numbers of copies of the microsatellite thus double each time the process is repeated. This technique is becoming semi-automated. On rare occasions PCR may modify the region between the primers and this modification may be amplified, known as 'false alleles'. To overcome the slight risk of this problem the DNA sample may be divided, with separate PCR amplifications carried out for each sample (Taberlet *et al.* 1996).

As a result of PCR it is possible to analyse very small amounts of DNA such as that extracted from hairs (Morin *et al.* 1992), feathers (Taberlet & Bouvet 1991), excrement (Höss *et al.* 1992) and even museum specimens of extinct taxa (Higuchi *et al.* 1984). Microsatellites can be used to carry out population estimates when it is possible to collect biological material and so identify individuals but not catch and mark (or even see) the species (see Fig. 5.12). A problem with using very small quantities of DNA is that for heterozygote individuals only one of the alleles may be amplified, a process known as allelic dropout, and thus the individual will be scored as a homozygote (Taberlet & Liukart 1999). The solution is to repeat the PCR a number of times to check that the results are consistent but this is obviously time consuming and expensive.

5.15.2 Identifying species and populations

For examining the relationship between divergent populations it is usually useful to study DNA that has not been subject to reshuffling during meiosis and so has an unambiguous line of descent. Mito-

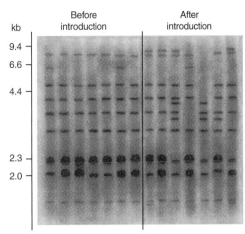

Fig. 5.11 A small population of Adders *Vipera berus* (a snake) at Smygehuk, Sweden, were in decline and exhibited inbreeding depression, with a high fraction of deformed or stillborn offspring (Madsen *et al.* 1999). The gell (on left) shows the lengths of DNA fragments cut by restriction enzymes (with a scale showing the lengths in kilobases). There is very little variation between individuals, showing that the population is highly inbred. In 1992 they added 20 males from another population but in 1996 removed all of these introduced males that had survived. The right gell shows the restriction-fragment length polymorphism from seven of the recently mature males in 1999, demonstrating that the visitors had increased genetic variation. Associated with this increase in variability the population also showed higher offspring survival and started to increase dramatically. This study illustrates the importance of preserving genetic variability to maintain wild populations.

chondrial DNA (mtDNA) is usually passed entirely from mother to egg. Although it can sometimes also be passed through males (Zouros *et al.* 1992) there is no evidence that the DNA then recombines (Hoeh *et al.* 1991). Furthermore, as only females transmit the mitochondria, the effective population is smaller, so speeding the rate at which mutations become extinct or fixed (possessed by the entire population) which simplifies comparisons between populations; mitochondrial DNA evolves five to ten times faster than non-repeating nuclear DNA (Brown *et al.* 1979). Which section of the mitochondrial DNA is used depends upon the expected divergence, for example, a section of the mitochondrial DNA is purely structural (called the 'D loop region' for vertebrates and the 'A-T rich region' for invertebrates) so the sequence of bases is unimportant and not subject to natural selection and DNA in this section changes much faster than other regions of the mitochondrial DNA making it useful for separating recently divided lineages (Taberlet 1996).

Variability in mitochondrial DNA is sometimes studied by cutting the DNA with restriction enzymes and determining the length of the cut sections. Cuts are mapped using single enzymes alone and combinations of enzymes. For example, cutting the previous sentence before every z gives three lengths of 26, 29, and 5 letters long, cutting before every b gives lengths of 42 and 18, while using both gives four lengths of 26, 16, 13 and 5. The location of each b and z can then be worked out. However, with automated techniques it is becoming increasingly easier to determine the complete sequence of lengths of DNA which provides much more information than this method.

Mitochondrial DNA in plants is surprisingly different to that in animals, as there is frequent recombination. Plant geneticists thus prefer chloroplast DNA even though it mutates relatively slowly. It is studied in much the same way as for mitochondrial DNA. Chloroplast DNA is usually transmitted maternally but may be transmitted by both sexes or even, in some gymnosperms, only paternally (Szmidt *et al.* 1987).

Mitochondrial and chloroplast DNA is often used to consider the pattern of relatedness between populations, how closely populations are related and how far they have diverged. As such it has revolutionised taxonomy, often with important consequences for conservation (Mace *et al.* 1996). This is useful evidence for helping to consider the justification of species, subspecies, races, evolutionary significant units and management units, although differences often need to be combined with other evidence (Section 3.2.3).

It can sometimes be very useful to be able to relate a sample or individual to a population or species. Mitochondrial DNA has been used to show that most of the Loggerhead Turtles *Caretta caretta* killed

in the North Pacific driftnet fisheries come from the Japanese breeding grounds and not from the Australian grounds (Bowen & Avise 1996). Variable sections of mitochondrial DNA can be amplified by polymerase chain reaction (Section 5.15.1) which has been used to identify the meat of Humpback Whale *Megaptera novaeangliae*, a protected species, for sale in Japan (Baker & Palumbi 1994). The same approach has been used to amplify mitochondrial DNA from canid faeces using the polymerase chain reaction to distinguish the faeces of the endangered San Joaquin Kit Foxes *Vulpes macrotis mutica* from those of another four species, thus allowing surveys to take place (Paxinos *et al.* 1997).

5.16 Ten major statistical errors in conservation

I believe practical conservation is in a sorry state because it is based largely on anecdote with very little convincing knowledge. Much of this lack of progress results from the following 10 statistical errors. As explained below, an experiment has a control, is replicated, unbiased, monitored and analysed. Hitherto most 'experiments' in conservation have been no such thing and often lack all of these properties.

1 *Not having a control.* Populations vary between years for many reasons including variation in the weather, numbers of predators, numbers of competitors or the abundance of food. Simply altering the management and recording the population changes is therefore usually insufficient information to be able to isolate the effect of the new management. The solution is to have controls that are similar and treated in exactly the same way except for the experimental difference. If experimental areas are electrofished and one species is removed then the control areas should be electrofished (but the species released). If experimental areas are visited weekly then controls should be similarly visited. This acting out control treatments may seem silly but the electrofishing may affect the predatory fish or the weekly visits may disturb herbivores or cause trampling. Controls are every bit as important as the experimental treatment.

2 *Not replicating.* A group of biologists decided to add ammonium nitrate to a shallow lake in England in the spring of 1987. By the summer the biology and chemistry of the lake had completely changed. This would seem overwhelming evidence that nitrates radically alter the ecosystem – except that the ammonium nitrate only got as far as the boat shed on the edge of the lake and was never added. The spectacular changes were for other reasons (Moss *et al.* 1986). This example illustrates the point that changing the management in a site and observing an ecological response, even a dramatic one, is not necessarily sufficient evidence that the treatment caused the change.

Unfortunately almost all 'experimental' management is carried out in single blocks with the treatment applied over the entire area so the results cannot be treated with any confidence. In lake ecology in England it is unrealistic to have many replicates but in most other situations replication is reasonably practical, for example, by dividing the area into blocks and treating some blocks and leaving others as controls. Controls should also be replicated as many times as the experimental treatment.

3 *Pseudoreplication.* Pseudoreplication is pretending to have replication when you have not. Suppose researchers wanted to study the growth of plants resulting from different fertilisers. They apply one fertiliser to one seedling and a different fertiliser to a second and after a month measure the height of each plant 20 times (repeated measures are likely to vary slightly due to measurement error) and show that the averages of each of these 20 measurements differ between the two seedlings and the two sets of measurements are highly statistically significant. Such an experiment would be rightly ridiculed. Each measurement is not an independent replicate but a pseudoreplicate (Hurlbert 1984) for the experiment obviously needs to be carried out with many seedlings, obtaining each fertiliser treatment with each seedling providing a single data point in the analysis.

Although the idea of measuring the same replicate repeatedly is clearly ridiculous, this is the experimental design used for many, if not most, management experiments. For example, suppose researchers were trying to establish the impact of grazing management by comparing two meadows, one grazed and the other not; they might be tempted to measure a series of quadrats in a grazed field, measure another series in the ungrazed field, then consider the quadrats as replicates. This is equally flawed as they are also pseudoreplicates. Any difference between the two sets of measurements may be due to physical differences between the meadows or past management, rather than due to the grazing management.

Pseudoreplication can be quite subtle. Consider an experiment to determine whether, prior to release, it is better to keep captive individuals in a cage with natural or artificial surroundings. Half are kept in a natural cage and half in one with artificial surroundings and their fate recorded once released. This too suffers from pseudoreplication. Any differences may be due to an undetected disease prevalent in one of the cages or due to one cage happening to be wetter or hotter. To be scientifically rigorous a number of cages should be used and each cage used as a replicate. The usual justification for pseudoreplication is that all other factors have been kept constant but in reality it is never possible to say there are no differences.

In practice, there will be occasions when there are only very few sites available and pseudoreplication is the only available option. The important points are, firstly, to realise that such a study is deeply flawed and, secondly, and most importantly, to realise the importance of creating properly replicated experiments wherever possible.

4 *Biased selection of experimental and control treatments.* It is essential to be ruthless in ensuring that there is no bias in selecting controls and experimental treatments. A common problem is to believe that the experimental treatment will be beneficial and thus select the better individuals or areas for this treatment. This seems a particular problem when the person in charge of the experiment is not in charge of selecting individuals or sites. Analysis of the data will then show that the experimental individuals or areas fare better, but it is impossible to tell whether this is a consequence of the treatment or the biased selection.

Random assignment of sites to experimental or control treatments works well with large sample sizes but for small samples there is a real chance that there will be a bias. For example, most of the experimental plots may be at one side of a reserve with most controls at the other or most of the first individuals to be released in a site may be randomly selected for the treatment and the later releases may be largely controls. A regular pattern may be more suitable (Hurlbert 1984) such as a checkerboard of control and experimental areas, or alternating control and experimental releases or having pairs of individuals and tossing a coin to determine which is the treatment and which is the control (Milinski 1997). Furthermore, with a regular pattern it is easier to analyse the data to test if there are other influences such as variation across the reserve or during the season. A problem with regular sampling is that most statistics assume random sampling, but there are usually sensible equivalents such as paired 't' tests or blocked Analysis of Variance which overcome this objection.

5 *Biased data collection.* Population estimates are often biased (Section 4.2). This is not necessarily a serious problem, for example, if half the individuals are missed in every count then this need not greatly affect the comparison of treatments. The real problem is if the bias differs between treatments. The treatment could easily affect the census, for example habitat management may change the visibility or activity of the species. Bias in relation to treatment could also result from one observer collecting vegetation data from one treatment and another studying the controls so that it is impossible to tell if any difference is due to the habitat management or observers. Another example would be if the same observer counted butterflies at all replicates of one treatment on one day and the controls on another day. Any difference could be due to the inevitable weather differences between the days.

Bias can also result from an over-zealous observer who is keen to support (or even refute) the expected conclusion and thus consciously or subconsciously biases the results. The ideal is to have the data collected by someone who does not know the experimental design.

6 *Confusing significance with meaning.* The aim of statistics is to determine what is important and believable. Most statistical tests are a two-stage process. Firstly determining the strength of the difference or relationship (e.g. r^2, χ^2 or t) and secondly the probability, P, that this difference or relationship could have arisen by chance. It is common to believe that a value of P below 0.05 announces a truth that must be acted upon while one above this means that there is no effect and it should be ignored (Yoccoz 1991). However, with a huge sample size, a small difference or weak relationship may be statistically highly significant although not worth acting upon. For example, I would guess that the growth rate of fish is likely to vary considerably with diet, to show some relationship with tank size and quite possibly to show a minute but genuine relationship with the colour of the walls. With a small sample size none of these would differ significantly, whereas with a huge sample all may be significant and a moderate sample may suggest that only diet is significantly different. It is thus necessary to decide how large an effect is worth acting upon; this is a conservation rather than statistical decision.

On occasions where it is particularly important that the science is correct it is appropriate to demand particularly strong evidence. Incidentally it is disconcerting that official safety standards in the nuclear industry are often based on $P < 0.05$ (Hall & Selinger 1986).

7 *Falsely believing there is no effect.* You are on a sinking ship. Two passengers each take a lifejacket from one cupboard, jump in the sea and float. Another two take lifejackets from a second cupboard, jump into the sea, and sink. What should you do? One solution is to calculate the χ^2 with Yates correction ($\chi^2 = 0.5$), discover that there is no statistically significant difference between the two cupboards ($P > 0.05$), and conclude that it does not matter which you use. I would use the limited information and take one from the cupboard which had the two floating life jackets.

We do not want to base conservation on false ideas and thus should design observations and experiments with sufficient samples and replicates to ensure we can be confident in the conclusions (i.e. any important differences or relationships will be detected and have a low value of P). However, for rare species or habitats it may be necessary to depend on few samples or replicates and if a decision has to be made on the available evidence, as in the sinking ship example, then it is reasonable to consider making decisions on weak evidence, say $P = 0.1$. It should however always be borne in mind that there is a reasonable chance that the conclusion on which the decision is based is wrong and any opportunity to extend the analysis should be taken.

It is much more straightforward to show an effect than a lack of an effect. The usual approach is to simply use lack of a statistically significant difference as evidence for a lack of an effect. However, there may be considerable differences between treatments or a correlation but the result is not significant, simply because the sample size is small. Furthermore, almost any process is likely to produce a significant effect if the sample size is sufficiently enormous, even though the effect may be minuscule.

To overcome the problem of defining the lack of a difference it is necessary to consider the statistical power, which is the likelihood of detecting a given effect. To do so it is necessary to make four decisions: (1) What effect size (w) is biologically important? (2) What level of statistical significance is being used? This is often set at $P < 0.05$ but need not be. (3) What is the acceptable likelihood (the power) of detecting an effect of agreed size if there actually is one? It is common to set the power at 0.9, a 90% chance of detecting the effect. While a higher value is obviously more satisfactory, it may require a huge sample size to obtain sufficient confidence. (4) What sample size can be collected? Kraemer & Thiemann (1987) and Cohen (1988) explain how to carry out power analysis.

Three of these decisions have to be made before anything can be said about the remaining fourth. For example, Thompson & Neill (1991, 1993) carried out an experiment to determine whether House Wrens

Troglodytes aedon prefer to use boxes that have been cleaned and failed to find any significant difference. They calculated that they were only 29% confident (power = 0.29) that the difference in occupancy was less than 10% ($w = 0.1$) but 98% confident that the effect was less than 30% ($w = 0.3$).

Alternatively, if there is already some information on the expected variability, then once the power, significance and effect size have been decided upon it is then possible to determine the required sample size. This leads to options of increasing the power by obtaining a larger sample size, redesigning the project or deciding that for the given resources there is such a low chance of obtaining a clear-cut result (i.e. the power is too low) that it is more useful to do something else (Greenwood 1993).

8 *Correlating trends over time.* An introduced aquatic plant has been spreading through a reserve and a frog population has declined by 90% since the plants were first introduced a decade ago. Plotting the graph of plant abundance against the frog abundance is likely to show a convincing correlation leading some to conclude that the plant resulted in the frog declines. The problem is that the frog population will also correlate with anything else that has increased over that period, such as ultraviolet light levels or global temperatures and local changes such as drainage, pollutants or ecological changes. A more detailed analysis of the timing and spread can make this more convincing. Were the frog's declines later in the pools that the plants were last to reach? Are the frog populations stable in those pools where the plant has yet to reach? Did the frogs recover in the couple of years when the plant was controlled in certain ponds? Was the decline greatest in years when the plant spread rapidly?

9 *Not thinking about the analysis until the data is collected.* A depressing number of studies cannot be analysed in a sensible manner as this process has not been considered earlier. Statisticians are usually much happier to be consulted over a flawed proposal than a flawed data set.

10 *Not writing it down.* Most practical conservation is based on anecdotes. Without basic information

being collected and made available to others it is difficult for conservation science to develop. Even if the experimental design is far from perfect then it is still extremely useful to know what was done and what the results were. This is even true for a non-replicated, non-randomised, uncontrolled study. Undocumented examples, spread verbally, cannot be distinguished from myths (see also Section 6.2).

6 Diagnosis and prediction

6.1 Why diagnose problems?

A failure to diagnose what is wrong is at the heart of much unsuccessful conservation. As described in Section 6.2, land managers and practising conservationists often believe they know the solution to a problem, although research then frequently shows they are mistaken. This is illustrated by the common situation of two experts holding confident but contradictory opinions: at least one of them must be wrong. One consequence of this misplaced confidence is that it suppresses research to determine the real solution.

A problem is that researchers appear to be showing the obvious, even when they are not. Consider the common situation of being told a species decline is due to, say, predation, eutrophication, human disturbance, an introduced species and a change in hydrology. After detailed work, researchers show the deterioration is actually largely due to hydrological change, partly caused by eutrophication, but the other components are unimportant. Although such a discovery is a major advance the usual response is 'I told you so'. A related problem is the tendency in species action plans and Red Data Books to list all the possible threats without identifying those that are critically important.

It is also useful to be able to predict the consequences of proposed changes and conservation biologists are increasingly being expected to do this, for example as part of the process of environmental impact assessment. The methodology and information required for understanding declines and predicting the consequences of changes are very similar.

6.2 A need for evidence-based conservation?

Some patients with heart blockages have additional ventricle beats and it has been shown that these can be suppressed by various drugs; this treatment thus became common practice (Morganroth *et al.* 1990). This follows the conventional methodology of identifying a problem and finding ways of alleviating it. However, analysis using randomised trials showed that patients given these drugs were more likely to die (Echt *et al.* 1991). Similarly, there are over thirty documented ways in which doctors can detect whether patients are obtaining insufficient air to survive, but when these are tested for precision and accuracy only a few both bear a relation to the actual airflow and can be confirmed by repeat measures (some were not repeatable even with the same clinician) (Sackett *et al.* 1998). In another study of general physicians in America, the researchers interviewed the physicians after each consultation with a patient and asked if the physicians had any questions they would like the answer to. The researchers consulted the literature to answer these questions and showed that in a typical day, eight clinical decisions would have been altered (Covell *et al.* 1985). Finally, the advice of clinical experts regularly differs from the consensus of already published studies (Antman *et al.* 1992, Warren & Mosteller 1993).

Concerns over a discrepancy between practice and evidence led to the development of evidence-based medicine (e.g. Friedland 1998) which is revolutionising clinical practice. Clinicians often base their decisions on intuition and experience such that it is often difficult for others to learn from them except through mimicry. Evidence-based medicine involves unravelling, understanding and making objective the logic of experts so that it is possible to replace mimicry and tradition with understanding. It builds on, rather than replaces, clinical skills, clinical judgement and clinical experience.

Traditional medical practice is, of course, heavily based on scientific evidence, but often largely derived from secondary information obtained from textbooks, from discussion with colleagues or from information obtained during training, with little use of scientific journals or reviews (Kanouse *et al.* 1995). Thus a study of hypertension showed that the main factor determining whether doctors decided to prescribe antihypertension drugs was not the severity of organ damage, as would be expected, but the number of years since the doctor graduated from medical school (Sackett *et al.* 1977, 1998). The real change in methodology of evidence-based medicine is for doctors to have the training to interpret studies and access to information so that they can review primary studies and continually update their methods.

Do we need a similar revolution to produce evidence-based conservation? At the University of East Anglia we have been carrying out a series of research projects to determine the consequences of various management practices and it is striking that repeatedly the conventional dogma turns out to be mistaken. For example, it is widely accepted that burning reed beds (a traditional practice disliked by many conservationists) kills many soil invertebrates, but a series of replicated, randomised and controlled experiments showed that this was not true; but flooding (a standard practice approved of by most conservationists) did kill them (Cowie *et al.* 1993, Ditlhago *et al.* 1993). As a second example, a UK government conservation scheme paid farmers to flood fields in winter to enhance populations of breeding waders but the flooding actually kills the earthworms on which the waders feed (Ausden 1996b). The waders are largely dependent upon shallow pools in summer but with the government scheme all areas receiving subsidies must be dry by the spring. As a final example, it was thought that the low survival of wader chicks on saline lagoons was largely due to a decline in fertility and predation was a major problem. Research showed that fertility has little effect but the major problem is chicks starving due to high salinities killing the invertebrates (Robertson 1993).

Conservationists are continually using methods on species and habitats similar to those that have been used elsewhere and would surely benefit from

being able to review past experiences. In practice, information is obtained partly from reviews and books (but rarely from primary sources) and largely by talking to others. In practice, even wardens managing similar habitats often do not exchange information, especially if working for different organisations or in different regions or countries. Similarly, Wright (1999) states that research results are rarely shared between parks in North America.

The following are the main techniques (Sackett *et al.* 1998) of evidence-based medicine which I have illustrated with a conservation example:

1 Convert information into answerable questions (e.g. do motorboats affect waterweed populations?).

2 Efficiently track down the best evidence with which to answer the question. This may include comparisons within parts of the site varying in boat activity, published papers or evidence from other sites.

3 Critically appraise evidence both for its validity (for example, does it commit any of the statistical sins described in Section 5.16?) and usefulness (is the water depth or plant community so different in the other sites that comparisons are not useful?).

4 Apply results of this appraisal.

5 Evaluate performance. A critical need is to find ways in which conservationists can record the exact problem (e.g. the spread of a particular problem plant), what was done (e.g. exactly how treated, when treated, exactly which equipment was used) and whether successful (ideally by quantifying, but at least by recording if the species increased, decreased or was equally abundant). A further need is for practising conservationists to have the training to be able to interpret studies in a critical manner. The essence of my point is that, for example, in Britain there must be dozens of wardens trying to control introduced Rhododendron *Rhododendron ponticum*, and dozens more trying to prevent damage to regenerating woods by introduced Muntjac Deer *Muntiacus reevesi*, attracting wading birds to wet grasslands, deciding when to allow grazing on fens or trying to restore damaged raised bogs. If each had an account of everyone else's methods, successes and failures, and those of wardens in other countries, then surely conservation would

be progressing at a more rapid pace. The internet provides the ideal medium for this.

One problem is that conservationists have too much to do already and documenting responses to management is yet another boring administrative task. Just as I am glad that doctors are taking time to learn whether a treatment works, I believe conservationists need to find ways of assimilating their collective experiences. Rather than waste time, the opportunity to learn from others and avoid repeating mistakes should actually provide more time for conservation that works.

6.3 Diagnosing why species have declined

Caughley (1994) and Caughley & Gunn (1995) argue convincingly that conservationists have concentrated too much on the problems facing small populations, diagnosing and reversing population declines usually being a more important problem. While it is often necessary to act before the reasons for the decline are certain, the decisions should be made on the best available evidence while still carrying out the work that confirms or refutes the ideas. However, Caughley & Gunn (1995) show that in many cases conservation programmes have failed to discover why a species has declined by being deflected by short-term actions and thus have not been successful in restoring the population even in the long term.

Most successful diagnoses are dependent upon a sensible monitoring programme (Chapter 4). It is often also necessary to census the prey species, competitors, predators or herbivores as well as measure some environmental factors.

It is usually best to take a systematic approach in determining why a species has declined:

1 Review the evidence to see if the species has declined. Is there evidence for decline on a large scale? Is the localised decline due to a shift in distribution with increases elsewhere? Has the range contracted? When did the decline take place? Is the decline within the typical fluctuations or cycles, or as a result of a return to typical levels after an atypically high period?

2 Collate the available information on the natural history and collect more if necessary. The key to understanding why a species has declined usually lies in the detail of the natural history. What habitat and microhabitat does it usually occupy? What are the predators, parasites, pathogens and competitors? What are the main causes of death? What are the main causes of breeding failure? If a plant, does it have vegetative growth through stolons, rhizomes or bulbs? What are the pollinators and seed dispersers? In what size and type of patches does successful germination and recruitment occur and when does this happen? Does germination and recruitment occur in pulses related to atypical weather or local events such as fires, floods or disturbance? If an animal, what does it eat and does this change through the year or with age? What are its usual habitat preferences and does this change through the year, especially in times of shortage, for example, where does it feed at the end of winter or find water in dry periods? Where does it breed?

3 Review the evidence for ecological changes. Has the management changed? Was there a history of grazing, burning or disturbance which has now stopped? Has grazing increased recently? Has the water regime or chemistry changed? Have introduced predators, competitors or parasites appeared in the area? Has exploitation increased?

4 List all the possible reasons for the decline. Keep an open mind as experience shows that the real reason for the decline may not be immediately obvious.

5 Test the hypotheses for the decline listed in stage 4. It is important that all the hypotheses are considered. Studies in single sites often fail to unravel patterns that are revealed when many sites are studied (Green 1995). The reason is that many factors will be changing simultaneously on a site and it is difficult to separate important from irrelevant factors. By comparing a range of sites it is easier to identify factors that are consistently important. For this reason it is often difficult for reserve managers to unravel the problems on their site if they view them in isolation. The main approaches for testing reasons for decline are:

(a) Carry out a population study, measuring breeding success and survival at all stages and

CASE STUDY

The Lord Howe Woodhen: diagnosis and recovery

The Lord Howe Woodhen *Tricholimnas sylvestris* is a flightless rail found only on Lord Howe Island between Australia and New Zealand (Miller & Mullette 1985, Caughley & Gunn 1996). It seems that this species was widespread on the island and was hunted by visiting sailors in the late 18th and early 19th centuries. I have structured this case study to conform with the method of diagnosis in Section 6.3.

1 When annual monitoring started in 1969 the population was only 20–25 individuals with six to 10 breeding pairs and productivity was low. It was restricted to two mountainous and precipitous summits (Mounts Gower and Lidgbird). By 1980 the species was reduced to only three to six pairs on Mount Gower.

2 Initial field studies of colour-marked birds determined territory size, and how to sex them (males are larger) and age them (adults have a bright red iris). Observations showed they nest in Providence Petrel *Pterodroma solandri* burrows and they feed on invertebrates, especially earthworms, by digging through the soil and leaf litter on the forest floor.

3 Humans settled on Lord Howe Island in 1834, and goats, dogs, cats and pigs were introduced but it was thought the environment had otherwise changed little.

4 Possible explanations for the decline and restriction to Mount Gower included:

(i) Insufficient petrel burrows for nest sites.

(ii) Insufficient food.

(iii) Black Rats *Rattus rattus* which Recher & Clark (1974) recorded as being abundant on the summit of Mount Gower and suggested this contributed to the woodhen's decline. Black Rats have been responsible for numerous extinctions of island birds.

(iv) Feral cats were present on the island but not on the summit of Mount Gower and so could be responsible for their constricted range.

(v) Two of the three known nests appeared to have been destroyed by the Currawong *Strepera graculino crissalis*, another endangered endemic.

(vi) The introduced Tasmanian Masked Owl *Tyto novaehollondiae* has been recorded taking domestic hens and taking a woodhen from a captive breeding pen.

(vii) Pigs were abundant, including on Lidgbird but not on the summit of Mount Gower. Pigs can compete for food and take eggs, chicks and incubating adults.

5 Each of the above hypotheses was examined.

(i) Nest sites (petrel burrows) and roosting sites were sufficiently abundant. The breeding success was higher for lower altitude pairs.

(ii) Invertebrates were sampled at eight points along six transects in 10 sites and were at least as abundant in the lowland areas indicating food shortage was not restricting the distribution. When a young adult male woodhen established a territory in the lowlands it became one of the heaviest birds.

(iii) Black rats were more common on Mount Gower than elsewhere and in field experiments they ignored eggs of domestic hens *Gallus gallus*. It was thus considered unlikely that egg predation by rats was the problem.

(iv) Stomach contents and scats of cats contained black rats and seabirds but no woodhens. They rarely ranged into the highlands and were considered too scarce to be a major problem.

(v) Currawongs were scarce and it was unlikely that their numbers had increased.

(vi) As the masked owls are nocturnal and woodhens in the wild roost in dense cover, the owls are unlikely to be a problem.

(vii) The pig and woodhen distribution did not overlap. This clear demarcation suggested the pigs could

(Continued)

(*Continued*)

be the major factor restricting the woodhen's distribution. The pigs took large numbers of earthworms but earthworm densities did not differ between sites with and without pigs. Many petrels were found amongst the stomach contents of the pigs and their range also was similarly discontinuous with the pigs. The implication is that any woodhens nesting in the pig range may well be eaten.

6 With the pigs considered the most likely reason for the woodhens' restricted range, a control programme was started with a bounty for each tail provided. About 180 pigs were shot leaving only one boar. The hypothesis that the pigs were the main problem was supported by an increase in the woodhen population with areas previously occupied by pigs rapidly being reoccupied.

It was decided that the low productivity on Mount Gower meant spread was likely to be slow. A captive breeding programme was thus started in 1980 in the lowlands and within 4 years the initial three pairs and descendants had produced 82 young. The high productivity of captive birds contrasted with those in the wild, implying the problem was with the habitat.

Little Slope was chosen as a site for experimental release, as it has abundant petrels and thus plenty of nest holes. It is surrounded by cliffs and the sea, restricting dispersal and making monitoring easier. Experimental release showed the survival of those raised in captivity was the same as those raised in the wild. Following the success of this experiment, further releases were carried out at a number of other sites.

7 The population is now stable at about 50–60 pairs, which is thought to probably be the maximum the habitat can sustain.

where possible determining the cause of death or breeding failure. It is easy to rush into wrong conclusions at this stage. For example, fire was considered damaging to Giant Sequoia *Sequoiadendron giganteum* and thus fires were extinguished, but regeneration only occurred after accidental burning. Similarly, in my experience native predators are often accused of being a major problem for bird populations as predation is obvious, however more detailed research then often shows the decline is more a result of subtle changes in land management.

(b) Compare breeding success or survival in sites or parts of sites where it has declined with sites or parts of sites where it has not. It can be very illuminating to carry out the population study described in stage 5a in both declining and healthy populations.

(c) Compare the environment in areas where the species has declined or gone extinct with areas where it has not. Quantify as far as possible both the management (e.g. whether grazed, burnt, enriched or subject to predator control) and the habitat (e.g. sward height, percentage of bare ground, phytoplankton density or density of predator faeces). Concentrate particularly on those aspects of natural history shown to be important in stage 5a. For example, the Silver-spotted Skipper *Hesperia comma* butterfly has disappeared from numerous sites in Britain, including a quarter of the populations in nature reserves (Thomas *et al*. 1986). Highest egg densities were on short cropped Sheep's Fescue grass *Festuca ovina* above 1.2 cm tall and next to bare ground. A comparison of sites showed that the population was much more likely to have persisted on sites with stock grazing, which produces the required habitat areas, than sites without. Following this research, grazing has been reinstated to a number of sites and the butterfly numbers have increased.

(d) Experiment within a site to change one of the environmental conditions and monitor the population. Wherever possible management should be designed in a way that also tests a hypothesis. For example, a comparison of summer and winter cattle grazing showed that winter grazing was more beneficial for the Silver-spotted Skipper. This method means that it is possible to take an informed and inspired guess as to the main problem and act upon this guess. The advantage of this logical and experimental approach is that it allows action while examining other possibilities and reveals if the guess is wrong.

(e) Relate the changes in population size within an area to the environmental changes. This is the most common but weakest approach. In practice, there is usually insufficient useful historical data on the magnitude and timing of changes of either the environment or the population, so attempts to relate them are difficult. Furthermore, too many factors usually change simultaneously so the role of each cannot be distinguished. It is easy to use this method to believe anecdotal information, draw the wrong conclusions and use this to justify failing to start the research that would reveal the true answer. In my experience the main reason for failing to diagnose problems is the myopic concentration on single sites, where factors cannot be unravelled, rather than looking for general patterns across sites. This approach is, however, very useful for generating hypotheses that can be tested with methods (a)–(d).

6 Link the reasons for decline discovered in stage 5 to the ecological changes documented in stage 3. Find ways of reversing the changes or removing their effect. This may involve habitat management, education or changing government policy. These may also be experimental by comparing population changes in areas with or without an education programme, or change in habitat management. It may be necessary to re-establish or reinforce the population. Continue to monitor the population.

7 As knowledge accumulates go back to stage 1 and repeat the process. This is particularly important if the population has failed to recover. Even if recovery is taking place this is sensible; there are often multiple reasons for declines and identifying and removing a major problem does not mean that there are not other serious problems to be identified and tackled. Continue until the population is back at the desired level.

8 Celebrate.

This general approach was used to rescue the Lord Howe Woodhen from extinction (see Case Study).

6.4 Predicting the ecological consequences of changes

It is often necessary to decide how seriously a proposed change will affect the conservation importance of a site. The process described below may reveal that the change is likely to have negligible consequences in which case it is a waste of effort to oppose such changes and overreaction also gives conservation a bad name. However, if the change is likely to have deleterious consequences, then a methodical, transparent and defensible prediction is more likely to be considered seriously by decision makers.

Ecology is not rocket science. It is much harder. We can fly rockets routinely and with confidence but making ecological prediction is more challenging. Paradoxically, ecology looks easy because it is so difficult. Although the main techniques seem simple and often involve string, buckets and mud, ecology is actually enormously complicated due to the web of predatory, competitive and mutualistic interactions between species. Furthermore there is often considerable environment variability and it is often necessary to consider behavioural responses. As a result of this complexity it is often difficult to make accurate predictions.

1 Determine the exact nature of the threats including where and when. A proposed golf course may not only change the habitat but may also add nitrates to water courses, require a road and extract water.

2 Survey the species and communities within the area that may be affected, paying particular attention to those considered of high priority (Chapter 3). The area may vary according to different threats. Thus it may be necessary to survey the freshwater communities downstream and the mammals over the wide area that may be affected by the road but the effects on the invertebrates or plants may be localised.

3 List all the possible environmental changes resulting from these threats. Thus water extraction could result in a lower water table in a marsh, lower water in a lake, reduced flows in a stream. If possible calculate the likely magnitude of these changes.

4 Use comparative information to predict how the communities or species of conservation interest are likely to respond to the environmental changes. For example, examine the vegetation in those areas of the marsh which already have a lower water table or compare the fish populations in streams that are similar but have lower flows. By measuring the environmental variables, such as water table or flow, over a range of sites, including those where the species or community is absent, it is possible to determine the relationship between these variables and the population or community. From an understanding of why species or communities are where they are, it is possible to predict the consequences of the expected environmental change. In some cases such studies will already have been completed while in others the data may be available to analyse. A considerably weaker approach is to use 'expert opinion' on say the requirements of different fish species but it is then important to determine the logical basis of their opinions as experts are often wrong.

It may be necessary to piece together indirect effects in stages using the same technique. Thus one study may show how reduced flows may result in higher water temperatures, another study show higher temperatures may benefit competitor or predator species and a third study show high densities of competitors or predators may result in reduced populations of the species of conservation interest. If the species of interest uses a range of habitats then it is necessary to split the analysis. Thus changes in the lake may alter the prey population and feeding conditions while changes to the marsh may affect where they breed.

5 Consider why the predictions from the previous section might be wrong. Do other factors operate at the threatened site? Will there be a long time lag? Will the absence of a certain species prevent the changes? It is probably usually easier to predict species declines than species increases. For example, J. Lubchenko (personal communication) attempted to predict the consequence of raising sea temperature next to a power station output. Those species predicted to decline usually did so, but those predicted to increase often did not, due to complex interactions of food, predators and parasites.

Such uncertainties justify the precautionary principle, that if there are doubts as to whether or not an action is ecologically damaging, the benefit of that doubt should rest with the environment (Wynne 1998).

6.5 Environmental impact assessment

Environmental assessment, environmental statement, environmental impact statement and environmental impact assessment all refer to a similar process of determining the likely consequences of a potentially damaging project. Environmental impact assessments are usually carried out because they are required by government or another decision-making body before they will give permission for a project to proceed. They are similarly required before some organisations, such as the World Bank, will fund projects. They are also sometimes carried out voluntarily by organisations that wish to minimise environmental damage, minimise the financial costs of damage compensation and rectifying damage, or prevent acquiring a reputation for being environmentally irresponsible. Environmental impact assessments are almost always paid for by those proposing the project and are usually carried out by consultants. Conservation bodies occasionally produce or fund complete or partial environmental impact assessments on projects that they have concerns about and these can be very useful for influencing policy makers.

Knowing that a project will face an environmental impact assessment may make the developers more conscientious in planning the project but environmental impact assessments are often an ineffective means of bringing about change, as by the time the assessment is complete the project planning is so well developed that any major modifications are deeply resented. One solution is to identify likely problems as early as possible and report them to the developers so changes can be considered. Strategic environmental assessment (Section 6.5.1) would also reduce this problem.

It is usually sensible to divide the project into four stages and consider the impacts of each stage separately. Not all stages apply to every project.

Preconstruction and planning: simply discussing the project may influence or delay the decisions of others considering investing in development or conservation.

Construction: activities, such as the need to temporarily drain an area or the presence of construction teams, will usually be brief compared with the lifespan of the whole project but can still have important long-term ecological consequences.

Operational: the obvious stage, when the project is functioning.

Post-operational: once the project is no longer functioning there may still be environmental impacts.

Most environmental impact assessments (Canter 1996, Treweek 1996, 1999, Harrop & Nixon 1999) consider not just ecological issues including damage to ecological processes and economically important resources but also issues such as landscape, noise, air quality, water quality, health, visual impact, disruption of traditional lifestyles, archaeological and cultural impacts. Wood (1995) describes how the approaches vary between different countries. Ridgeway *et al.* (1996) provide the materials for running an environmental impact assessment training course. Most assessments are variants of a similar procedure.

1 Is an environmental impact assessment required?
This step, called screening, identifies those projects needing an assessment. The seriousness of the impacts will depend upon the project's size, nature and location. For unclear cases an initial environmental assessment can be used to decide whether to proceed to a full assessment. Some countries list the types of projects that routinely require assessment.

2 Agree on terms of reference. The terms of reference state the key issues, the decisions required and the options that have to be evaluated. If the environmental impact assessment is required for legal reasons then this is likely to determine some of the procedures.

3 Quantify what the project involves. As well as data on area, size, location and intended project lifespan, there is likely to be technical data on the project, such as the amount of water consumed or the pollutants produced. It may also be necessary to collect such data from equivalent projects for those aspects for which data is not provided.

4 Identify the main issues. This step, known as scoping, entails consulting individuals, consulting organisations and reviewing the literature, in order to decide on the issues of greatest importance. Methods such as checklists or matrices (step 7) can be used here as a preliminary assessment of important issues prior to more detailed predictions. It is necessary to ensure that both the entire area affected (not just the area of the development) and the long-term cumulative effects are considered.

5 Collect baseline environmental data. The aim of this step is to determine the current situation. Ecological studies usually consist of collating information on the distribution and status of important communities and species. Pertinent data is also collated on the physical and chemical environment. In many cases there will be very little data. This step usually involves a combination of literature reviews and field surveys (see methods in Chapters 2 and 4) to fill gaps in the available information and data analysis. The data used will tend to be dependent upon availability, priorities (e.g. endangered or flagship species) and ease of surveying. As a result, surveys usually concentrate on birds and plants but often include some work on butterflies and mammals. In practice, most surveys are so superficial that many species will be missed (Treweek 1999) and many are underfunded and have to be carried out in an unsuitable season. Ethical consultants should either refuse contracts which cannot produce an adequate survey or make the limitations clear. If there are standard methods for surveying a group or habitat in a country then these should usually be used to aid comparisons. Some measure of the extent of population variability is also useful so that any observed change can be placed in context. This step also needs to identify those species and habitats that are of priority (see Chapter 3). A quantifiable index of the importance of a species or habitat is usually considered more acceptable than a simple impression of rarity or priority. Surveying and identifying priority species and habitats is of course essential, but many environmental impact assessments concentrate on this rather than on the central but harder task of predicting impacts.

6 Establish a process of consultation. Consultation is now accepted as being important and should be carried out at all stages, but the legal requirement for it varies enormously between countries. Consultation may improve the environmental impact assessment by correcting facts, raising issues that would otherwise have been missed, identifying public concerns (which may differ from the issues raised by experts), producing sensible modifications, suggesting alternatives and allowing involvement that may reduce opposition to the final report.

Public consultation can consist of inviting written responses or holding public meetings, in which case dividing people into discussion groups allows a greater range of opinions, encourages dialogue and reduces the probability of it being dominated by a few individuals with strong views.

7 Predict impact and its significance. It is much harder to assess ecological impacts than, say predict changes in noise level or air pollution. The main method for predicting ecological consequences is the framework described in Section 6.4. The main technique is to understand the requirements of species and communities so that the consequences of changes can be predicted. Assessing non-ecological impacts uses comparable approaches, as described in Canter (1996) and Harrop & Nixon (1999). As well as considering the likely consequences, it is important to assess alternative project locations, timing or methods. In almost all cases there will be some uncertainty over the impact and broad terms are often used (e.g. unlikely, possibly, likely, probably, almost certainly). The predicted impacts should be placed in context. Risk assessment involves determining the likelihood of various outcomes and the consequences of each (Royal Society Study Group 1992). This is easiest for projects for which there is abundant data on their consequences, such as the human mortality rate associated with hazards such as dam failure, railways and aeroplanes. In principle this could be applied to produce ecological risk assessment, for example by estimating extinction risks, although in practice it is often too difficult.

Impact assessments thus usually consist of a general expression of the likely main problems, using the techniques outlined below, and some more detailed analysis of the major problems, using the methods in Section 6.4.

Checklist: This is a simple list of possible physical/chemical, socio-economic, biological and cultural changes. Expected impacts may be described under each and can be given a score of the likely magnitude. They are a useful step towards encouraging comprehensive coverage, and providing a structure for subsequent analysis and discussion.

Leopold matrix: This is a table with rows listing the activities (divided into project stages) including potential problems (Leopold *et al.* 1971). Thus a road may lead to habitat loss, noise, the risk of spillages, associated development and increased hunting and so each of these is given its own row. The columns list the components, e.g. all the communities, species or ecological processes that may be effected. Each grid cell is given a symbol according to whether effects are likely to be a negligible, beneficial, detrimental, or a combination of these. In one version, a score is given to each potential impact, such as from -10 (expected considerable deleterious impact) to $+10$ (considerable benefit). One technique is to divide each grid with a diagonal line and state the magnitude of the impact (-10 to $+10$) in the top left corner of each cell and the importance (from $1 =$ least to $10 =$ highest) in the bottom right. Other elaborations include distinguishing whether impacts are direct or indirect, unknown impacts, impacts for which mitigation measures or design solutions are known. Leopold matrices are useful for assessing and presenting the main issues to be considered but are poor for analysing indirect consequences. The results of Leopold matrices are obviously subjective so the justification (such as using methods in Section 6.4) should be documented.

Descriptive matrix: The rows are aspects, such as air quality, fisheries, recreation, a key plant community or a species of concern, and the columns consist of impact characteristics, such as adverse(A)/beneficial(B); strategic(S)/local(L);

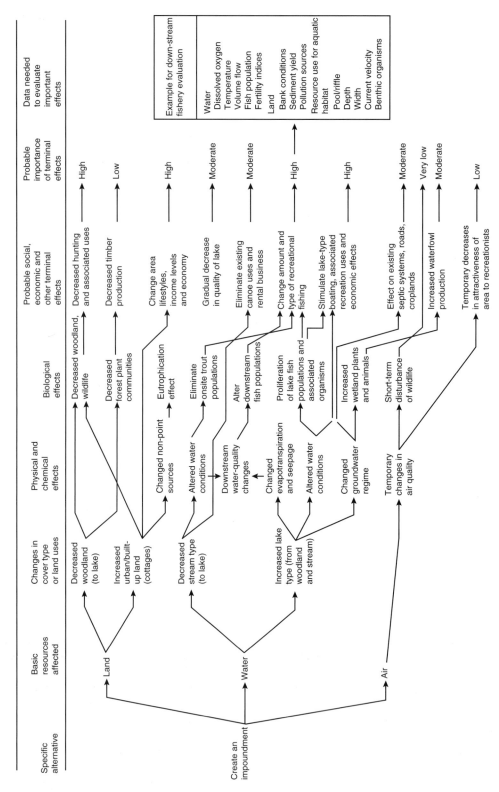

Fig. 6.1 Example of a network diagram to assess the consequences of creating an impoundment (from Canter 1996).

Table 6.1 Proposed general structure for an environmental impact assessment report (modified from Canter 1996).

Non-technical summary	
Chapter 1	*Introduction*
Chapter 2	*Description of the need for the project*
Chapter 3	*Description of the proposed project*
	Description of the actual project and how it will function
	Timing of construction, operation and dismantling of project
	Effectiveness of the project in meeting stated need
Chapter 4	*Description of affected environment*
	Boundaries of study area
	Baseline conditions
	Interpretation of existing quality of components
Chapter 5	*Impacts of proposed project*
	Identification, description and ideally quantification of impacts on environmental components
	Interpretation of significance of impacts
	Mitigation measures for adverse impacts
Chapter 6	*Evaluation of alternatives*
	Description of alternative proposals
	Selection method and results leading to proposed action
Chapter 7	*Planned environmental monitoring*
	Need for monitoring
	Description of sampling regime and monitoring programmes
	Outputs from monitoring and decision points
References	
Glossary of terms	
List of abbreviations	
Index	
Appendices	Technical description of project
	Pertinent laws
	Description of scoping programme
	Environmental factors considered but not considered relevant
	Description of public participation programme
	Species list and distribution maps
	Impact calculations
	Construction specifications of mitigation measures

long-term(L)/short-term(S); intermittent(I)/continuous(C); direct(D)/indirect(I); and irreversible(I)/reversible(R). The final column gives comments such as mitigation measures and suggestions. Thus a project to drain a lake for a day to mend a dam might read 'fisheries, A, L, S, I, D, R, potentially serious but can be ameliorated by catching fish and keeping in aerated pools next to the lake and by carrying out in winter when fish are more tolerant'.

Networks: These are designed for analysing indirect effects by examining the connections. As shown in Fig. 6.1, the consequences of creating an impoundment can be examined and the likely consequences evaluated. A problem with networks is that they can become very complex and difficult to interpret.

Geographical Information Systems: These are used to identify the areas in which problems arise. By adding the data to a geographical information system it is possible to identify the areas of conflict and this can be used to examine the consequences of different proposed schemes. A cheaper, easier but often less satisfactory (and much less impressive) approach is to simply draw maps on acetate sheets and lay them over each other to identify issues.

8 Devise mitigation methods and design solutions. Design solutions should reduce the damage. Mitigation measures compensate for the damage done, for example by restoration (Section 12.9), in order to reduce the impact on species and habitats that are affected by the proposal. The proposals need to be specific and realistic.

9 Devise monitoring methods. It is necessary to know if mitigation methods and design solutions work. This will require monitoring (methods in Chapter 4) targeted at the key species, habitats or processes which may be impacted. It is necessary to ensure the monitoring procedure has sufficient statistical power (see Section 5.16) to detect trends of a given magnitude within a reasonable period.

10 Prepare environmental statement. The report is usually open to public scrutiny. Canter (1996) stresses the need to use straightforward language, intelligible to an intelligent but uninformed reader, concentrate on the main message which is the environmental consequences of the project, ensure the report is sensibly structured and ensure that it looks attractive and professional. Get comments from a number of colleagues. Table 6.1 gives a proposed report structure.

11 Review procedure. As environmental impact assessments are usually produced or funded by the project proponents, the decision-making body will often carry out an independent review, for example by giving the assessment to a panel of external experts, to ensure that it is of sufficient quality and that it has followed agreed procedures and terms of reference. There is also the problem that

consultants employed to write the environmental impact assessment may also be too biased towards the perspective of their employers. Open independent review by known experts both reduces the likelihood of this bias and makes it less likely that the accusation will be made. External reviews are also likely to improve standards. Conservation organisations sometimes organise reviews if they believe the environmental impact assessment is substandard.

12 Carry out post-project analysis. This is necessary to determine whether the agreed conditions are being followed, whether predictions are accurate and whether impacts are being managed to acceptable levels. This will also help reveal if action is required to correct for unexpected consequences.

13 Review and publish effectiveness. Documenting the success of predictions, recommendations and mitigations would greatly improve future assessments.

6.5.1 Strategic environmental assessment

Environmental impact assessment is a reasonable method for examining the consequences of a solitary major project. It is, however, unsatisfactory where there are numerous projects, each of which are likely to have minor consequences, but collectively the cumulative or synergistic effects may be considerable. Strategic environmental assessment consist of reviewing the consequences of plans and policies rather individual cases (Therival *et al.* 1992). It has been used for planning strategies for oil exploitation, windfarms and transport (Treweek 1999). Strategic environmental assessment also has the considerable advantage of making early strategic decisions rather than using environmental impact assessment to prevent or modify proposals that others have invested time, money and effort into.

7 Conservation planning

7.1 Why plan?

Plans are boring. Planning is tedious. Good conservationists know what they want to do and should be getting on with it. Good reserve managers should be wearing out the soles of their shoes before the seats of their trousers. Why then is it necessary to plan, especially as time and resources will be diverted from direct conservation to paperwork and meetings? The answer is that many conservation projects flounder through poor planning. This is particularly true for those that are complex and involve numerous individuals and organisations. However, even conservation projects with one capable person in charge often lose direction due to poor or insufficient planning. Even excellent conservationists may be committed to the wrong solution or have conflicting opinions on the best approach. It is thus sensible to review all the evidence and opinions and ensure that the approach adopted is tested. Furthermore, if the analysis is presented in an explicit manner then it can be questioned and reassessed. Planning has the advantage of providing the opportunity to agree priorities for action, organise complex programmes in a sensible sequence, assign responsibilities for action, determine budgets both for the entire plan and for components within in it, determine work programmes, agree on how to monitor and evaluate the success of the work and allow the project to continue even if staff change.

This planning approach is fundamental to many areas of conservation although the terminology and structure may vary. It underpins species action plans, site management plans and integrated conservation development plans. It is also an important component to running successful organisations, education programmes or political campaigns. The planning structure derives from the observation that many businesses, like many conservation projects, are not sure what they are trying to achieve, how to make progress and whether or not they are succeeding.

There are, however, innumerable plans sitting on shelves around the world accumulating dust. There is no point in writing a plan unless someone will implement it, so any plan should have a clear purpose and readership. The implementors usually need to be involved in preparation of the plan, such that they feel a sense of ownership, for there to be any likelihood of action. This is true for all plans, whether a site management plan, a species action plan, plans to change exploitation, an integrated conservation development plan or a plan for changing an organisation. Consultation requires sensitivity and should take into consideration the culture of the relevant communities; nationalities and local communities differ as do groups such as politicians, landowners, farmers and other conservationists. Social sensitivities are important in deciding who should be consulted either as part of a planning meeting or an open meeting. Plans are often more likely to be implemented if in a local language. Some communities (including most conservationists) have an aversion to planning jargon, in which case avoid using terms like 'mission' or 'strategy'. One frequent mistake is to overpackage documents intended for discussion so they are perceived as complete and final. If plans are too long or too detailed they will be ignored.

Even with full enthusiastic participation, plans are readily forgotten and promises of actions superseded by other commitments. There needs either to be a mechanism to ensure the plan proceeds, for example by integrating within current procedures (Wilson 1994), or a real persistence to see the plan enacted. It can often require doggedness to implement changes and ensure that agreed actions are actually carried out.

Unless the planning process can influence budgets or time allocation it is unlikely to be successful and the separation of planning and budgeting results in the failure of most plans (Gray 1986). Plans regularly assume additional funding, people or equipment, but lack a realistic mechanism for providing it. It is often better to write a plan that is practical with the available resources, with a separated section outlining further possibilities should extra resources become available. However, in some cases the plans can be used to justify extra funding by both identifying what is needed and convincing others that the proposals are well considered.

Both the strength and weakness of planning is that it formulates thinking. The strength is that it encourages the collection of evidence and the analysis of major issues that might not otherwise be considered in depth. The weakness is that planning meetings are often tiresome, with individuals arriving unprepared and collectively making decisions without the detailed consideration that such important issues warrant. In reality, the experience of day-to-day decisions largely drive strategy rather than strategy largely determining decisions (Mintzberg *et al.* 1998). Innovative strategies usually do not come from logical analysis, but from the intuition and experience of leaders and employees involved in solving specific problems (Cambell 1999). The planning process may then deny or diminish their role (Wilson 1994). An extreme perspective is that the term strategic planning is an oxymoron, as planning often suppresses the creation of strategy (Mintzberg 1994).

A common failing of planning is to regard the production of a plan as an end in itself. In reality the plan should only be the start of a process of innovation, experimentation, analysis of accomplishments and approaches, and review. Analysis and action are then parts of the same activity, not isolated procedures (Ketelhöhn 1997). Progress is likely to be made much more rapidly in an open non-judgemental environment where mistakes and problems can be discussed frankly without blame and where individuals can readily change their opinions when presented with new evidence. Detailed plans have the problem that the situation is likely to change

and thus the original plan no longer applies. For this reason a clear mission can be useful as the basis for deciding how to respond to unforeseen events.

Plans usually go through several drafts and are updated periodically. Unless given a version number or date, there is the opportunity for considerable confusion.

To summarise, it is usually a mistake not to plan at all but elaborate deeply prescriptive plans usually take too much time and are rarely consulted. A good plan outlines a vision, reviews the available evidence, suggests directions and priorities and provides a basis for further analysis and learning.

7.2 The planning process

A plan is most likely to be successful if: (a) the need for it is understood by those involved and they are fully consulted during the preparation of the plan; (b) there is a process of monitoring and review to adjust the plans as necessary; and (c) the resource implications have been considered so that the plan is realistic.

The modern approach to strategic planning is to consider the following stages.

Mission. What is your vision? This seems easy. In reality this can involve painful debate as it requires agreeing on exactly what you really want to achieve. It should be positive, visionary and give a strong sense of purpose. The rest of the plan should flow from this mission. It may be omitted when the mission is obvious, such as in recovering a species, and is particularly useful for complex issues such as running an organisation.

Current position. Collate relevant information that helps you understand the current situation and the expected response to change. This includes developing hypotheses, gathering facts and analysing them. Thus this might involve considering whether and why a species has declined and what the most effective means are of restoring the population or whether the membership is as expected for such an organisation, how recruitment, retention and age structure compare with similar organisations and a review of the success of methods that other organisations use to increase membership.

Objectives. Decide exactly what you want to achieve, giving specific objectives. These should be realistic, measurable and specify a timescale. Self-imposed objectives are more likely to be achieved than those that are imposed from outside. Measurable objectives can help avoid the common problem of the plan being completed but never implemented, yet considered a success. After the mission, determining the objectives is often the hardest part of the plan. Ensure that there is enough time for completion. It is wise to distinguish between ends objectives, which show whether the project is achieving its mission (e.g. 'restore the population of breeding green turtles at the site to 250 by 2005') and means objectives, which show whether the proposed activities are reducing the problems (e.g. 'reduce the incidence of illegal turtle hunting to less than two cases a year by 2003'). Objectives should relate to results ('ensure at least 90% of shrimp netters have fitted and use a turtle excluder device within two years') rather than the activity ('provide free advice on turtle exclusion devices to all fishers and a publicity campaign to 80% of local schools). Within this framework the activities on advice and publicity would be included within actions. In reality the process of deciding on realistic objectives will also involve some consideration as to what actions are practical and thus the planning process is not always the linear sequence portrayed here.

Strategy. Outline the main broad policies for achieving the objectives.

Actions. Specify the actions that need to be taken, who will do them, how long they will take and how much they will cost. Either a priority should be allocated to each action (e.g. high, medium or low) or they can be ranked in order of priority. State the experiments, such as comparing different educational approaches, that will be carried out to improve the operations.

Monitoring. Devise realistic procedures for monitoring the work done and especially for monitoring whether the objectives are being achieved. Any unexpected events should also be documented.

Reviewing. A timetable of periodic review should be built into the plan. The results from monitoring and from experiments should be used to adjust the actions, and sometimes the strategy, to ensure the objectives are reached; this process is often called adaptive management. By also monitoring the actions it is possible to distinguish theory failure (the actions took place but did not have the anticipated result) from programme failure (the planned actions did not take place).

7.3 The species action plan process

The typical structure of a species action plan is given in Box 7.1 and an example is given as a Case Study. Writing a species action plan usually starts with a meeting of all those involved (a 'pathfinder meeting') which identifies the threats and limiting factors, defines targets and identifies actions. These actions should be ranked or classified according to priority. The meeting should agree objectives towards which to focus conservation action and against which the success of action can be monitored (Williams *et al.* 1995). In a plan it is often helpful to separate actions into categories, such as research and monitoring, habitat management, species management, education and public awareness, or action to change government policy and legislation. The plan is usually written by one person but circulated widely for consultation before being approved by the organisation or organisations involved. To implement the plan, work outlined in the actions should be incorporated into the work programmes of the individuals and organisations involved and budgeted for. It is usually the case that there are some indications as to which approaches will be most effective but insufficient evidence to be confident. Most action plans should thus involve experiments and the process of diagnosis described in Chapter 6.

Snyder (1994) is critical of elaborate action plans as he suggests they are rarely consulted. He recommends brief accounts of the factors limiting populations, broad research needs and potential conservation strategies.

A team or organisation (the 'lead organisation') may take control of the development and running of a species action plan. Conservation bodies often assume that they should take this lead but in reality it may sometimes be more effective for the

CASE STUDY

The UK Corncrake species action plan

The Corncrake *Crex crex*, a bird of Northern Eurasian damp grasslands, has declined enormously over the last century. The actions resulting from this plan have helped stop the decline and resulted in local increases. This is a summary of the UK 1989 plan (from Williams *et al.* 1995) that I have restructured slightly to fit the format of Box 7.1.

A European species action plan for the Corncrake was also written following a similar format with a pathfinder meeting in Poland attended by 23 of the 34 countries which hold Corncrake (Crockford *et al.* 1996, Sutherland 1994).

Current position

Sustained decline and contraction of United Kingdom range for at least 100 years. The main threats are loss of traditional grassland habitat mosaics and changes in grass management and cutting techniques, especially loss of hay meadows through abandonment or conversion to sheep-grazed permanent pasture, and adoption of earlier cutting dates.

Legal status

A globally threatened species: Annex 1 of the EC Birds Directive; Schedule 1 of the Wildlife and Countryside Act 1981 (in Britain) and Wildlife Order 1985 (Northern Ireland).

Priority

High

Objectives

1 To maintain numbers of Corncrakes in Britain at or above the 1983 level (479 singing males).
2 By 1998 increase the total numbers of singing males in Britain to at least 600.
3 To maintain the range of Corncrake in Britain at or above the 1983 level (82 occupied 10 km squares).

4 By 1998 increase the range of the Corncrake in Britain to at least the same number of 10 km squares occupied in total in 1988 (90 occupied 10 km squares).
5 In the long term to re-establish Corncrakes in parts of their former range in the United Kingdom.

Broad policies

1 Protect the nests, adults and young from destruction due to cutting of meadows for hay or silage, through advisory and species protection work with farmers and crofters; seek provision of grant aid to minimise the financial impact of such measures on farmers and crofters.
2 Encourage continued production of hay, and its sympathetic management for Corncrake, through special designations such as Environmentally Sensitive Areas in the core areas of its range in the United Kingdom.
3 Develop and promote ways of supporting traditional land management practices associated with crofting, which appear to benefit the Corncrake.
4 Develop and promote an European Union-wide hay premium scheme, sustain and encourage hay production at European level for grassland species, including the Corncrake.
5 Outside core areas of current Corncrake distribution, conduct experiments to assess whether new centres can be established at grassland nature reserves in the United Kingdom through appropriate habitat management and use of tape lures.
6 Investigate levels of mortality due to feral and domestic cat and Mink *Mustela vison* predation and assess the possibilities of reducing such mortality.
7 Reduce birdwatching pressure on Corncrakes and their habitats in the Western Isles through promotion of a 'showing Corncrake to people' project.

(Continued)

(Continued)

8 Investigate the level of mortality in Quail *Coturnix coturnix* nets in Egypt and assess the potential for reducing such mortality.

9 Institute a population-monitoring scheme and undertake a complete United Kingdom census at 5-year intervals; to continue to research Corncrake ecology and the factors influencing its numbers and distribution.

Actions

1 Policy and legislative

(a) Promote and support Corncrake grant schemes in Scotland and Northern Ireland.

(b) Ensure designation of core Corncrake areas as Environmentally Sensitive areas.

(c) Design and promote a 'Hay Premium Scheme' and a new system of support for crofting.

2 Site safeguard

Ensure that high-density Corncrake areas are designated as Sites of Special Scientific Interest in Britain or Areas of Special Scientific Interest in Northern Ireland and/or as European Union Special Protection Areas.

3 Land acquisition and reserve management

(a) Purchase, when opportunities arise, strategic core Corncrake areas.

(b) Ensure that appropriate active management for Corncrakes is carried out on all nature reserves in Corncrake areas.

4 Species management and protection

(a) Reduce mortality of adults and young from mowing operations by direct wardening and the promotion of Corncrake-friendly techniques.

(b) Advise crofters of potential risks to Corncrakes from large numbers of feral/domestic cats.

(c) Support localised programmes to control feral Mink and Ferret *Mustela furo*.

(d) Prevent disturbance by birdwatchers by advice and direct intervention.

5 Advisory

Provide written and verbal advice for all those involved in the management and conservation of Corncrake areas.

6 International

(a) Assess the nature and scale of Corncrake mortality in Egypt.

(b) Liaison with Birdlife International in Europe on international efforts and needs for Corncrake conservation.

(c) Promote research and conservation measures for Corncrake in the Republic of Ireland.

7 Future research and monitoring

(a) Conduct full surveys of the breeding population in Britain and Ireland every 5 years.

(b) Carry out annual monitoring of breeding numbers at key sites.

(c) Conduct periodic habitat surveys in key sites.

(d) Undertake an economic, technical and agronomic study of grassland management in key areas.

(e) Experimentally test the value of tape lures in attracting Corncrakes to unoccupied areas of suitable habitat.

8 Communications and publicity

(a) Publicise the conservation needs of the Corncrake in local media in core areas.

(b) Promote responsible behaviour by birdwatchers through articles in birdwatching magazines.

(c) Conduct surveys after mowing to assess the density and timing of nests and broods and incidents of mortality.

(d) Investigate the responses of Corncrakes to the approach of mowing machinery.

Review

A review meeting is held every autumn to discuss the action for the next year. The plan will be fully reviewed and rewritten every 5 years.

Objectives with clear biodiversity targets, sometimes referred to as ends objectives, have the merit of encouraging concentration on the real purpose of the action plan. By 1998 the success in delivering each of the above objectives was as follows:

1 This was easily met in each year from 1993.

2 This target was met in 1997 (637) but declined to 589 in 1998. The reason for this decline is not known.

3 Comfortably exceeded, as has increased to 93 squares.

4 Exceeded with 93 squares.

BOX 7.1

Writing a species action plan

This format is modified from that used by the Royal Society for the Protection of Birds (Crockford *et al.* 1993).

Current position

Conservation status of the species: reasons for concern and inclusion in the action plan process, population size, geographical distribution, trends, level of population and conservation knowledge, ecology, limiting factors and threats thought to affect population numbers and distribution with classification (high, medium, low, unknown) as to their relative importance. Summary of conservation action undertaken and in process.

Legal status

The international and national legal status.

Priority statement

The importance attached to overall action for the species in terms of high-, medium- or low-priority categories.

Objectives

Goals at which to aim in taking conservation action.

Broad policies

The general strategy that will be used to achieve the objectives.

Actions

These are described with priority (high, medium or low) and responsibilities for implementation and a broad estimate of timescale, staff time and costs.

Proposed actions are listed in the categories:

1 Policy and legislative
2 Site safeguard
3 Land acquisition and reserve management
4 Species management and protection
5 Advisory
6 International
7 Future research and monitoring
8 Communication and publicity.

For quick reference by those implementing the plan, each action is given a priority rating according to its cost effectiveness and achievability, and is assigned to a particular work programme (or work programmes), specifying who does what and when.

Review

Timetable and procedures for monitoring effectiveness and for reviewing and updating the plan.

---

organisation that has the greatest power to determine the fate of the species (such as a forestry company or water authority) to take responsibility with the conservation body acting as advisers and commentators.

7.4 The site management plan process

Almost all reserve management is based on management plans. The scale of this process will differ enormously but even for very small sites visited occasionally by volunteers it is usually worth creating a very brief plan. Start with a meeting of all involved to discuss the plan and agree who will do what. It is usual for one person to write the plan but that person should consult widely. A detailed prescription for a management plan (based on Hirons *et al.* 1995) is given in Box 7.2.

It is important to first collate the information on the importance of the site and then use this to establish priorities and decide on management. This is essential as almost any management will be detrimental to one set of species while benefiting another set. The common problem of starting management without considering the priorities has frequently resulted in damaging some important habitats or species.

BOX 7.2

Writing a management plan

Summary

Briefly describe the importance of the site, the objectives and strategy. This is valuable for managers, organisation directors, media, the public, and fund-raisers.

Current position

This section should collate the published and unpublished information available locally and nationally. Information documenting changes in the site, such as changes in grazing pressure, and the consequences of change, are particularly useful. Much of this information is best given as maps in appendices.

General information

1 Location
2 Site status (e.g. protected area designations)
3 Tenure
4 Site definition and boundaries
5 Legal and other official constraints and permissions required
6 Main fixed assets

Physical information

1 Climate
2 Hydrology
3 Geology
4 Soils

Biological information

1 Habitats
2 Flora
3 Fauna

Cultural information

1 Commercial use
2 Educational use
3 Recreational use
4 Archaeological interest
5 Aesthetic considerations, such as landscape
6 Research, survey, monitoring
7 Past and current conservation management

Evaluation

This section describes the importance of the site in terms of its international, national, regional and local status.

1 Size
2 Diversity
3 Naturalness
4 Rarity
5 Fragility
6 Typicalness
7 Recorded history
8 Position in ecological/geographical units
9 Potential value for future development:
   Role in habitat and species action plans
   Visitor services
   Education
   Marketing, retail
   Marketing, recruitment of members
   Public affairs
   Demonstration and advisory work
10 Intrinsic appeal
11 Other criteria
12 Identification/confirmation of important features
13 Operations likely to damage the special interests
14 Main factors influencing the management of the site
15 Land of conservation or strategic importance in the vicinity of the reserve.

Objectives

List the main priorities for the area. These could include protecting a particular plant community, increasing the numbers of a particular species, increasing the number of visitors or ensuring sustainable exploitation of particular species.

Broad policies

By reference to the site's main features, describe the general policy for management.

(Continued)

(*Continued*)

Actions

This section lists what needs to be done over the next 5 years in order to meet the objectives. Each action should be prioritised, considering the time and resources required and available.

1 *Habitat and species management* This could include management prescriptions, for example for grazing, burning, patrolling to maintain refuges, creation or predator management.

2 *Visitor services, interpretation and education* This could include prescriptions for laying out nature trails, walks, open days for local people, guided walks, provision of displays and leaflets, facilities for school educational programmes, membership recruitment or retail sales.

3 *Estate services and major machinery* This could include prescriptions for positioning and erecting observational hides, work stations and other buildings, maintaining or erecting fences, gates, paths, roads, buildings, sluices, cattle-handling units, acquisition or maintenance of tractors and other agricultural and forestry equipment and boats.

4 *Public relations and administration* This could include prescriptions for renewal of leases, formal meetings with neighbouring landowners and local authorities, action to combat nearby developments, information to be sent to other organisations, volunteers and their tasks, minimum staffing requirements, identification of possible reserve extensions or liaison meetings with other conservation organisations.

5 *Research, monitoring and survey* This will provide the information to say whether the other management prescriptions are resulting in the desired objectives.

Monitoring

This could include prescriptions for surveys and monitoring of biotic groups, contributions to national monitoring programmes, assistance to on-site biologists, monitoring effects of management programmes, studies of particular species or attitude surveys amongst visitors.

Review

Information should be updated continuously, for example, as new species for the area are discovered or there are changes in ownership. The entire plan should be reviewed at regular intervals, say every 5 years. This section should outline the plan for updating and reviewing.

Many site management plans are completed but not executed (Caldecott 1996). One major reason is that a site management plan can be written from an office, with perhaps a brief visit to the site, but implementing the plan requires considerable time in the field with marked consequences both for the personnel and agency involved. Unused site management plans may not only waste resources but may even be counter-productive (Caldecott 1996) as the plan itself may provoke action by other users of the land, such as logging companies, mining companies or local people, or result in unnecessary public relations problems.

8 Organisational management and fund raising

8.1 Why is organisational management important?

Conservation is usually thought of as consisting of scientific research or action to alleviate problems and these are the interesting, enjoyable and satisfying aspects of conservation. I am sure that many will be surprised, or even shocked, to find a chapter on organisational management in this book. However, as Clark *et al.* (1994a) argue, organisational problems are the major weakness of many conservation programmes and the book by Clark *et al.* (1994b) gives numerous examples of such organisational difficulties hindering effective conservation.

8.2 Leadership and management

Much of the success of conservation organisations depends upon the management and leadership skills of the director. The function of management is to make it easier for the workers to work and not, as some managers think, for workers to enable managers to manage. The skill of leadership is to provide direction, inspiration, confidence and enthusiasm.

Conservation bodies often consider themselves as being very different from companies, but in reality they are often similar except the purpose is to raise money for conservation objectives rather than to make a profit. One way to attain more of the conservation objectives is then to find ways of raising income and controlling expenditure, just like any company. Financial management is often harder for charities as the objective is to break even, which is a difficult balancing act, rather than simply make a profit. It can even be argued that financial management of charities has to be even more rigorous as there is not the constraint of the marketplace to impose discipline.

There are, however, some critical differences between charities and companies. In business there is a clear link between competence and income: if the goods or services are poor or expensive the customers go elsewhere. For charities the link is often tenuous (Bryson 1995), as the numbers of members and donations may have little relation to the actual effectiveness at conservation. Too much emphasis may be on gaining money and pleasing members rather than achieving conservation aims. At the other extreme too little investment in fund raising means that the organisation achieves less than is possible. Most businesses have the simple objective of making money but charities have more complex aims. It is for this reason that the mission needs to be carefully defined.

Conservationists are often highly motivated, committed, hard working and underpaid and the collective values of the organisation have to be cherished by the management. The crude application of management ideas from business can result in dissatisfaction through increased administration and a feeling of lack of trust, without increased efficiency.

Managers need to create simple systems, straightforward measures for monitoring success of those factors that are of real importance (e.g. population size of species of concern, number of visitors or the percentage of members renewing), clear individual or team responsibilities, and the capacity for individuals from different parts of the organisation to work together on projects (Waterman 1994).

8.2.1 Leadership

A good manager can plan and ensure the smooth running of a project. A good leader can motivate others so as to gain the most from the staff. Effective leaders provide vision, direction, gain commitment to the conservation mission and motivate others. There is no simple recipe for successful leadership as much depends upon the nature of the organisation. Flourishing organisations and very successful projects are almost invariably those with good leaders. However, good leadership but poor management can lead to high but unsatisfied expectations.

Leaders have three main roles:

1 Provide direction. The direction should be in terms of conservation objectives rather than empire building. It needs to be absolutely clear what the mission and objectives are. Direction is set through behaviour and example rather than through words and paperwork as behaviour reveals what the leader really considers important. Visionary leaders can imagine what an organisation could be like, what it could achieve and how this vision can become reality.

2 Motivate to gain commitment and co-operation. The main role is to encourage and inspire by setting an example. Determination, hard work, a predilection for action and the ability to inspire enthusiasm are all important. Celebrate success.

3 Make the best use of the skills and energies. Much of the skill of leadership is to look after the needs of each individual and of the group while ensuring the task is achieved. Show support and provide confidence. Criticise in private but praise in public.

8.2.2 Delegation

It is now well appreciated that although the traditional strictly hierarchical management may be effective in a static world it is often too sluggish to respond to change as the management decision makers are too remote from the realities. Modern successful companies thus tend to have fewer layers with greater autonomy, flexibility and decision-making power at lower levels (Heller 1997). Similarly in conservation programmes, Clark *et al.* (1994a) suggest that extensive bureaucracies can result in narrowly defined professional roles, intelligence failures, weak science, slow responses and information being distorted as it moves up and down the hierarchy.

Most people work best when given independence and responsibility. Delegation usually increases morale, productivity and abilities of the staff. Delegating responsibilities also means that outsiders can be dealt with efficiently rather than waiting for authority from above. Delegation often works best when responsibilities are not fragmented, for example, it is often sensible to entrust entire tasks to single individuals. Ensuring others are informed of the delegates new role usually reduces confusion and can increase a delegate's pride and confidence. Delegation is not, however a panacea to solve all problems; the four main problems of delegation are: (1) poor decisions on important issues may be made by less experienced staff; (2) reduced control; (3) loss of morale amongst those who do not desire greater responsibility and (4) greater potential for chaos as individuals make different incompatible decisions or give conflicting messages to others, for example on the amount of support the organisation will provide (Peiperi 1997). A good independent team can be extremely effective with high morale and a freedom to act but a bad independent team can be a disaster as they proceed undetected with poor decisions that could have been prevented by guidance. Balancing the level of delegation with the appropriate supervision requires skill and a good knowledge of individual's strengths and weaknesses.

'Hard' delegation entails dictating exactly what is required and by when. 'Soft' delegation entails outlining the problem, when and how progress will be monitored and the precise conditions under which reporting back is necessary (for example, if the population declines by 5%, the budget is going to be exceeded or there is any bad publicity) and the limits to authority. Such 'management by exception' allows the manager to concentrate on other issues, knowing that there will be regular reports and any problems will be reported. The staff member is free to work on the project, knowing any significant deviations have to be reported. It may be useful to start this process of soft delegation with a discussion of the approach in order to provide initial guidance as otherwise subsequent guidance may be construed as interference.

8.3 Types of conservation organisations and their problems

There needs to be a balance between consultation and the ability to make decisions. One approach adopted by most organisations is to have an individual responsible for day-to-day decisions and have a governing council that decides the strategy. The members of these should be selected for their skills in a wide range of relevant technical fields such as the law, finance or publicity but they also need to have commitment and understanding of the aims of the organisation. Appointment for a fixed term, e.g. 4 years with a year's gap before re-election is possible, encourages fresh views and provides a tactful way of losing less useful members.

Hudson (1995) lists a series of common faults of directors.
• Talking too much in meetings.
• Consistently siding with one interest group within the organisation while claiming independence.
• Blaming the governing council.
• Always seeing people's weaknesses rather than strengths.
• Not delivering on tasks they committed themselves to.
• Not delegating explicitly.
• Criticising colleagues behind their backs.

As Hudson says, everyone makes mistakes, but successful directors learn from these and find their own formula for effective leadership. There is usually too little celebration of success, yet this is one of the most effective ways of generating and maintaining direction and enthusiasm.

Conservation organisations can usually be divided into governmental or non-governmental organisations (NGOs). Non-governmental organisations follow a continuum from entirely voluntary to entirely professional but I have tried to divide this continuum.

Entirely voluntary organisations

Voluntary organisations are typically run by a committee of chair, secretary, treasurer, organisers of specific tasks and a few other ordinary members. The committee will often discuss practically all

decisions. They can be very effective, especially when applied to single issues. They are often directed by one or two highly motivated individuals but if these leave, or lose interest, the organisation may become ineffective. Whether such organisations are effective usually depends upon whether committee members can devote sufficient time and whether they possess the range of required skills. Such organisations often lack the basic skills of running meetings (see Section 8.5).

Small organisations often face crises of confidence. It is often the case that all the work is done by a few overstretched individuals, attendance at some events is poor and there are periods when almost nothing is achieved. In my experience some very successful small organisations have these confidence crises, even when they achieve a remarkable amount. Dwelling in public on failures, such as poorly attended events, by complaining at the event itself or in subsequent newsletters probably only discourages further attendance. A positive tone is more likely to attract others as well as attract funding.

Mainly voluntary organisations with a few paid staff

These often develop from entirely voluntary organisations. If the key motivating individual is employed the organisation may become a sometimes rather curious, but often highly effective, one-person show. If this person leaves or loses interest the organisation often crumbles. Alternatively, the organisation may employ staff from outside with the strategy and direction largely directed by the committee of volunteers.

Largely professional organisations

Professional organisations often develop from mainly voluntary organisations with paid staff. Typically a full quota of staff are employed full-time, but the governing council is voluntary. The governing council should avoid routinely reconsidering details or decisions already delegated to staff members, as this is unnecessary and demoralising to the staff. A common problem is that the staff often develop considerable expertise and confidence such that they no longer look to the governing council for guidance and may even view them as an irritating irrelevance. It is thus particularly important that the governing council members have both sufficient skills to provide a real role, such as experience in running a similar sized company, public relations, financial management or conservation biology, and that they appreciate the ethos of the organisation. It can also be useful to ensure that governing council members are trained, for example by initially being given an induction into the working of the organisation and regular opportunities to meet staff members.

Even with a highly skilled staff, the governing council still has three important roles.

1 Determining the mission, objectives and strategy, monitoring the success in achieving these objectives, allocating resources to different purposes and ensuring long-term financial security.

2 Mediating when interests conflict. A strong governing council is necessary to ensure that the interests of particularly powerful staff members do not dominate. It may also be necessary to resolve differences between staff, the membership and funding bodies.

3 Appointing, supervising, monitoring and, in exceptional cases, sacking the director. The chair of the governing council also has the important functions of managing the director (not the reverse) and providing the director with support and encouragement. It is not generally appreciated that directors can feel isolated and even lonely, especially when dealing with issues that cannot be discussed with the staff. Directors usually only hear criticism and problems, even when they are doing a very good job, and sometimes need reassurance. The chair and governing council also need to ensure that they are appropriately consulted and fully informed.

The governing council of a charity are usually responsible for the actions of the paid staff and for the finances and have to be made aware at the start of their legal responsibilities.

Senior managers, especially those whose roles have developed within the organisation, tend to become involved in decisions that are the responsibility of others below them. They must be made aware that this leads to confusion for those outside the organisation as to who makes the decisions. This can lead to demotivated staff upset with the interference and stressed but inefficient managers who, by interfering with the work of others, lack the time to do their real job of managing and leading.

Government organisations

In many countries government organisations are the only organisations taking conservation action. Government organisations sometimes have great opportunities as they may be relatively well connected and securely funded and it may be easier for them to bring about political change. Against this, there may be a perception that they should not question policies that affect other, more powerful, areas of government such as development, agriculture, water use, fishing and forestry. There is often less freedom for individuals to make decisions so the staff may be less well motivated. They are often also tied up in bureaucracy. Government organisations often suffer from similar problems to the largely professional organisations discussed above.

8.4 Collaboration between organisations

Conservation organisations and individuals often collaborate to share expertise, pool resources and prevent counterproductive competition. Several organisations speaking with one voice can often make for a more powerful message. Such collaboration is often highly successful but can also lead to bitter disagreements and a breakdown in the collaboration, as shown in projects to conserve the Black-footed Ferret *Mustela nigripes* (Clark 1997) and Giant Panda *Ailuropoda melanoleuco* (Schaller 1993).

It is essential to appreciate that organisations and individuals will often have objectives in addition to the straightforward conservation objective. These could be to gain publicity, increase membership, carry out research, obtain overheads, obtain employment from grants or keep animals or plants in captivity. A failure to understand and accept why others are involved can lead to problems. If conflicting objectives are likely to be a problem it should be discussed at the start: for example what are the main objectives? Who is responsible for making decisions?

Who will do the work? Who will provide the funding and equipment and how much? Who takes the credit? Who gains the publicity? Collaboration is often easier between organisations that are very different (such as a botanic garden and an education charity), as they are dependent upon each other for the project and are usually not competing for the same funds or opportunities.

8.5 Meetings

Meetings are an excellent way of bringing together individuals with a variety of views and expertise to make difficult decisions requiring a variety of perspectives. Section 10.6 describes how to be effective within meetings. Meetings can also promote co-ordination and save time by bringing relevant people together to exchange information. However, it can often be more efficient for a single person to make decisions and consult where necessary. An individual with personal responsibility for a decision will often consider it more carefully than will a committee. Furthermore, an individual can usually make decisions much more quickly than having to wait for the next committee meeting.

The basic rules for running meetings include:
- Ensure that the meeting has a clear function, a timetable and the participants know the limits of their authority.
- Ensure the meeting is sufficiently small to be effective (ideally less than 10 people) while including those that need to be consulted and a range of skills and perspectives.
- Seat all participants so that they can easily see each other.
- Prevent distractions. Either start with coffee or stop for coffee, as serving drinks or food during meetings usually breaks concentration.
- Organise the agenda with the important items first, as otherwise important decisions tend to be rushed.
- Ensure that all papers are circulated well in advance and both explain each issue and provide sufficient background information that members can form questions and opinions prior to the meeting. The paperwork should make clear the

decisions that have to be made. If the chairperson asks individuals questions or asks for their opinion then it gives the clear message that they should come prepared.
- Allow discussion and the expression of a range of views but prevent digressions.
- Ensure debate on important issues. Trivial items often attract more thought and debate than fundamental decisions that affect the organisation's future. In my experience, a lack of discussion on fundamental issues does not mean that participants have read, understood, thought through the consequences and decided the decisions are right: it usually means they have no idea what is happening.
- Have a chairperson who ensures decisions are made and not just discussed.
- Make a decision on each issue and then move on.
- The chairperson should summarise the decisions after each item and at the end gives a quick summary reminding everyone who will do what.
- Produce and circulate minutes soon after the meeting, listing the decisions and who will implement them.
- Do not allow major items to be included in 'any other business', as this allows issues to be unexpectedly introduced at the end of the meeting that should be on the agenda with background papers and can lead to poorly considered decisions. If an important item occurs at short notice then details can often be sent by email beforehand.
- Emphasise the work between meetings, which is usually what matters, rather than the meeting itself. Ensure the promised actions are carried out. Contacting those named on the minutes prior to the deadline to remind them of their obligations is often effective. The deadline is often the next meeting but this often means the work is carried out in short pulses a few days before each meeting and if it cannot be done, for example because the person to be contacted is away, or the committee member who should do it is ill or unexpectedly busy, then it can slip away undone or be delayed until just before the next meeting. One possibility is create a committee enforcer to remind others of their tasks and check on progress. This could be the chair's main responsibility.

• Many committees repeatedly return to the same issue without making progress. This can be overcome by ensuring decisions are clear and actions implemented, by the chair ensuring the committee has a work plan that is compatible with individual member's agendas and the chair ensuring that discussion does not return to issues on which progress is unlikely.

8.5.1 Generating ideas in meetings

The interaction between individuals in meetings can often generate ideas. Approaches for encouraging this include:

Brainstorming. Ideas comprising a few words are shouted out to a facilitator who then writes them on a board or large sheet of paper so they can be easily read. No comments or criticisms are allowed, regardless of how silly an idea is. It is usually necessary to explain the process and rules beforehand and if this is a novel technique it is often best to practise first, for example, by asking for suggestions on possible uses for paper clips. Once complete, the best ideas are then selected. This is useful for producing novel and imaginative ideas but is less suitable for solving complex problems.

Division into subgroups. Each subgroup (also known as breakout groups) is given a specific task and asked to report back to the entire group. This is useful when there are too many technical issues to resolve in the time available or when the group is too large for useful discussions.

Cards. These can be circulated and completed then the facilitator collects them. For example, in a Nature Kenya meeting to determine which sites were of highest priority, the first stage was to agree on the criteria used for prioritising. Each participant was asked to consider a site they believed should be conserved and write three reasons, one on each card, why it is important. The facilitator collected these cards as they were being produced and grouped similar ideas on the board. This entire process took less than five minutes, showed that there was general agreement and led to the rapid creation of criteria. This technique is useful in producing a rapid consensus from a large group.

Borrowing ideas from others. This easy technique is greatly under-used. Most problems have already been repeatedly solved. Many individuals and organisations are very happy to provide or exchange ideas. Approach non-conservation organisations; they share many similar problems and are often willing to provide considerable expertise and advice when they would not be able to provide direct funding.

8.6 Crisis management

A reserve floods, a disease spreads through a population of an endangered species, an organisation faces financial ruin or damning stories appear in the media: such crises test managerial skills. The accepted approach to crisis management is to view it as being no different from ordinary management, just at a faster pace. It tends to be a badly planned, instant response that turns a crisis into a disaster.

1 Stay cool and calm, both to give an impression of control, and to ensure bad decisions are not made in undue haste. Make it clear to those concerned that the problem is being treated seriously.

2 Consider the urgency of each aspect of the problem.

3 Ensure there is the capacity for yourself and others to plan by delaying, delegating or cancelling other work. Bring together those necessary to plan the response.

4 If appropriate, set up a crisis management centre and a communication system to provide instant intelligence and to communicate with the team, press and others.

5 Run through exactly the same decision-making process as described in Section 7.2. If urgent decisions have to be made then ensure that the decision-making process is as thorough as possible in the available time. This may entail developing short-term plans, contingency plans to respond to developments or emergencies and consideration of when long-term solutions should be planned and introduced.

6 If there are media interests then address the concerns as much as the facts. Consider having a senior representative present to show the issue is being taken seriously. If responsible for a problem, show sympathy, concern, reassurance, competence and ideally show that you have made changes as a consequence (Bland 1998). Reassurance is often best from a credible and authoritative third party. Contact the journalists and give them a story, as unhelpfullness is unlikely to stop the issue being published and it is almost always better that it is you, rather than a critic, that provides the material. Media stories usually develop quickly and often in unpredictable ways and, once established, it is difficult to refute false accusations or reverse negative images. A rapid response can often be effective in preventing wrong or distorted stories but ensuring someone is always available to respond to the media requires planning.

Most problems develop gradually and the real skill is to recognise and prevent developing crises. It can be useful to anticipate future problems by asking 'what-if' questions (if a particular situation occurred, what would be the impact and what should the response be) (Sherden 1998). It can also be useful to consider what is needed to deal with a crisis, for example, would the telephone system be sufficient?

Some become good at dealing with crises as a result of running a disaster-prone group. A more responsible way of gaining experience is to enact a crisis, ideally including other relevant organisations, with participants playing the role they would carry out in reality, while some play roles such as journalists or politicians. One person is responsible for developing the crisis 'it is 3 a.m. on the third day and the fire has spread to the adjacent woodland and burnt down two holiday homes at x'. An observer can be used for documenting the process and identifying problems. The exercise ends with a review of procedures and suggestions for changes. This can be very good at revealing gaps and weaknesses in the system.

8.7 Fund raising

Funds should be sought only to meet the agreed objectives of the organisation: an organisation that changes its objectives to follow easy sources of money easily loses its way. It is too often assumed that fund raising is the priority when other activities may be better for achieving the objectives.

It is usual to estimate how much money is expected to be raised from fund raising from past experience of the same or other organisations. This provides some basis for estimating projected income and reasonable expenditure. Create both an annual budget of projected income and expenditure to plan the year's activities and a monthly version to detect likely cash flow problems. Consider the likely uncertainty in the income, for example it may depend on the weather at the main event or upon a few large donations or legacies; some conservation organisations have had severe problems from depending on such unpredictable funding. The solutions are to ensure that there are capital reserves to overcome such shortfalls (for example by paying legacies into a separate fund from which a fraction is removed each year), ensure flexibility within the spending plans so that expenditure can balance income or take out insurance to cover major risks, such as bad weather at the main fund-raising event. It is usually wise to diversify the sources of funding to spread the risks.

Short-term pulses of money are particularly difficult to manage as they may involve acquiring new staff, buildings and equipment and it is enormously easier to spend more money than less. Suddenly increasing the work load to deal with a project can affect the whole organisation. Furthermore, a sudden pulse of money can take the pressure off fund raising so that as the money runs out there are extra staff, a yawning gap in the finances and little effort and expertise in finding continuing funds. Similarly, money to help buy land is often the easiest to obtain but the long-term costs of maintaining the nature reserve need to be considered.

Fund raising needs to match the image of the organisation, as organising a black tie dinner or a sponsored 24-hour bath in baked beans will alter how the organisation is perceived. This is particularly important as fund-raising events often attract more interest than conservation events; similarly a sales catalogue may be read more thoroughly than a newsletter. Thus, if the fund raising appears extravagant or the goods for sale seem low quality or inappropriate then this may alter the public perception. Similarly, providing incentives for buying goods in bulk, or giving expensive prizes to particularly successful fund raisers, may seem a sensible way of raising net profit but may be counterproductive by creating the impression of wasting money that should be spent on charitable objectives. Certain types of organisations attract people because they want to be involved and they may be disillusioned by repeated requests for money, especially if there are no other opportunities to be involved.

Most organisations use just a few means of fund raising, although in practice there are hundreds of methods, such as providing a sales catalogue, collection tins, offering advice on how to complete a will (hoping to be included), sponsored activities, sale of food or drink, sales of goods, sales of second-hand goods, tax-efficient donations, auctions of gifts or promises, running training courses, sponsoring a species, running holidays, corporate members, government grants, discos, exclusive meals with famous guests, sponsored walk/bird race/swim/silence, lottery for donated goods, appeals to buy plots of land, direct appeals for money for a specific project, sponsorship for specific items, and appeals for regular automatic contribution from bank account or salary. The list is far from comprehensive. My point is that each organisation tends to restrict itself to a narrow range of methods and it is useful to examine the methods used by both similar and very different organisations.

Attracting money is easier if you can identify, create and make known a set of unique qualities. For example, being the most effective conservation campaigning group in the country, the group that is dedicated to conserve a local site, the society that provides the best care for elderly members on field trips, or the consultancy that has the most comprehensive biodiversity data.

Practically every stage of marketing can be improved by experiments. For example, calculate the response to a given advertisement in a set place in terms of each of the number of responses, replies to the brochure sent out, renewal rate, amount each spent on other goods or time given as a volunteer. To do this it is essential that the responses can be linked

to the method (for example by marking the form that has to be completed or asking people who reply where they heard of the offer). This background level of responses can be compared with experimental improvements at each stage, such as a different advertisement, different location, different brochure, whether a second letter is sent if there is no response to the brochure, whether a reply envelope is provided or whether return postage is provided. By carrying out these experiments on subsets the risk is minimised.

Income from membership has the advantage of providing a reasonably predictable source of funding which makes planning easier. Having a membership has other advantages such as volunteers providing assistance and backing. As with fund raising, most organisations only use a few of the host of methods for attracting members including: having a membership form in a catalogue, having a membership form in a catalogue of a company selling related goods, leaving leaflet dispensers in public places, providing a gift for joining, providing a gift to an individual who gets a friend to join, agreeing with another organisation that you both advertise each other or leaving cards on cars at reserves thanking them for visiting and giving a membership form.

Surveys can be useful for increasing membership, attendance at meetings, visitors to sites or numbers buying goods. They can provide information on the type of people who participate, where they live, how they heard of it and what they like or dislike about it. However, this does not provide the really interesting information which is why people did not join, attend, visit or buy. Surveys of those who do not participate but are clearly interested, for example by being a member of a similar organisation, are useful. It can be especially illuminating to survey those who show signs of being dissatisfied, for example those that resign or ask for details but do not respond. Identifying the type of individual that will make donations is invaluable for targeting future appeals (Wilberforce 1998).

Direct mailings appealing for money can be a very effective means of fund raising. Sending to randomly selected names is unlikely to be successful, hence many organisations spend considerable effort in collecting names, addresses and information on interests and spending patterns. For example, to encourage people to supply their contact addresses, some organisations offer to send further information, offer to send a gift, ask for a petition to be signed or have a competition. These can be designed to attract people with particular interests. Lists of people buying particular products, members of related organisations or subscribers to magazines are also widely used. The current fashion for questionnaires is usually linked to such name gathering. Lists can also be bought, swapped or donated. Some lists are out of date or inaccurate and thus will have a low success rate. Although highly effective, many conservationists find such unsolicited mail offensive. If providing a list to someone else it is sensible to agree firm conditions on its use and add a fictitious name to a correct address to see what that version of the list is used for. Direct mailings are often directed at particular segments of the population. This is particularly useful if the mailing list is detailed. For example, mailings can be sent to those that are full members of an organisation and that replied to one of the last two appeals before, or those that subscribe to a particular magazine, or those that said on a questionnaire that they are under 30 and care about animal welfare. Different mailings can be sent to different segments. Selling or swapping mailing lists can be resented by the membership list, especially if they consider the use inappropriate. In many countries there are now strict data protection laws which restrict the way in which names can be gathered, exchanged or used.

Advertisements can often be expensive but can be a good way of raising members or funds. In selecting locations it can be effective to contact those with similar objectives that have already advertised there. Advertisements are usually more effective with a headline. The golden advertising rule is to focus on what the customer wants to buy rather than what you want to sell. Thus emphasise the actual benefits to the purchaser, for example 'attract woodpeckers to your garden' or 'help save the turtles', not the features or the organisation's name. Be brief and ensure the advertisement does not look cluttered.

Free advertising by attracting publicity to a story is not only cheaper but, as the source appears independent, is often more credible and effective. For this to work there must be a genuine story (see Section 10.3).

Businesses and organisations can be major sources of funding. Find out the name of the key relevant person (for example by telephoning) and contact them directly using their name. There are a number of possible contacts within any organisation such as the Chief Executive, Marketing Director, Public Relations Director, Personnel Director and any proposal will have to be modified accordingly to be compatible with their concerns. Identify the person or organisation who is a potential funder and call or visit to see if they might be interested. Building a strong relationship can be as important as the contents of the proposal. Companies are often more willing to provide goods (e.g. providing the computing system) or services (e.g. printing the newsletter or training staff), than money as they feel more involved and can often provide considerable assistance at little cost to themselves. This may also lead to direct sponsorship.

Consider the interests of donor individuals or companies and repay their generosity and ensure that they are thanked. This could be by keeping them informed about the project, giving publicity, inviting them to see the work in operation, inviting them to social events or providing a special class of membership. This takes time but is essential. Some individuals may not be at all interested in the regular working and not even wish to be a member but may still respond to appeals or be prepared to give a legacy.

It can be useful to record the time spent on different fundraising approaches and their effectiveness to assess the efficiencies of different methods, learn from mistakes and provide feedback to the target-setting process. Performance indicators, such as membership enrolments per event or responses per 100 brochures can be very useful for deciding future strategies.

Consultancy work can be a lucrative way of funding the organisation. Many conservation organisations have acquired considerable skill and information on conservation and habitat management. It is, however, essential to ensure such work is fully costed to ensure it is actually benefiting the organisation. Either it should be to do work that is a priority for the organisation or it should make a genuine surplus. Beware that it is easy for such work with short-term deadlines to monopolise the organisation and thus hinder the real objectives.

8.8 Grants

The rules for obtaining grants are reasonably standard whether the project is large or small. There is often an application form. The most important consideration is to view the proposal from the perspective of the grant giver. Most grant-giving bodies usually have far more high-quality projects than they can fund and so are likely to be highly selective. They may also fund less than the requested amount. A grant is more likely to be funded if easy to understand, short, straightforward, divided into logical sections and presented in an attractive manner. Grant-giving bodies usually want to ensure that the project is within the subject area they fund, that the project is worthwhile and that it will be carried out satisfactorily. The major concern of most donors is whether the expertise is available to carry out the project and if there is value for money. The proposal should thus make clear the expertise of the group and previous successes. It is much easier to get money for a new project than for the core funding of the organisation. Funding is often easier if specific items or projects are listed for funding rather than a general request for money.

It is a waste of both your time and that of the grant givers to write a proposal that is completely inappropriate to that grant giver or to ask for a sum that is larger (or even smaller!) than they are prepared to donate. Most grant givers publish guidelines. It is often possible to speak to the person responsible for allocating grants who can give you guidance on what type of projects they fund, the criteria they use in selecting grants, the approximate typical value of grants they give away and the time they take to make decisions. Some organisations are keen to discuss potential projects to direct them towards

the type of work they wish to fund. It is useful to show your grant application to someone else with experience of obtaining grants and especially to someone who has experience of that particular grant-giving body.

Most grant proposals should include the following sections:

Objectives. Make sure that these are clear, for example 'to understand why plant species x is declining' or 'to set up a school education programme to discourage the killing of species y'. The objectives should also be clearly achievable, as described in the subsequent sections.

Background. A summary of previous work and the known information to place the proposed work in context.

Outcomes. These may also be referred to as targets. Specify these clearly, for example 'to determine the role of grazing and the addition of fertiliser on the decline of species x' or 'to provide an education package consisting of a teachers pack, poster and a quiz which will be circulated to all 78 schools in the area'. It is increasingly fashionable to be expected to set precise targets, for example describing what will be achieved at the end of each year in a three-year project.

Methods. The committee or person evaluating this project will want to be sure that you have the experience, knowledge and abilities to carry out this work. Be precise about exactly how the work will be carried out. State the number and size of exclosures and quadrats, the levels of fertiliser that will be added, how the results will be analysed, or state how the poster will be designed, printed and distributed and what the main message will be. Show you can do it.

Participants. It is often necessary to include details of the main participants describing their training and experience in this type of work. Make it clear that you, or the person who is being employed, have the experience in experimental design and botanical recording or in graphic design and education to carry out the work. State if you are collaborating with others or getting outside advice.

Applications. This work is obviously being done for a purpose. Make it clear how these results will be used. This might be in the form of practical work, scientific papers, popular articles, reports, talks or advice to others. Outline additional advantages such as any training that will be provided.

Budget. Again be detailed. Grant-giving bodies will often cut vague items, those that are clearly over-priced or those that should already be available. However, make sure it is sufficiently funded as it is not usually possible to go back for more. Some grant-giving bodies allow overheads to be charged to pay for the running costs such as basic facilities. This is often fixed as a set percentage (say 40%) which can then be added to the grant. If overheads are not charged but staff and facilities are needed to run the project then this can cause financial problems. It is common to underestimate the time and costs involved in running projects, especially considering there will probably be occasional disasters in which the staff have to be heavily involved. Unless salary costs include everyone involved, including secretaries and administration, then an overhead on salaries of 100% is typically required for a project to break even. If the funding is to do something the organisation wants to do anyway then a loss may be acceptable while if the work is undertaken purely for profit, it is essential to ensure that the overheads are sufficient. Some grant-giving organisations expect their funding to be supplemented by funding from other sources. Some allow the cost of staff also working on the project, but not included in the grant, to be included as the supplementary contribution of the organisation. It is important to try to find out the rules and preferences of a given funding body. If awarded a grant, remember to claim the money on time and to give good publicity if this is welcomed.

Permission and permits. The project may need permission from the landowner for access or from the government as a result of working on a protected species or using techniques such as collection, trapping or marking or sometimes simply for carrying out research. It is often necessary to include copies of these permits or make a convincing case that it would be straightforward to obtain the necessary permits. Obtaining support from other bodies may further increase your credibility. It is sometimes necessary for the head of an organisation to sign a form or include a letter saying that they are prepared to allow the work to be based in their organisation.

9 Education and ecotourism

9.1 Why educate?

Education is one of the major techniques available to biological conservationists. Any proposal which depends upon a change in behaviour or compliance with new legislation relies on education if the change is to occur. Five common objectives are to encourage a general interest in wildlife, generate greater awareness of conservation issues, bring about a specific change in opinion, disseminate specific information or provide training.

Education is often considered separately from other aspects of conservation, but in practice most conservation programmes require some education and it is essential to consider the conservation objectives of a relevant education project. For example, maintaining protected areas is easier if there is public support, which often leads to political and financial support and greater adherence to rules and regulations (Shepard & McNeely 1998). Following an education project in Morro do Diabo Park in Brazil, there was a clear change in attitude to the park, for example, the community helped to extinguish a forest fire that threatened the park, pressurised the local government to relocate the city's garbage dump away from the park and there was some evidence for reduced hunting and timber cutting in the region (Jacobson & Padua 1995).

It has been suggested that the greatest and most depressing problem in conservation is not habitat loss or overexploitation but the human indifference to such problems (Balmford 1999). Overcoming such indifference is likely to depend on providing both the opportunities to appreciate areas and species, and education to highlight the ecological, aesthetic, cultural, spiritual, recreational and economic importance. Conservation education often concentrates on scientific arguments while a wider perspective is likely to generate greater support.

9.2 Planning and running an education programme

The planning process described in Section 7.2 applies equally to education projects. Successful projects are usually a consequence of well thought out research, targeting, planning, execution and evaluation (Jacobson & McDuff 1998).

- *Evaluate environmental problems and solutions.* Consider the best approach to tackling conservation problems (see Chapter 6) and the role, if any, of education.
- *Have clear goals.* Many conservation education programmes drift in an aimless manner without clear conservation benefits and sometimes even without real educational benefits either. It is necessary to have a clear vision of the purpose of the education. An example of a clear goal was when conservationists in Australia decided it was important to improve support for the native species and so promoted chocolate Easter Bilbies, based on the Greater Bilby *Macrotis lagotis*, to replace the Easter bunny. This also made it easier to carry out control programmes on the introduced European Rabbit *Oryctolagus cuniculus*. Another example followed from data showing that seabird populations were declining along the Gulf of St. Lawrence, Canada, as a result of illegal exploitation, but surveys showed that most residents were unaware of the law. A youth education project was created with the objective of reducing exploitation, both through changing the views of the younger generation but also as this was considered the best means of introducing these ideas into the community (Blanchard 1995).
- *Identify objectives.* With clear objectives it becomes obvious what the purpose of the teaching is. One approach is to list the main messages. A useful technique is to state objectives under five categories: *awareness* (e.g. after the education

programme the participants should be able to list the main threats to local marine resources in the community), *knowledge* (should be able to identify the most important fish species and give their life histories), *participation* (should be able to plan and implement a programme to teach tourists how to snorkel without damaging the reef), *skills* (should be able to survey fish populations) and *attitudes* (should be able to describe how people's opinion of the reef differ) (Braus & Wood 1993).

With quantifiable objectives (such as over half the children should be able to identify the five commonest trees) it is then possible to determine whether the programme has been successful. In 1975 an opinion poll in Rwanda showed over half were hostile to Gorillas *Gorilla gorilla* yet 10 years later, after a concerted education programme, a large majority had acquired positive feelings towards them and pride in Rwanda's role in their conservation (Kingdon 1990).

Where the education programme may have a real impact on the problem it is best to have objectives that determine whether the conservation aims are being achieved. Many West Indian Manatees *Trichechus manatus* are injured or killed by boats in Florida but guidelines to 'Save the Manatee' in Florida and associated education led to significant declines in mortality (Blangy & Wood 1993).

It can be useful to have objectives that vary from ends objectives (what is trying to be achieved, e.g. 'reverse the declines in orchid populations') to means objectives (steps that need to be taken to achieve the ends objective, e.g. 'reduce the percentage of orchids picked to less than 2%' or 'ensure that 80% of school children understand the importance of the orchids').

• *Use existing organisations, groups and structures.* It is usually more efficient to work through the organisations and groups already present than attempt to start new ones. For example, others may already distribute newsletters or visit schools and may be prepared to incorporate additional material.

• *Determine target audiences.* Identify the exact audience you wish to educate. Those with the ability

to improve the situation are not necessarily those that readily attend conservation programmes.

• *Understand the target audience.* Specific groups, such as farmers, students or politicians, usually require different approaches. It can be useful to consult members of the target audience and ideally involve them in designing the programme. South African farmers, who are almost all Afrikaners, used to kill jackals by placing poisoned baits which also killed many predatory birds. A programme to convince them not to do so became much more successful after it was run by Afrikaners. The education message should be given by someone who understands the community and preferably by a figure whom they respect. Similarly if preparing materials for schools, speak to local teachers to see if the

The collection from caves of bat guano for fertiliser was a major source of income in several villages in Thailand, but in the early 1980s, Tuttle visited and discovered that guano sales had recently halved. This problem was due to reduced guano production, attributed to the population declines of these insectivorous bats as they were accidentally caught in the nets of poachers catching fruit and nectar bats for restaurants (Tuttle 1990, Morton & Murphy 1995). Tuttle gave educational materials to both the local major and the Buddhist monks who owned the cave and also suggested that it would be economic for them to hire a warden to protect bats near the cave. Eight years later, as a result of successful protection, guano sales had increased from US$12 000 to $88 600 and busloads of tourists were now visiting to see the bats with local people profiting from selling refreshments and souvenirs. Schools still teach about the bats using materials modified from the original donation and the uniform of one school incorporates a bat T-shirt. As a result of the interest and education efforts the region has been designated a non-hunting area by the Thai government (photo: M.D. Tuttle, Bat Conservation International).

CASE STUDY

Conservation stickers on Sumba

In August 1994, Nicola and I became engaged on the beautiful island of Sumba in Indonesia. To celebrate we commissioned a set of stickers from an artist friend, Greg Poole (who painted the cover of this book), with captions provided by Paul Jepson of Birdlife International Indonesia Programme and invited our wedding guests to buy sets of stickers at £10 each. We raised enough for 10 000 stickers, which have been distributed by Birdlife International Indonesia. This has been such a cost effective means of spreading simple conservation messages that similar sets of stickers have since been prepared for other islands.

The captions are: 'let us conserve Sumba's beautiful and unique nature', 'protect nature on Sumba' (Pied Bushchat *Saxicola caprata*), 'protect birds' (Paradise Flycatcher *Terpsiphone paradisi*), 'only on Sumba' (the Sumba Hornbill *Aceros everetti* is one of the country's six endemics and most people on Sumba are surprised to hear that these species are not found throughout the world), 'more beautiful in the wild' (this subspecies of Yellow-crested Cockatoo *Cacatua sulphurea citrinocristata* is only found on Sumba and is frequently captured for the pet trade), 'conserve Sumba's forests' (there is very little forest left).

MARILAH KITA LESTARIKAN ALAM SUMBA YANG INDAH DAN UNIK

JAGALAH ALAM SUMBA

JAGALAH BURUNG

HANYA ADA DI SUMBA

LEBIH INDAH DI ALAM

SELAMATKAN HUTAN DI SUMBA

Hasil Kerjasama

Departemen Kehutanan

BirdLife
INTERNATIONAL
INDONESIA PROGRAMME

untuk menunjang kegiatan
KELOMPOK KERJA KSDA
KONSORSIUM PENGEMBANGAN DATARAN TINGGI
NUSA TENGGARA

Disponsori oleh teman-teman Nicola & Bill Sutherland sebagai hadiah pertunangan mereka di Sumba

CASE STUDY

Public involvement in the conservation of Tiritiri Matangi Island, New Zealand

Public involvement can help reduce the costs of glamorous restoration projects, provide goodwill, support and education (Craig 1994, Galbraith & Hayson 1995). Tiritiri Matangi Island near Auckland, New Zealand, used to be farmed and regularly burnt. The last stock had been removed in 1972 and a decade later a plan was made to enhance the conservation interest whilst still allowing free access along the trails. Public finances helped run a tree nursery and the public paid for the privilege to plant trees that they helped raise. This resulted in the planting of 300 000 trees in 8 years, so increasing the proportion of bush from 6% to 60%. Seven species of locally or nationally rare bird species were introduced. A charge has been made to attend their releases which were greatly oversubscribed. The release of the South Island Takehe *Porphyrio mantelli hochstetteri* was attended by over 500 people and attracted considerable media coverage with funds for the release being paid by a sponsor. For the release of Brown Teal *Anas aucklandica chlorotis* a raffle was held to personally release one of the birds. All the seven released species have survived and bred better here than elsewhere. Unlike many other New Zealand islands the only introduced predator was the Pacific Rat *Rattus exulans* which was eradicated in 1993, leading to the possibility of releasing invertebrates, reptiles and amphibians. A supporters' group has the major objective of funding this project from wharf charges, donations, sponsorship and sale of merchandise and there is considerable help from volunteers for monitoring and habitat management. The important lesson from this project is that, when practical, public participation may be highly beneficial.

Those students who had been involved in the conservation programmes on Tiritiri Matangi help the conservation officers release stichbirds *Notiomystis cincta* from their carry cage (photo: M. Dimond).

CASE STUDY

Global Rivers Environmental Education Network (GREEN)

This project encourages school teachers to take their teenage students to the local river to monitor water quality then analyse watershed usage, identify socio-economic factors causing river degradation and present their findings and recommendations to local officials (Stapp *et al.* 1995, Jacobson & McDuff 1998). Water quality is reasonably straightforward to assess and this provides the opportunity for students to participate in data collection and analysis with obvious conservation relevance. The field kits measure dissolved oxygen, faecal coliforms, pH, biochemical oxygen demand, temperature, total phosphate, nitrates, turbidity and total solids. These field test kits can be bought in many countries (most are accurate to 2%) or made using standard laboratory equipment. Surveys for specific sensitive invertebrates can also be carried out. The accuracy of the data is critical if it is to be used for making recommendations so there may be a need to reconfirm evidence of pollution and ideally obtain independent confirmation. This programme has acted as a network linking schools in over 55 countries so they can exchange data and ideas. It also has the advantage of stimulating greater international awareness. Schools within a region are encouraged to collaborate to study the same watershed.

The standard two-week programme consists of:

Lesson 1   Discuss water quality issues and view slide presentation.

Lessons 2–3   Learn the nine water quality measures (dissolved oxygen, faecal coliforms, pH, biological oxygen demand, temperature, total phosphate, nitrate, turbidity and total solids).

Lesson 4   Test the quality of the water.

Lesson 5   Analyse the data and interpret results including calculating the water quality index. Understand the results in terms of use of the river by humans and wildlife.

Lessons 6–10   Write an action plan and act on it.

In practice this standard programme is creatively extended to adapt to local conditions and interests.

Copies of the *Field Manual* (Mitchell & Stapp 1997) can be obtained from William Stapp, 2050 Delaware, Ann Arbor, Michigan 48103, USA. It has been translated into Bengali, Chinese, Czech, German, Hebrew, Hungarian, Italian, Japanese, Russian and Spanish. There is also a manual for low-cost monitoring that includes instructions for making inexpensive equipment.

approach and level is appropriate. Individuals also differ as to whether they prefer to learn from formal lectures, by being given the opportunity to think through the problems and solution or by participating in exercises to illustrate the required approach. A range of such techniques is often most appropriate.

• *Develop an organisational plan and budget.* Ensure that the responsibilities, timescale and resources have been considered and are understood by other participants in the programme. Generous funding can result in an ambitious project that cannot

be sustained when that source stops providing. It has been suggested that, rather than programmes based on foreign aid, it can be better to finance the project by linking local sponsors to a popular cause (Butler 1995).

• *Devise a relevant programme.* The programme should be directed at people's interests and concerns. One approach is to appeal to self-interests such as economic or cultural values, for example by linking waterbirds to water quality or forests to water conservation. Increasing national pride in the national or global importance of species or areas can generate grassroots support for conservation. One approach is to build a programme around an attractive, readily identifiable species which is linked to the major conservation issue (such as loss of a particular habitat), but does not have negative associations, such as being considered a pest.

• *Build necessary support.* Assistance from government, education authorities, the community, industry or other conservation bodies may be helpful or essential. For example, in the Bahamas, conservationists aiming to conserve the Bahama Parrot *Amazona leucocephala bahamensis* gained donations of bumper stickers, church support, funding from businesses, 2.5 million bags with conservation messages printed by a grocery store, an eight-page supplement included in a major newspaper, 10 million letters franked with a parrot stamp and a rap song (Butler 1995).

• *Plan for potential problems and resolution of conflicts.* For example, at an event: What if it rains? What if someone falls ill? What if double the expected of number of people appear? What is the procedure if an individual objects strongly to the message? How will conflicts be dealt with?

• *Encourage participation.* Individuals learn more from doing than watching. As just one example, nationwide surveys of conspicuous species can attract interest and can also produce useful results.

• *Provide direct contact with the environment or resource.* Making wildlife accessible greatly strengthens the interest and concern. This is particularly important with the increasing isolation of modern people from the natural world. Whale watching has greatly increased the support for whales. A nocturnal torch safari for invertebrates can give children as much excitement as seeing distant spectacular mammals from a vehicle. Hands-on experience obviously has to be compatible with the welfare, dignity and conservation of species and habitats.

• *Select appropriate educational media.* Decide whether talks, television programmes, World Wide Web pages, posters, T-shirts, touring vehicles, videos, leaflets or stickers are the most appropriate

for the particular situation and how they should be presented. As examples: Scottish Natural Heritage's guide to sea turtles and their protection is laminated with a hole in one corner so it can be tied to a boat; Birdlife's Mount Kilum project in Cameroon commissioned Prince Afo-Akom to sing about their project and the tape reached number two in the hit parade; posters describing the forest on the Comores and its endemic fruit bat were often kept out of sight as a treasured item but stickers were widely displayed (Hurst 1998); the most popular soap opera in St. Lucia is written by a conservationist and has an environmental theme.

• *Consider the problems of teachers.* Most teachers are very busy and have little time to develop new material. Conversely, if provided with suitable material that can be used with minimum modification, they will often do so. The ideal is material that is compatible with their curriculum and can be directly used in the classroom yet still provides the flexibility for teachers to extend, reduce or modify the materials as appropriate for their requirements. Successful options include lesson plans, full details of practical exercises, project suggestions or activity sheets that can be photocopied. If providing visits for school groups, consider creating a teacher's pack with suggestions for introductory classes, background information and follow-up exercises.

A major problem with conservation education is that it often requires identification skills that many teachers feel they lack. One solution is to appreciate the environment without demanding identification of everything, for example by searching for specific textures or colours. My personal experience of such exercises is that they really awaken the senses. Another solution is to concentrate on processes such as food webs, succession, competition or adaptation rather than identification. A third solution is to provide teachers with information packs or training courses to provide the required knowledge. Often guidance on the species likely to be seen on a particular route at that time of year and how to identify them will solve most problems. A combination of all three methods is often sensible.

• *Test out the material.* It can be useful to examine the response of the target audience to a range of similar materials. Do they listen to talks, read the posters, watch the videos to the end or follow the guided walk? At what stage do they lose interest? Are they absorbing the messages? It can similarly be very useful to test material while developing it. It is important to be realistic about the amount of material that can be retained. For example concentration drops markedly for talks over 10 minutes without visual aids and display captions with more than about 20 words are unlikely to be read.

• *Ensure the programme is enjoyable.* Any programme is likely to be more successful if it is enjoyable or entertaining. The educational talks on parrots on Dominica start with the project team dressed as a parrots entering the school (Butler 1995). Charles Vatu, who runs the Big Bay conservation project in Vanuatu, is also well known as the captain of the Vanuatu football team and can provide football training to attract interest before introducing conservation. Other possibilities include cartoons, puppets, drama, debates, art, jokes, games, quizzes and songs.

• *Monitor the programme.* As with other aspects of conservation, it is extremely valuable to assess the success of projects in relation to the measurable objectives. Measures can vary from the number of people present, the results of questionnaires

Conservation education should also be enjoyable. Charles Vatu (right) of Big Bay Conservation Area is also captain of the Vanuatu football team and often combines football training with conservation. The existence of the forest at Big Bay depends entirely upon the goodwill of the two local villages and especially their chiefs Solomon (left) and Moses (photo: William J. Sutherland).

asking participants to evaluate the programme, or the changes in behaviour such as poaching. Such monitoring allows programme modification or creation of new programmes. An educational project in St. Lucia, tested in one school, showed that teachers needed thorough training before they were confident to use the material, thus a teachers' programme was introduced (Jules & Leal Filho 1998).

Following the terminology and example introduced earlier it can be useful to monitor the ends objectives, which show whether the project is achieving its goal ('reverse the declines in orchid populations'), the means objectives, which show whether the proposed activities are reducing the problems (e.g. 'reduce the percentage of orchids picked to less than 2%') and the activities carried out (e.g. education display created, daily visits by reserve warden in flowering season). It is then possible to see if the conservation goal is being achieved, and if not, whether it is because of programme failure (the planned actions did not take place) or theory failure (the actions took place but did not produce the anticipated changes).

• *Review the programme.* Ensure that there are regular opportunities to consider changes. The facts may no longer be true, the material may look old or the message may no longer be appropriate. It is easy to ignore outdated material, especially if seen daily.

• *Enhance local control and support of programme.* It is often an objective for externally organised programmes to devolve responsibility to the local community. This may require training.

• *Develop long-term plans for sustaining the programme.* Projects are often established on short-term funding or with a burst of enthusiasm. It is also often easier to get the funding to set up a new project than maintain an existing one. It is thus sensible to plan carefully, perhaps by linking with institutions that will give long-term funding, so that the project can persist in the long term.

• *Disseminate programme results.* As with most other branches of conservation there is little reliable information on the successes and problems of specific approaches. Ensuring the results are widely available is invaluable.

9.3 Identification guides

The production of identification guides is perhaps the most effective means of conservation education. Yet some countries lack even a bird field guide and even more lack one in their national language. Basic field guides of the most conspicuous species in the local language could have a considerable conservation impact and are an obvious funding priority. The production of the first sensible field guide to British dragonflies (Hammond 1983) initiated enormous interest in the group which has led to considerable advances in ecological knowledge, massively greater effort to conserve them and has created a market for a number of better field guides. I am sure many other groups could be much more popular, better understood and better protected should sensible guides be produced.

Considering that reasonable field guides have been produced for 60 years (Petersen 1934), it is remarkable how dreadful so many recent guides are. A secret to producing good identification guides is to find critical features that distinguish the species from all others. For example 'the Hawksbill Turtle is the only sea turtle with overlapping plates along the back' or 'Sirenians, are large sea mammals which differ from dolphins by the lack of a dorsal fin and from seals and sea lions in lacking hind flippers. The Dugong *Dugong dugon* differs from other Sirenians in its concave rather than rounded convex rear edge to the tail.' It is of course useful to give other information to confirm the identification, in case the critical feature could not be seen or to identify from a distance as the observer becomes more expert. The norm is to just list a mass of features, only some of which can usually be seen and most of which are shared with other species: it is then difficult to be confident in the identification and can lead to incorrect identifications. Identification is harder when the characters are continuous i.e. X is larger, hairier and greener than Y. The ideal is still to try and find a critical character e.g. 'X always has leaves at least 3 times longer than wide which Y never does' or 'only X has hairs on the leaf margin'.

Bad layout is a common problem of field guides. Field guides often have an intimidating mass of similar looking pictures. Often further reading shows many are of the same species, many are never found in the country, some are restricted to very different habitats and the few remaining species have clear characteristics that distinguish them. The skill is to make this process as straightforward as possible. The ideal is to have the text and distribution map opposite the figures or at least have a page opposite the figures outlining the main identification features. Cluster different pictures of the same species together (it can also be successful to draw a line around them) and ideally give the species name on the figure. If this is impractical use one number per species and list opposite in the same order as on the plate. Some guides have a large block of empty space opposite each plate which they could fill with key identification features. The layout should be designed to ensure it is easy to determine what the options are (i.e. what species of that appearance are present in the area) and then how they can be distinguished. Paintings are usually much better than photographs. If possible give a range of ages, sexes and views. It is useful to state distribution if helpful for identification. While using distribution for identification is not ideal, I believe that this will reduce the number of records that are obviously wrong and by making identification easier will increase the confidence of identifying species outside the known range. The pictures should be a sensible size as large pictures mean few per page so comparisons are difficult and often also means the book, being larger, may be less convenient to take in the field.

In the current year two superb field guides have been produced covering the birds of Southern Africa (Sinclair *et al.* 1999) and Europe (Mullarney *et al.* 1999). These both show the way forward: sensible layout, maps and text opposite the illustrations and especially masses of critical features for each species. Mullarney *et al.* (1999) outline the critical features on the plates which make this a sheer joy to use. Sinclair *et al.* (1999) start each species account with a terse statement of how to separate the species under consideration from any other similar species. Anyone preparing a guide to any taxa would gain by examining these guides.

Keys, in which identification proceeds by means of a set of branched questions, are much cheaper to produce and are the standard and invaluable means of identification for many taxonomic groups. They often require considerable experience before it is possible to get the right answer. Unless the key ends with a critical feature or illustration or unless it is possible to check with specimens then it is difficult for a newcomer to the key to know if the identification is correct which can result in some spectacularly wrong identifications.

9.4 Ecotourism

People have been travelling to see wildlife for at least a century. The current interest in ecotourism is partly a realisation of the ability of tourism to be used as a positive conservation tool by increasing funding or awareness. However in part it results from operators exploiting a new commercial market with little concern either for the conservation benefit or even whether they are causing damage. Ecotourism encompasses two ideas: one is experiencing wildlife and the other is causing minimal environmental damage (Boo 1993, Lindberg & Hawkins 1993).

Ecotourism can have considerable benefits for conservation by providing economic incentives for protecting a habitat or population (Tobias & Mendelsohn 1991). The benefits can be local, in terms of a contribution from park fees to the local community, increased local employment or the opportunity to provide accommodation. Sometimes the benefits can be largely regional or national, as little of the profit goes to the conservation site but there are considerable benefits to hotels, restaurants, car hire companies or governments. A problem is often then that the people who benefit differ from those that bear the costs. Only US$7 million of the US$350 million made from nature reserves in Kenya has gone back to the parks (Olindo 1991). Calculating the value of ecotourism for a region can be useful for justifying state investment.

Ecotourism is likely to increase concern for conservation. Thus in both Maluku, Indonesia, and Southern Luzon, Philippines, divers register and pay a fee before visiting local coral reefs with the result that the villagers now supervise the area and prevent fishing using explosives (Caldecott 1996). In Costa Rica and Namibia private nature reserves are created for profit.

Ecotourism can be important in changing perspectives. Wolves *Canis lupus* are unpopular amongst farmers but wolf-howling trips for tourists at Algonquin Provincial Park, Ontario, Canada attracts large numbers of participants (Carbyn 1979). It is likely that the enormous worldwide interest in whale watching is the most important factor affecting the acceptability of future exploitation. Furthermore, the fact that people will visit and pay to see species and habitats can alter local attitudes.

There are many examples of successful ecotourism but also others where tourism is damaging to conservation. As ecotourism may bring people to the most important and sensitive sites there is the real potential to cause more damage than benefit. Individuals may benefit from actions, such as using anchors carelessly on coral or taking tourists too close to sensitive species, that destroy the resource they all depend upon (the Tragedy of the Commons described in Section 13.1). Much will depend upon the extent to which there are strong controls as a result of land ownership, social pressures or regulations. It can be unwise to encourage ecotourism unless there is a clear mechanism of control including the motivation, physical ability and authority to restrict damaging activities.

Planning a sustainable ecotourism project often involves the following procedure.

1 *Analyse the current position.* Review the biodiversity. What are the components of biodiversity that attract tourists or are sensitive to tourism? What is the distribution and status of each component? Is it changing and what are the threats?

Review the current tourism. How many visitors are there? When do they come? How long do they stay? Where do they visit? How much do they spend and on what? What are the reasons for these patterns? Analyse the attractions of the conservation site in relation to those available locally, nationally or internationally. Consider the location in relation to tourist routes, transport and accommodation. Review the infrastructure, facilities, staff, tourist

CASE STUDY

Managing tourism in the Antarctic

The number of tourists in the Antarctic is increasing rapidly. The International Association of Antarctic Tour Operators aims to reduce interference between groups of visitors and minimise environmental damage (Splettstoesser & Folks 1994). Most tourism is by cruise ships which take passengers ashore in inflatable dinghies.

The main Antarctica tour operator guidelines are:
• Enforce the visitor guidelines in a consistent manner, bearing in mind that guidelines must be adapted to individual circumstances. For example, fur seals with pups may be more aggressive than without pups so visitors should stay further away.
• Hire a professional team. Place an emphasis on lecturers and naturalists who will not only talk about the wildlife, history and geology, but also guide passengers when ashore.
• Ensure that for at least every 20–25 passengers there is one qualified naturalist lecturer/guide to conduct and supervise small groups ashore.
• Limit the number of passengers ashore to 100 at any one place and time.
• It is the responsibility of the tour operator to ensure that no evidence of visits remains behind. This includes garbage of any kind.
• Respect historic huts, scientific markers and monitoring devices.

The main Antarctica visitor guidelines are:
• Do not disturb, harass, or interfere with the wildlife.
 – Never touch the animals.
 – Maintain a distance of at least 15 ft (4.5 m) from penguins, all nesting birds and true seals, and 50 ft (15 m) from fur seals.
 – Give animals right of way.
 – Do not position yourself between a marine animal and its path to the water, nor between a parent and its young.
 – Stay outside the periphery of bird rookeries and seal colonies.
 – Keep noise to a minimum.
 – Do not feed the animals, either ashore or from the ship.
• Do not walk on or otherwise damage the fragile plants, i.e. lichens, mosses and grasses. Damage from human activity among the moss beds can last for decades.
• Leave nothing behind and take only memories and photographs.
 – Leave no litter ashore.
 – Do not take souvenirs, including whale and seal bones, rocks, fossils, plants, or other organic material, or anything that may be of historical or scientific value.
• Do not interfere with protected areas or scientific research.
• Do not smoke during shore excursions (fire hazard).
• Stay with your group or with one of the ship's leaders when ashore.

Creches of Adelie Penguin *Pygoscelis adeliae* chicks (foreground) and tourists (background) on Coronation Island (Photo: R. Coggan, British Antarctic Survey).

services, educational service and consider their use, quality and finances.

Consider what is currently preventing tourists from visiting. Speak to potential visitors to discover what they are looking for. Where are the potential visitors? What types of people are they? What will they expect? How much will they spend?

2 *Analyse costs and benefits of changes.* Increased tourism will almost inevitably also cause some conservation problems and it is thus important to consider whether there is a net benefit. The damage can be direct, such as increased disturbance or trampling, or indirect, such as the demands for water, firewood, polluting activities or hotels and restaurants. Section 6.4 describes how to predict the consequences of changes. Tourism may also result in deleterious local economic or cultural change including increases in crime, gambling, prostitution and HIV (Shaw & Williams 1994). Tourists may weaken cultures but can also strengthen them through interests in arts, crafts, and culture which may retain meaning for local people. Thus on Malakulu, Vanuatu, Chief Emil wanted to rediscover his community's dormant cultural traditions. The elders could thankfully still remember them, and tourism provided the funding that gave them the time to relearn their customs and reawaken their culture.

Sites that were once visited only by naturalists, scientists or backpackers who were thankful for basic facilities are now becoming standard tourist locations. This wave of tourists can be astonishingly oblivious to their surroundings and expect the same standard of food and facilities in an isolated mountain lodge as in a city hotel. It is thus important to consider the likely expectations of sorts of tourists that will be attracted and the time, expense and frustration spent meeting them.

3 *Decide upon objectives.* The objectives could, for example be to make a living so your conservation work can continue, expand local employment, ensure an area is profitable and so protected, change the attitude to a species or provide education facilities to increase general conservation awareness.

4 *Devise a strategy.* The strategy should decide how many tourists are wanted and of what type. This is easiest to achieve by collecting information from other sites and deciding where you wish to be relative to them. The most successful approach is often initially to bring together the participants in a workshop to agree upon general policies for establishing the required infrastructure, legislation and training.

Create a business plan, calculating the investment and likely number of participants. Building slowly allows the experience gained to be used. The most useful technique is to visit elsewhere and observe service standards, facilities and products.

Much of the secret of a successful ecotourism project is to identify a niche in the tourism market and target a segment or segments of it (Cooper *et al.* 1998). The segments may vary according to economy, origin, age and behaviour (Youell 1998). It might be to attract dedicated naturalists with a specific objective, tourists breaking their journey to another site, those seeking a wilderness experience, a budget or luxury version of an attraction elsewhere or workers from a nearby town wanting a weekend break. In each case the expected facilities, experience and price will differ, as will the most effective marketing and publicity.

5 *Plan zoning and regulations.* Zoning (see Section 12.10.1) is often a key element of ecotourism but this is dependent upon sufficient control mechanisms. The zones may include areas for extensive use with facilities, areas of less intensive use and sensitive areas where access is forbidden. Consider where the problems will be greatest and how they may be resolved. Consider if there is sufficient control to regulate the tourism if it becomes successful. Regulations are usually much easier to set up at the start of a project than after operations have started. Consider whether voluntary agreements or a permit system will work. Avoid conflicting activities in the same area.

6 *Provide required infrastructure and facilities.* Often a change in the extent of ecotourism is only possible with an investment in accommodation, transport, footpaths or signs.

7 *Review procedures*. Much successful tourism results from a persistent attention to detail such as good advertising (usually offering a brochure), an effective brochure, good personal communication, efficient and friendly procedures for sending out details and confirming bookings and attention to tourist needs. Decide beforehand if you are committed to this level of service.

8 *Train*. It can be useful to provide training for local guides in languages, expectations of tourists, field skills and guiding. It is sometimes appropriate to establish certification schemes by which an individual is tested and evaluated on his or her knowledge (Fennell 1999). It may also be necessary to devise ways of ensuring tourists adopt acceptable behaviour in relation to collecting, access, behaviour to other people, or distances to animals.

9 *Provide education*. The layout of visitor centres, trails and notice boards needs to be considered within the general facilities and infrastructure, such as car parks. Directions are more likely to be followed if the necessity is explained, they are positive, expressed simply and provide helpful information at the same time, i.e. 'please keep on the path so as not to disturb the breeding birds' rather than 'do not leave the path'. Guidelines on responsible behaviour are more likely to be respected if accompanied by useful information such as routes, species present or ecology. Interpretation can also help foster civic pride and ownership amongst the host community.

10 *Consider profit sharing*. Consider how benefits can be shared so that all gain. A share of park fees may go to the local community. In developing countries where tourists are prepared to pay substantial fees, it may be appropriate to have different fees for locals and visitors.

For some reason the most beautiful, natural and diverse areas usually have few working lawyers, accountants or civil servants. Many such areas do not have the money, technical knowledge or experience to manage tourism. External assistance can then be beneficial.

While being a tourist I am sure it is invaluable to be generous to the people who are actually responsible for deciding the fate of an area or species. Consider also taking spare binoculars or field guides to give to such people.

11 *Monitor the successes and failures*. Ecological monitoring should also be devised to detect any adverse consequences. It is also useful to collect data on the origins of visitors, how they heard of the site and their reaction to the experience and facilities. Listening to tourists and providing what they want seems obvious but this stage is a critical one in separating successful and unsuccessful tourist operations. Monitor the success of different advertisements or brochures.

12 *Reviewing*. As ecotourism develops and the situation changes there will be new opportunities and problems. A process of continual review is important.

10 Bringing about political and policy changes

10.1 Why enter politics?

The main means of bringing about changes are education (described in Chapter 9), persuasion, economic incentives and changes in the law. It is often better to work within existing institutions than to attempt to create new competing institutions. In bringing about changes in other countries it is important to help develop the local organisation.

Conservationists usually bring about change by some combination of campaigning and negotiation and these are described below. The balance depends on the circumstances. Hard campaigning may result in the desired policy changes but they may be unworkable due to the resentment of other parties and may produce antagonism which may hinder future discussions on policy. Negotiation is often effective but may result in unnecessary loss of biodiversity due to compromise or the length of time it takes to complete the negotiations.

It can be very effective for different individuals or organisations working on the same issue to vary in their emphasis on campaigning or negotiation. Those emphasising campaigning may create public concern and the demand for change while those emphasising negotiation are more likely to be consulted by decision makers in bringing about this change.

Suggestions for change should be made as early in the decision-making process as possible and it is thus useful to understand the details of the process in order to be effective. Lying down in front of bulldozers is dramatic but rarely effective on its own. Bringing about change often requires tenacity and a rather visionary view of the possible eventual solution.

Individuals seeking to bring about conservation change often start by creating a non-governmental organisation. This is often an excellent way for furthering the cause. Such organisations need clear objectives and writing one or two short sentences that state these objectives can be useful in focusing the organisation and enabling others to understand its function. Any organisation needs to unite those who share the objectives, usually as a membership. The membership can then help raise money and help advance the conservation objectives, can provide a reasonably reliable and stable income and is more democratic. Simply possessing a reasonable sized membership shows policy makers that there is interest and concern.

10.2 Campaigning

1 Decide if campaigning is really worthwhile. Campaigns usually involve much time and effort and can be demoralising if lost. Conversely, a failure to be involved in some issues may give the wrong signals. Consider whether campaigning for a change in legislation is better than an education programme.
2 Decide exactly what change in policy or legislation is sought. Have a text available.
3 Collect evidence. It is usually essential to show that there really is a problem. Evidence may be data on the decline in a species or habitat, or proof that a problem, such as persecution, exists. Photographs and videos can be effective evidence. The best evidence is: (a) independent, meaning it is provided by someone who has no personal interest in the issue; (b) precise, giving sufficient details to be convincing that the witness was not mistaken; and (c) contemporaneous, meaning that it was created when the event took place rather than later (Day 1998). The need for the campaign has to be easily understood by non-experts. Campaigns using untrue, questionable or even insupportable facts are easy to dismiss and may increase people's conviction that the entire

Water extraction in Mono Lake

Mono Lake, a saline lake in eastern California, was the subject of a protracted legal battle that was resoundingly won by conservationists (Hart 1996). The issue started with the extension of the Los Angeles aqueduct into the Mono Lake basin in 1941. Four of the creeks leading into Mono Lake had been diverted to provide water to the Owens River, which in turn now provided water for Los Angeles. In 1970 the capacity to extract water from the Owens River was greatly increased, thus more was diverted away from Mono Lake.

Biological surveys, started in 1976, showed that the ecological importance of the lake had been unappreciated. It holds the second largest colony of California Gulls *Larus californius*, migrant birds include 1.5–2 million Eared Grebes *Podiceps nigricollis*, 30–70 000 Wilson's Phalaropes *Phalaropus tricolor* and 30–70 000 Red-necked Phalaropes *Phalaropus lobatus*. The numbers of wintering ducks had declined from an estimated one million in 1948 to about 14 000.

In its natural state, water only leaves Mono Lake through evaporation, hence a reduction in input will result in a smaller lake. A water budget (see Section 12.7) was developed and it was calculated that at existing extraction rates, the area of the lake would be halved, the volume reduced by a third and the salinity increased four-fold. The ecology of Mono Lake is simple and consists largely of algae, alkali flies *Ephydra hians* and a brine shrimp *Artemia monica* only found in Mono Lake. The birds feed largely on the flies and shrimps. Laboratory studies showed that at salinities double the 1976 levels the shrimps had very high mortality and the flies failed to pupate. It thus seemed likely that much of the ecological interest of the lake would show a further dramatic deterioration.

Considerable publicity about Mono Lake and its problems was obtained through talks, press interviews and events. Each year since 1979 a group of concerned people have removed water from the now diverted

inlet and walked 8 km to pour it into Mono Lake. Similarly, an annual cycle ride took water from the pool in front of the Water and Power building in Los Angeles to Mono Lake.

The conservationists also argued that this water source was not essential. They calculated that small-scale projects to reduce water wastage in Los Angeles could easily compensate for the loss. Furthermore, in some years there was so little rain that no water was taken from the inflows to Mono Lake, yet this caused no problem in Los Angeles.

A series of legal disputes took place. In 1979 a combination of the National Audobon Society, Friends of the Earth and the Mono Lake Committee filed a suit against the Los Angeles Department of Water and Power to reduce extraction, arguing that it did not comply with California's public trust doctrine. This doctrine, derived from ancient Roman codes, specifies that the government has the duty to protect navigable bodies of water for everyone's use and benefit. There then followed a complex series of court actions with the California Supreme Court concluding in 1983 that "The human and environmental uses of Mono Lake – uses protected by the public trust doctrine – deserve to be taken into account". Although a landmark decision, it did not, however, result in any changes.

Ironically, the next major step forward came through the protection of non-native Brown Trout *Salmo trutta* and Rainbow Trout *Salmo gairdneri*. During wet conditions in autumn 1984, water overtopped a reservoir and carried tens of thousands of fish into an inlet creek where they thrived. However, the water was due to be diverted from the creek on the 1st November. Conservationists discovered that the Fish and Game Code forbids removing water from creeks below dams and persuaded the Assistant District Attorney to state that anyone touching the valve would be arrested. Subsequently, a minimum flow rate was agreed.

(*Continued*)

(*Continued*)

Eventually, following many legal victories for the conservationists, widespread support by the public and politicians and a new Mayor in Los Angeles, a solution was found by which water conservation and reclamation would replace much of the removed Mono Lake water. After 15 years and tens of millions of dollars, in 1994 a plan was agreed to restore the water levels so that Mono Lake could revert to its natural state and a restoration plan was agreed in 1998.

As well as resulting in improved conditions at Mono Lake, the legal decisions have helped change attitudes towards water use. The Mono Lake committee was actively involved in helping obtain US$80 million for water conservation programmes in Los Angeles which have considerably reduced the water use per person and created more reliable and economically valuable water programmes.

With water likely to be an increasingly important issue, this shows solutions are possible. Although largely fought in the courts, this success was due to a combination of research, education, publicity and persistence as well as skilled legal representatives.

As part of the campaign to reduce the divertion of water to Los Angeles that would otherwise go into Mono Lake, an annual event was organised to collect vials of water from Los Angeles, cycle over the mountains and pour them into Mono Lake. Such publicity events, combined with political pressure and scientific research, resulted in an agreed restoration plan (photo: Mono Lake Committee).

campaign is mistaken. This is particularly true if the campaign is organised from outside the country or from a different cultural group that it is easy to be critical of.

4 Understand the real objections to the suggested change. Who is likely to object and why?

5 Decide upon the most effective means of producing change. To what extent is it through direct discussion with policy makers and to what extent through raising public support? Find out who holds what influence within the organisation whose policy you are trying to change. Who can produce change? Who is likely to block change? Decide upon the appropriate level at which to lobby for that particular issue. Is it better to influence the person carrying out the work on the ground or someone at a high level in an office? Identify the important deadlines and determine what needs to be done for each.

6 Attract wide support. Change is more likely if a range of individuals and organisations are making similar demands.

7 Use publicity (see Section 10.3). The process of producing bad publicity for problems and good publicity for solutions is extremely effective. Publicity is excellent in generating pressure and creating awareness. Decision makers are more likely to dismiss issues if they have never heard them raised before. Publicity also helps fund raising and recruiting.

8 Seek a solution that is acceptable to both you and the opposition. Concessions or capitulation are more likely if there is an easy way out that does not humiliate the opposition.

10.3 Publicity

Press releases should be actual news, brief (often less than a page), have a terse but interesting headline, a first paragraph that tells the entire story and give an available contact person with telephone numbers. Describe what you have done, who you are, why you did it, when you did it and where you did it. Notes for editors can give supporting information, for

CASE STUDY

International collaboration to reduce pesticide poisoning

There was evidence that certain populations of Swainson's Hawk *Buteo swainsoni* had declined over parts of their North American breeding range. Satellite tracking showed that nearly the entire world population wintered in a restricted region in the pampas of central Argentina. In the winter of 1994/5 biologists found 700 dead Swainson's Hawks in this region. Interviews with farmers implicated the pesticide monocrotophos, a powerful organophosphate. Farmers were using it to kill large grasshoppers, the main prey of the Swainson's Hawk (Krapovickas & Lyons de Perez 1997). In the austral summer of 1995/6, researchers discovered more deaths of the hawks and estimated perhaps 20 000 were killed. Laboratory analysis confirmed that most of the deaths were caused by monocrotophos. This insecticide, widely used throughout the globe, is acutely toxic to birds and can kill birds in nearly all legal uses. It is not recommended for use for grasshoppers or locusts as there are readily available and safer alternatives. The American Bird Conservancy, a U.S.-based non-governmental organisation, then the Birdlife partner, interceded with the largest manufacturer, Ciba-Geigy (now Novartis), and alerted its Argentinian partner, the Asociacisn Ornitolgica del Plata. A committee was formed in Argentina including INTA, the national agriculture agency, SENASA, which manages pesticide registration and control, the Secretariat of Natural Resources, and several provisional agencies. The Asociacisn Ornitolgica del Plata and the American Bird Conservancy worked with the committee.

The American Bird Conservancy documented the killing of birds around the world from the use of monocrotophos and demanded that Ciba-Geigy stop its global manufacture and sale. Monocrotophos was no longer allowed to be used in the USA and more nations were banning it because of its toxicity to humans. Ciba-Geigy responded by suggesting a meeting with all parties represented. Such a meeting was held in the autumn of 1996 in Washington, DC. This meeting led to a landmark agreement in which Ciba-Geigy agreed to withdraw monocrotophos from the pampas region, buy back or exchange existing stocks, mount an advertising campaign with farmers and applicators and carry out further research. Ciba-Geigy also agreed to work with other manufacturers of the insecticide in Argentina to do the same and was successful in convincing the others to follow suit. Ciba-Geigy acted responsibly under the pressure of both the USA and Argentinian governments and non-governmental organisations. Some believe that Argentina and Ciba-Geigy were under great pressure to stop the hawk deaths as complaints could influence Argentina's sales of agricultural products and Ciba-Geigy was involved in a major corporate merger. Farmers were concerned about the possibility of an export ban and by the fact that monocrotophos is classified as a World Health Organisation 1b pesticide, meaning humans are also very vulnerable. Meetings continued in Washington with Ciba-Geigy assuming leadership in responsibly acting to stop the hawk deaths and re-evaluate the use of monocrotophos. Ciba-Geigy produced pamphlets, posters and radio and television advertisements to inform farmers of the problem. New labels were placed on all containers of monocrotophos in Argentina warning farmers that it was illegal to use it on alfalfa, where most of the hawk deaths occurred, and that it was illegal to use monocrotophos on grasshoppers. The American Bird Conservancy and the Asociacisn Ornitolgica del Plata suggested that the government ban monocrotophos and in response the national agriculture agency (SENASA) issued Resolution 396/96 which banned monocrotophos use on alfalfa and grasshoppers.

(Continued)

(*Continued*)

The Asociacisn Ornitolgica del Plata visited numerous communities, participated in radio and television programmes and met with school teachers, agricultural associations and agrochemical distributors. They also developed a network of Swainson's Hawk monitors.

Only 24 dead birds were reported in 1996/7, suggesting this problem is coming under control. As a result of this programme not only has the future of the Swainson's Hawk become more secure but there is now the awareness, monitoring, legislation and laboratories to make other pesticide poisoning less likely. The rapid success of this project appears to be a result of effective collaboration between organisations and the effective application of pressure.

example about the organisation, background to the issue or the biography of the spokesperson.

Most effective public relations campaigns are based on the following principles:

• Outline the problem in a way that matters to people. This may involve creating an emotional appeal or emphasising links with wider concerns, such as linking fish conservation to water quality or forestry to flood defence.
• Ensure the message is clear and focused. It should be possible to explain the message in a sentence.
• The story should be news. The media is likely to be more interested if it is linked to a report, meeting, statement or event. Using humour is often very effective as long as it does not trivialise the story.
• Make it easy for the media to use the material. Provide photographs or film or the opportunities for these to be taken. Give short striking quotes, 'sound bites'.
• Think carefully about the language. Coining the magical term 'ancient forest' instead of the rather pejorative term 'old growth forest' was a turning point in the preservation of the forest of the north western United States (Evans 1999). Similarly, whether cutting down the forest is called logging or harvesting greatly alters perceptions.
• Facilitate grassroots action. Make it easy for others to respond, for example, by providing fact sheets for people to use as a basis to write to politicians. Encourage involvement through open meetings.
• Identify an effective spokesperson. This person should be knowledgeable, confident, persuasive, reasonable, ideally charismatic and not make unrealistic demands. The most successful often illustrate general points with anecdotes from personal experience. The spokesperson must be easy to contact.
• If likely to be regularly involved with the media, it is often worth establishing contact with reporters as personal links make publicity much easier.

News stories develop rapidly and possess inertia so it is difficult to change the thrust of an established story, even if it is wrong or irrelevant. Being able to respond rapidly to refute misinformation or redirect the story can be extremely effective. It is also useful to collect press cuttings, internet coverage, and record radio and television coverage as a valuable basis for planning future coverage.

10.4 Negotiating and conflict resolution

Conservation policy making may result in disagreements and conflict, usually concerning exploitation, access, land ownership, pollution or land use (Lewis 1996). Conflict management and negotiation are very similar but in conflict management there is a history of disagreement and the re-establishment of trust is often a particularly important component.

If local interpreters are used, bear in mind that they often interpret information as well as the language. In my experience, they are often dismissive of the desires and beliefs of local people.

1 Decide what you wish to achieve and what you are prepared to concede. What is your ideal settlement and what is the minimum you will agree on? What concessions are you prepared to make? What arguments justify your position?

2 Research your opposition. This research should continue during the negotiation. What is the conflict about? What stance is the other party likely to take?

What are they likely to concede on? Are there any ambiguities or uncertainties that can be removed through the transfer of information or collecting additional information? What is the legal or constitutional context for the conflict? Are any resources available to help resolve the conflict?

Determine which individuals or organisations are, or should be, involved in the conflict resolution. Determine who speaks for them, their interests in the issue, the stance they have taken, their power, and whether they would gain or lose from the conflict being resolved and why.

3 Focus on underlying interests (the participant's fundamental needs and concerns) rather than positions (the proposals suggested to satisfy interests). For example, a group's position may be to demand access to a forest but the underlying interest is a source of heat, which they traditionally obtain by collecting wood. Once interests are recognised it is easier to discover other solutions.

4 Aim high. It is always easier to negotiate down than up.

5 It seems that a very successful strategy in tough negotiations is to trivialise any concession of the other party while emphasising the inconvenience of any concession you make.

6 Create rapport and trust. This is particularly important if there is a history of conflict. It may be sensible to carry out acts that make it clear you are committed to finding an acceptable solution. Individuals and representatives of groups are more likely to accept a solution if they believe they have been listened to, their ideas are valued and they have been respected as individuals. The solution is also likely to be better and viable. Others are more likely to consider your requests if you have shown yourself to be trustworthy. From a cynical perspective, friendly discussion often helps reveal the other party's stance.

7 Aim for a settlement which both sides find acceptable. Searching for such a 'win–win' solution is more likely to resolve the problem in the long term than one which only serves the interest of one party.

8 Remain true to your real objective but remain flexible and seek novel solutions as to the means through which this can be achieved.

9 Consider the whole package as otherwise, if making piecemeal agreements, it is possible to run out of acceptable concessions.

10 Try to ensure participants are happy at each stage and especially with the final agreement. The agreement is more likely to hold and work if the participants leave satisfied and a good relationship may be essential for further discussions on other issues.

11 If you lose, reflect on what happened, learn the lessons and move on. A lot of energy can be wasted wallowing in past disputes.

10.5 Changing legislation

Ensuring a change in legislation is often the most effective contribution that a conservationist can make. Most conservation efforts tend to be very local but a change in government policy may alter the entire landscape. This is often a gradual process. For example, it is often easy to introduce fairly vague statements about furthering conservation within say agricultural, forestry, fisheries or water use legislation. However, once these vague statements are within the legislation, they can be used as a lever to introduce more meaningful changes in legislation and policy. Bringing about change thus often requires tenacity and a clear vision of the possible solution.

1 Legislation is usually easiest to change when the precise way in which the law could be changed is understood. The constraint is often the need for someone who can draft the necessary legislation. It can thus sometimes be useful to employ the services of a lawyer or parliamentary draftsperson. Some legislation is only effective if it has an associated budget or means of generating income and this needs to be incorporated from the start. Examining similar legislation for other countries is useful, both for suggestions for content and to have a precedent, which politicians find greatly reassuring.

CASE STUDY

Reducing traffic damage to a roadside reserve

Due to intensive agriculture in England many strips alongside roads provide the best remaining areas of semi-natural habitat. One such area is Cherry Hill in Suffolk, which contains a remarkable number of rare British plants.

This site was being seriously damaged by vehicles pulling off the road and observations showed that this occurred when a lorry encountered another vehicle along this narrow section of road. The verge is only 3–4 m across, so this erosion of just 1–2 m was obviously important. The local council did not have the finances for a traffic survey which they considered essential information. In 1997 the local wildlife trust organised a traffic survey using volunteers with the standard data sheets used by the council. There was an easy solution – preventing lorries from using this road which involves a short diversion. Letters were written to the council about this problem by a range of governmental organisations, non-governmental organisations and individuals. In 1998 the suggested change was made, unannounced.

Although enormously less expensive or time consuming than the dispute over Mono Lake, the principles of researching and identifying the precise problem, finding a realistic solution, showing a commitment (the traffic survey), applying pressure and encouraging policy makers to make specific changes is the same in each case. In both cases success occurred suddenly and unexpectedly.

Lorries were destroying a narrow strip of roadside which contained a remarkable number of rare plants. A simple campaign of research into the problem, suggesting a solution of redirecting lorries and applying pressure on the local council, resulted in the recommendation being adopted (photo: William J. Surtherland).

2 It is usually best to find a sympathetic civil servant or politician who is prepared to push through the changes. This clearly has to be someone who is effective. If the proposed changes are likely to be popular in some circles then it may be appropriate to select someone with ambitions who would gain from being associated with the changes. If the proposed changes may be unpopular with some, then it may be best to select someone who is not frightened of being associated with such controversy. In selecting someone, get to know those involved in government and seek their advice. If voting behaviour is recorded, find out which people voted for similar cases.

3 Work out how it is best to get the change through the system. Be sensitive to the political balances and the political reality. Support is much more likely if it fits into the individual's or group's agenda so it can be useful to understand wider agendas. It is usually considerably better to have the legislation proposed by the party in power, but it may be more difficult to find someone who will promote such a change. Making life difficult for civil servants or ministers who want to reject your proposal through pressure from the public or other politicians often encourages progress.

4 Initiate or encourage a process of consultation with individuals and organisations to incorporate

improvements to the suggested changes and widen the support for the changes. Get support from opposition parties, which is usually easier for biodiversity legislation than others as it is usually not a party political issue.

5 Lobby all stages that are involved such as politicians, civil servants and advisers to ensure it does not fail or become diluted.

6 Compromise if necessary. It is common to be told by individuals or organisations that they will block the entire package unless a specific change is made. Do they mean it? Do they have the power? The skill is to distinguish genuine threats from bluff.

7 Once legislation has changed then it is often necessary to ensure the consequences are understood by those affected and responsible for implementation.

10.6 Meetings

Many meetings consume time without any progress being made. This is often due to inadequate preparation and follow up and people venting their grievances and opinions or just talking without suggesting practical or acceptable solutions. Some meetings do not have a clear function. I was made the chairperson of the conservation committee of one organisation and immediately disbanded it as the paperwork took up too much staff time and the committee generated too little of use (I closed down the monitoring committee at the same time). I recommend this satisfying activity! In this section I consider the issue of bringing about change within existing meetings while Section 8.5 considers how to run meetings. In some meetings there may be some people present whose objective is to stop the meeting coming to a conclusion which might threaten the status quo or with agendas that are counter to those of much of the rest of the committee. Such meetings are then best thought of as a process of negotiation (Section 10.4).

Some people are extremely effective in bringing about change in meetings. Much of this is due to preparation and considering exact changes. Many people turn up at meetings without having looked at the paperwork or thought about the issue.

In my experience, effectiveness on committees derives from one of four approaches. (1) Ideas may appear so sensible, and are presented so clearly, that they will be adopted. (2) Suggestions accompanied by the offer to do or fund the work are often readily accepted. (3) Ideas that involve work or resources from others are more likely to be adopted if the person suggesting them has assisted the committee, for example by carrying out some work or providing resources. (4) Some individuals acquire power by means of their status or unpleasantness. Such people are often extremely efficient at pushing through self-serving decisions. In the presence of such people, meetings will largely comprise negotiation, conflict resolution or undemocratic subversion.

A problem arises when external interests arise, such as conflicts or coalitions between individuals, organisations or countries that block sensible solutions. Understanding the political realities makes it easier to circumnavigate such problems.

If ideas are put forward and accepted then it is often important to agree a timetable and agree exactly who will do what. Ensure this is minuted and subsequently acted upon. Persistence in ensuring the agreed action points are carried out is often essential for ensuring ideas are not just agreed and forgotten.

10.7 Economic instruments

Most conservation problems are a result of some damaging operation providing a short-term profit to certain individuals even though they often result in a loss to others. Introducing changes to the economics to make damaging operations pay for the cost of the damage they cause can thus be highly effective.

Some damaging activities may be supported by 'perverse' incentives through grants or tax relief (James *et al.* 1999). For example, most fisheries are greatly overexploited and the fishers are poor. Providing financial incentives to support the fishers may seem sensible on social grounds but only encourages further overexploitation and makes the situation even worse. In many cases these incentives are provided as a result of pressure from lobbies and are usually not economically sensible, even before the environmental damage is considered.

In most cases the person causing the environmental damage gains the profits but does not pay for the costs. Thus pollution has to be cleaned up by someone else and others suffer the health consequences or the deteriorating environment. A response to this problem is the 'polluter pays' principle, which is that if the real costs of damage are paid by those causing the damage it is both fairer and will encourage more restrained use. For example, Denmark has a pesticide tax, the USA and Denmark have taxes on both chlorofluorocarbons and halogens, while Austria, Finland, Norway and Sweden tax fertilisers.

Economic incentives can be effective but can however easily lead to corruption. Thus schemes, say to reduce nitrate application, manage land in a particular manner or provide compensation for stock killed by carnivores, should be devised in a way that can be easily checked as otherwise the money can be claimed with no conservation benefit. Incentive schemes are often only successful if linked to education programmes or advisory services.

10.8 The importance of international agreements

Governments sign international agreements to gain prestige, be part of the international community and avoid criticism. International agreements often form the basis for national legislation. Once signed, they may be used to pressurise for change as international expectations may be harder for governments to ignore than local pressure. Thus ironically, although considered a government tool, they are often of greatest use to non-governmental organisations campaigning for change (this has been called the Convention Paradox). The potential embarrassment caused by allowing damage to protected sites can aid protection. For example, the South African government were giving serious consideration to allowing mining of the mineral-rich sand dunes near Lake St Lucia, which is a Ramsar site, but protected it following international protest and independent on-the-spot appraisal under the monitoring proce-

dure of the Ramsar Convention. Countries breaking the conditions of the convention have to explain their actions at regular meetings (conventions of the parties) and the threat of such embarrassment is often sufficient to bring action.

10.8.1 Convention on Global Biodiversity (1992)

This convention, usually referred to as the Biodiversity Convention, which resulted from the Rio de Janeiro Earth Summit, has been ratified by 183 countries. This is a framework convention which sets broad goals, leaving regions and countries to decide how they should be implemented. Glowka *et al.* (1994) gives a detailed account of the convention and its implementation.

The main biodiversity conservation requirements of signatories are to create protected areas; restore degraded ecosystems; legislate to protect threatened species; identify, regulate and manage damaging activities; introduce environmental impact assessments and develop national strategies, plans or programmes for the conservation and sustainable use of biodiversity. In practice many countries are much more likely to follow the conditions if urged to do so by local conservation organisations.

This convention has stimulated many countries to develop National Biodiversity Strategies where such policy documents did not exist, thus forcing themselves to consider positive conservation activities such as formulating conservation goals and integrating ecology into other parts of society. The Convention's secretariat is in Montreal, Canada.

10.8.2 Convention on International Trade in Endangered Species of Wild Fauna and Flora (1973) (CITES)

Usually known as CITES, this has been one of the more successful international conservation treaties. It has been successful in restricting international trade in endangered species. CITES' main function is to maintain three appendices of species, for each of which a different extent of trade is allowed. If a species is listed in Appendix I then international trade is forbidden except with special permission. Species on Appendix II have their international trade controlled. Appendix III lists species whose trade is forbidden by certain countries but are not listed on the other appendices.

Each CITES member has to create a national management authority and implement the treaty. CITES is co-ordinated by a secretariat in Switzerland.

10.8.3 Convention on the Conservation of Migratory Species of Wild Animals (1979) (Bonn Convention)

This is often referred to as the Bonn Convention (where the secretariat is based) or CMS. It provides strict protection for a list of species. It also provides a framework for collaborative conservation agreements between the states through which each species on a second list migrate. It has mainly been applied to birds, but also to bats and dolphins. The Bonn Convention also calls for research and surveys.

10.8.4 Convention on Wetlands of International Importance especially as Waterfowl Habitat (1971) (Ramsar Convention)

Usually known as the Ramsar Convention, after the place where the convention was agreed, this treaty provides the framework for international collaboration on wetland conservation which includes mangroves and coral reefs. Contracting countries have four obligations (Davis 1994). (1) Incorporate the consideration of wetlands conservation within their national land-use planning. (2) Designate at least one wetland of international importance ('Ramsar Sites') according to specified criteria. (3) Promote wetland conservation by creating nature reserves. (4) Train staff in wetland wardening, research and management and consult other countries especially for species or areas in common. There are currently 116 countries participating and over a thousand Ramsar sites.

Ramsar provides small grants from a fund, international expertise, and resources. For further details of Ramsar contact: Ramsar Convention Bureau, rue Mauverney 28, CH-1196 Gland, Switzerland. Email: ramsar@hq.iucn.org

11 Species management

11.1 Why manage species?

Species management is an admission of failure. It is obviously much more satisfactory to provide the correct habitat management or harvesting regime so that there is no need to resort to manipulating individual species. Species management is particularly important when a species is in rapid decline or the population is very small. The techniques for animals include providing supplementary food, providing breeding sites, removing predators, restricting disturbance and re-establishment. For plants the techniques include pollinating, weeding to remove competitors, fencing to exclude herbivores, creating new individuals from tissue culture, collecting seed and planting out seed or seedlings. Techniques for preventing excessive harvesting are described in Chapter 13.

It is usually sensible to combine management with the process of diagnosis (Chapter 6) to determine or confirm the problem. Many long-term programmes of intensive management have not incorporated diagnosis so it is unknown if past management was worthwhile or what should be done now. It is often useful to write a species recovery plan (see Section 7.3) before starting to manage a species.

11.2 Manipulating wild populations

11.2.1 Creating breeding sites

Artificial sites for nesting, resting or hibernating have been designed for a wide range of different species, including bats, lacewings, earwigs, bees, otters, birds and snakes. Artificial sites can have much higher success than natural sites (East & Perrins 1988). Artificial sites are more likely to be used in areas where natural holes are rare, such as short-rotation plantations. Artificial sites should produce the ideal conditions, thus successful boxes for most birds are dry and sufficiently insulated to keep inhabitants warm in cold weather and cool in hot. Some wood preservatives are potentially highly toxic to occupants (Racey & Swift 1986) and so should be selected and used with care. Artificial sites are often more conspicuous thus running the risk of considerably increased human disturbance or predation.

The habitat can also be manipulated to make it more suitable for breeding, for example by encouraging vegetation for species requiring thick cover or discouraging vegetation for those requiring open locations. Artificial islands tend to be safe from terrestrial predators. Examining and quantifying natural breeding sites and then considering how these can be created either individually or as a product of habitat management can be successful (Thomas 1991). The microclimate will vary considerably with the vegetation cover, slope and direction. Thus in the northern hemisphere species at the northern edge of their range are likely to prefer bare south-facing slopes while those at the southern edge are likely to prefer north-facing slopes and may benefit from cover.

11.2.2 Supplementary food

Supplementary food is usually only provided as a last resort. It can enhance populations by helping to reduce starvation or enhance breeding output when food shortage is a real problem and it is sometimes used when the usual feeding places are associated with risks such as predation, persecution or poisoning. Food is sometimes provided to improve visibility to visitors but this is often considered inappropriate, especially with the current move towards increasing naturalness (Section 12.1).

If huge amounts of food are provided then large populations can be attracted to one location which may make a spectacular sight. One concern is that populations may leave adjacent sites; these may then stop being managed for the species so that the population then becomes dependent on the artificial feeding.

Concentrations of food and individuals can spread viral, bacterial and parasitic diseases, with *Salmonella* infection being a particular problem (Wilson & Macdonald 1967). Diseases are a particular risk if the feeding or drinking areas become contaminated with faeces. Super-abundant artificial food can also increase the soil or water nutrient levels with possible deleterious health and ecological consequences.

11.2.3 Hand pollination

Hand pollination is used to increase the number of seeds produced when it is thought that this may be limited by the level of natural pollination. In practice, hand pollination usually involves erratic outcrossing of as many individuals as can be found. In the absence of information on the breeding system or pollinator behaviour it is likely that this will radically affect the genetic composition of the next generation and thus should be only adopted if consistent low seed production is a real problem (Cropper 1993). It is thus useful to know the breeding system of the plant being conserved, such as whether it is self-incompatible, obligate selfcross or apomictic and the species of pollinator (see Section 5.9.1).

Artificial pollination can involve moving pollen from the anther to stigma of the same flower (autogamous cross), between flowers on the same plant (geitonogamous cross) or to another plant (xenogamous cross). It is usually desirable to pollinate within the same population. Pollination across populations is usually only acceptable if one of the populations has reached critically low levels and either consists of just one sex or the remaining individuals are incompatible. It is obviously important to ensure that the donor and recipient are definitely the same species, particularly for those taxa that do not have physiological barriers to cross-pollination, such as many orchids, which thus readily hybridise if artificially pollinated (Cropper 1993).

The usual approach is to transfer pollen using a brush or toothpick. The pollen can be kept in a glass vial (a film canister works reasonably well) during the transfer or the brush or toothpick may be suspended through a hole in the vial lid. Flowers may be shaken above a Petri dish to collect pollen. The anthers or male flower can be picked and the anthers pushed against the stigma of another flower (Kearns & Inouye 1993).

11.2.4 Controlling parasites

Parasites may reduce survival and breeding output. Treatment of captive animals is routine but treatment in the wild is usually considered justifiable only in exceptional circumstances. Ectoparasitic repellents, as used to protect horses, can be used if it is clear they are not going to poison the species to be protected. The tropical nest fly *Passerimyia heterochaeta* has been a problem for the Echo Parakeet *Psittacula egues* on Mauritius, resulting in the death of nestlings (Jones & Duffy 1993). Adding 5% carbaryl dust to the nest at the start of the season and renewing frequently has removed this problem (Greenwood 1996). The same technique was used in artificial nest baskets of the Pink Pigeon *Columba mayeri* to control hippobosid flies *Ornithoctora plicata* which may cause anaemia and transmit diseases between nestlings.

11.3 Controlling predators, herbivores and competitors

An astonishing range and number of species have been introduced outside their natural range with parts of New Zealand seeming more English than England. As a result, introduced species are one of the major worldwide conservation problems as well as being a major economic problem. Introductions lead to extinctions through predation, competition, diseases and hybridisation (Godfray & Crawley 1998).

The African land snail (left) has been introduced to numerous Pacific islands for food. They became a pest and so a predatory species of snail *Euglandina rosea* was introduced which had little impact on the intended target but resulted in the mass extinction of numerous species in the genus *Partula*.

Some species, such as *Partula turneri* (right) survive on islands to which *Euglandina* has yet to be introduced. These are being bred at London Zoo as an insurance against extinction in the wild (photos: William J. Sutherland).

Introduced predators such as rats, cats and dogs have had disastrous consequences, especially on islands where the species have evolved in the absence of mammalian predators. Of the 110 bird species known to have become globally extinct since 1600, 34 (31%) can be attributed to introduced predators such as cats, rats, mustelids, mongooses and monkeys (Groombridge 1992). Other introduced predators can have surprisingly drastic effects. The introduction of the Brown Tree Snake *Boiga irregularis* to Guam resulted in the extinction of all the bird species on the island (Pimm 1987, Savidge 1987). The African Land Snail *Achatina fulica* has been introduced to various Pacific islands where it became a pest by eating crops. To reduce this problem a predatory snail *Euglandina rosea* was introduced which had a minor effect on the African Land Snail, but rapidly eradicated other snails, including innumerable species in the genus *Partula* (Hadfield 1986).

Introduced species may often out-compete native species. This is a particular problem with introduced plants, especially when the habitat has also changed (Pimm 1992).

There are surprisingly numerous cases in which the spread of an introduced species has coincided with the decline of a related native species and research has shown that it is not simple competition but that the introduced species carries a disease

to which it is immune but which is devastating to the native species. For example, the spread of the American Signal Crayfish *Pacifastacus lenuisculus* in Europe resulted in the decline of native White-clawed Crayfish *Austropatamobius pallipes* and Noble Crayfish *Astacus astacus* due to the fungal disease *Amphanomyces astoci*.

Another problem is that introduced species may hybridise with the native species (Simberloff 1998), either resulting in introgression as when the American Ruddy Duck *Oxyura jamaicensis* breeds with the endangered White-headed Duck *O. leucocephala* in Europe, or a loss of fecundity, as when male American Mink *Mustela vison* mate with female European Mink *M. lutreola* and the resulting foetus aborts (Rozhnov 1993).

As a result of these serious problems much conservation effort is spent on the depressing task of killing introduced species. To minimise future problems it is clearly important to be rigorous in preventing further introductions. However, the message not to introduce species has not been grasped even by those that should know best. The person responsible for the African Land Snail and *Euglandina* introduction programmes is probably responsible for more known extinctions than any other individual, yet has clearly not understood the obvious lesson that introductions are irresponsible. He has acknowledged that these introductions have caused problems but has suggested the solution is to introduce a further predatory species! Equally astonishingly, at an international ecological conference in 1998 all 1500 participants from around the world were given a 'green pen' which contained viable seeds, which have presumably germinated in rubbish tips around the world.

Conservationists are far too reluctant to eradicate small populations of recently introduced species, although it would often be reasonably straightforward to do so. Unfortunately the tendency is to consider them as entertaining quirks rather than potentially serious problems.

Prevention of introductions is clearly important and for isolated sites precautions against introductions should be routine. Many of the precautions listed by Merton *et al.* (1989) for Round Island apply widely.

• Ensure there is the legislation to prevent intentional introductions and the power to prosecute for infringements.

• No permanent mooring points should be provided to which boats could tie up for extended periods.

• Access should be by permit only and the permit should contain clear instructions about the precautions necessary to prevent introductions. The party leader is responsible for ensuring the group members read and act on the instructions.

• Seal boxes and ensure they are vermin proof. Plastic barrels or plastic crates with screw top or clip lids are recommended. Boxes can be sealed with masking tape. Bagged items such as tents should be tied tightly to ensure no animals can hide inside.

• Baggage should be stored in a rodent-proof building prior to transfer to the island.

• Baggage should be carefully checked prior to departure. Any items accidentally opened should be repacked and resealed.

• Boats used for transport should be carefully checked to ensure that they are rodent free. This is best achieved by adding poisoned bait prior to and during the transport.

• Baggage should be inspected carefully during unpacking.

• All visitors should ensure that footwear, socks and other clothing are free of seeds.

• Rodent-destruction kits and a contingency plan should be available in the case of either rodents being discovered or a shipwreck on or near the island.

• Visiting groups should look out for signs of introduced species. Early detection makes eradication considerably easier.

Section 5.4 gives further advice on hygienic fieldwork techniques.

Boats are a major source of introductions of rodents. One technique is to ensure all baggage is transferred by a second smaller boat. There are dock designs that are supposed to be rodent proof and designs to reduce the risk of introductions of the Brown Tree Snake.

Although introduced species are the major problem and cause of extinctions, there can be situations in which native predators, herbivores and competitors can be a problem, such as when protecting a critically endangered species.

11.3.1 Eradication of problem species

Complete eradication can be achieved through shooting, trapping or poisoning. Feral goats were eradicated from the 148 km^2 San Clemente Island, California, to protect the seven endemic plant species. Over a 17-year period 29 000 goats were shot, with the last goats being detected by the presence of radio-collared 'Judas' goats (Keegan *et al.* 1994). Feral populations of the Coypu *Myocaster coypus*, a South American rodent, in Europe, North America and Japan can cause marked changes in plant communities and weaken river banks. The population was eradicated in Britain using floating traps baited with carrots. The key to success was a trial eradication in one area to test the feasibility and methodology, followed by a programme with a considerable bonus to the trappers, equivalent to 3 years' wages if eradication took place within the estimated minimum possible time. With this considerable incentive the trappers exterminated the species (Gosling & Baker 1987).

With recent developments in poison baits it is increasingly realistic to eradicate species, such as

Fig. 11.1 Côté & Sutherland (1997) carried out a meta-analysis of those published studies which examined the effect of predator removal on bird populations. Removal had clear benefits in increasing the production of fledglings and juveniles but the increase in number of adults was less clear. One reason is that compensatory survival will reduce the effect. Although predator removal can clearly have a role, this study implies that other approaches such as habitat improvement may often be more useful. Photograph of the first author's grandfather, who was a Canadian gamewarden, and her uncles.

rats, from islands using bait placed on the ground or dropped from the air (Innes *et al*. 1995). Baits are usually placed in a grid at approximately 25 m intervals with higher densities in areas of particular abundance (Merton *et al*. 1989). The poisoned bait is usually covered to keep it dry and reduce consumption by other species. Plastic pipes are effective and relatively easy to store and transport but milk cartons, inverted ice cream containers with a hole in one side and even coconut shells with an end removed have all been used. Dying bait green may make it less attractive to birds (Caithness & Williams 1971). There is merit in using a range of bait covers, poisons and locations (e.g. under cover, next to objects or in the open), both in case individuals differ in preferences and to reduce learning. Follow-up surveys, for example by providing a different bait to detect survivors, followed by localised poisoning is sensible. It is likely that immunocontraception is likely to be a powerful technique in the near future (McCallum 1996).

Introduced plants are often cut down or pulled up but this is not always as straightforward as it seems. Cutting trees and shrubs often simply results in new growth from the cut stem. Either the roots need to be removed or the cut stem needs to be treated with a herbicide. Pulling plants from the ground may simply create bare patches exactly where the plant has dropped its seeds and so can actually increase the problem. It is thus useful to experiment in an area before starting a large programme.

A pilot study on a small area is often worthwhile to determine the practicalities, problems and to estimate the likely success. Prior to starting a programme to eradicate Rabbits *Oryctolagus cuniculus*, Merton (1987) provided non-toxic bait dyed with 0.05% concentration of tetra ethyl rhodamine (rhodamine B) which fluoresces orange under ultraviolet light; by shooting a sample of rabbits the uptake was estimated. It was then possible to show that the proposed technique was likely to work.

It is sensible to conduct a full risk assessment, for example by using the methods in Section 6.4, before an eradication scheme, especially one that involves poisoning or trapping. This should determine what other species are present, whether they will be exposed and the likely consequences. It may be necessary to move vulnerable individuals elsewhere or into captivity during the eradication and release them afterwards. The risk assessment should consider whether reinvasion of the problem species is likely and whether eradication of one species is likely to result in increased populations of others, so producing greater problems. An unsuccessful eradication programme not only wastes money but can cause increased problems if it causes a temporary but ecologically damaging population explosion (Bullock & North 1984). It is obviously essential to ensure there is no recolonisation after a successful control programme.

11.3.2 Control of problem species

It is often impossible to eradicate the problem species, either as it is not practical to eliminate the species or prevent recolonisation. In such cases it is usually better to plan a realistic long-term control strategy than carry out an unsuccessful eradication scheme with the species recovering. Such control schemes often concentrate on those areas where the species is the greatest problem, for example, on a small scale it is possible to clear vegetation surrounding individuals of a rare plant. In Madeira, Black Rats *Rattus rattus* have been controlled on cliffs used for nesting by Zino's Petrel *Pterodrama madeira* but not elsewhere and this has resulted in increased breeding success. Localised predator control has also been used before re-establishment to reduce mortality on newly released, naive animals (Short *et al.* 1992, Miller *et al.* 1994).

11.3.3 Exclusion of problem species

It is often considered impractical or undesirable to eradicate or control the problem species. In such cases one solution is to exclude them from the areas where they can cause damage.

It is often reasonably straightforward to exclude herbivores by fencing areas or even individual plants. However, many species of conservation concern gain in the long term from the presence of herbivores, for example by their disturbance creating bare ground for germination or by grazing competitors.

Predators may also be excluded. For example, wire mesh screens laid over individual nests of Green Turtles *Chelonia mydas* and Loggerhead Turtles *Caretta caretta* prevented Red Foxes *Vulpes vulpes* and Golden Jackals *Canis aureus* from eating the eggs, although predation occurred at 81% of unprotected nests (Yerli *et al.* 1997).

Predators can often be discouraged by simple physical means such as by creating islands or unsuitable habitat. There is considerable potential in excluding predators by the use of high-frequency noise, or synthetic or natural repellents. The effectiveness of these in the field has not been well studied and deserves much more research.

11.3.4 Changing the behaviour of problem species

Training predators, for example to associate eggs with unpleasant experiences, is potentially a powerful conservation technique but it is still unclear how practical or effective it will be in the field (Cowan *et al.* 2000). For example Red Foxes *Vulpes vulpes* given Pheasant *Phasianus colchicus* meat treated with 17α-ethinyloestradoil avoided unadulterated Pheasant meat when tested on eight occasions over the subsequent years.

Alternative feeding has been used to distract predators from the species of concern, for example dead white mice have been provided to a pair of Kestrels *Falco tinnunculus* to stop them removing chicks of Little Terns *Sterna albifrons* (Durdin 1993). One concern with alternative feeding is that, if carried out for long periods, it could result in an increased predator population and so it seems likely that its use should be restricted to critical periods.

Manipulating behaviour has the advantage of not needing to kill the predators and could be a major technique in the future so this subject justifies much further research.

11.4 Captive breeding

Conservation off site (*ex situ* conservation) can be a useful tool but it needs to be accompanied by conservation within the site (*in situ*). Captive breeding can attract publicity, education and concern which can help the *in situ* conservation. Alternatively, captive breeding can divert funds from *in situ* conservation and, worse still, reduce the concern for conserving wild populations.

Maintaining captive populations is expensive relative to other conservation investments (Balmford *et al.* 1997). Captive breeding can result in a loss of genetic diversity, problems with inbreeding (Frankham 1995), selection for the captive environment (Lachance & Mangan 1990, Frankham & Loebel 1992) and loss of cultural traits, such as predator avoidance (Curio 1993) or simply produce individuals that are too tame. Hybridisation is a particular problem when bringing together species which are not normally adjacent (as in botanic gardens). Captive animals may show physiological changes, for example the stomach of birds may shrink if given high-quality food in captivity (Piersma 1998). Cooper *et al.* (1988) suggest that many of the developmental abnormalities attributed to inbreeding in captive-reared birds are more likely to be a result of infection, poor nutrition or faulty artificial incubation.

The main requisites of successful captive breeding are usually as follows.

• An understanding of the natural history so that the diet, conditions for social behaviour, breeding habitat and conditions such as light and humidity are appropriate.

• Some knowledge of the tolerance to variability in, say, temperature, humidity and water chemistry. Those species living in variable environments, such as shallow pools, estuaries or bare rock usually have greater tolerance of changing conditions than those in stable conditions, such as deep lakes, deep seas or caves.

• Rigorous hygiene. Food and water must not be contaminated with faeces. Allocate a separate set of equipment, e.g. nets or brushes, for each container.

• Quarantining, inspecting and testing new individuals to ensure diseases are not introduced. The emphasis now is on prevention of diseases as treatment is often difficult.

• Screening and understanding what parasite load is normal. This is best done in the wild or as soon as individuals are first brought in.

• Regular monitoring of health, environmental conditions and maintenance techniques followed by analysis of data.

• Experimental changes to management. Either document the situation before and after change or, more rigorously, have treatments and controls, ideally with replicates. Document the results.

• Determine the cause of any illnesses or deaths.

• It is best to consider stock that might be used for re-establishment as being in a constant state of quarantine, especially avoiding related species from other areas. If there is any risk of infection then it is usually unwise to use the individual for re-establishments.

11.5 Plant propagation

Plant propagation either uses wild plants or those raised under artificial conditions. Gardening exper-

tise (e.g. Brickell 1996) is obviously useful. The commonest and least damaging technique is obviously to propagate from seed. For the majority of species, seeds germinate most readily as soon as they are mature. The methods for breaking seed dormancy vary between species with Ellis *et al.* (1985) and Baskin & Baskin (1998) giving guidance on each family. Dormancy occurs in two main ways: coat-imposed dormancy and embryo dormancy.

Coat-imposed dormancy arises because tissues prevent water uptake or gas exchange and in the wild the embryo will not start growing until the coat is damaged by fire, frost or microbial activity. Seeds with hard coats may need soaking or scarifying before they germinate. Soak by immersing in water that has recently been boiled until the seeds swell (this may take between 10 minutes and 72 hours) and sow straight away. Large seeds can be scarified by filing or nicking with a knife, while smaller seeds are rubbed with emery paper. Seed coats can be corroded with chemicals such as sulphuric acid.

Embryo dormancy usually arises from chemical growth inhibitors. Seeds needing cold stratification should be kept warm for a few days in moist compost within a plastic bag (absorbing water is an essential part of the process) and then placed in a refrigerator at 1–5°C until they germinate (usually 3–18 weeks, varying with the species). A deep freeze is ineffective as it is too cold. Warm stratification involves a similar treatment but at 20–35°C with 20–25°C often being ideal. Fluctuating the temperature within each day usually leads to greater germination (Thompson & Grime 1983). Some embryo dormancy can be broken by soaking in a growth stimulator such as giberellic acid.

The best way of understanding germination and dormancy requirements is to first sow seeds outside at the time of seed fall and then count seedling emergence regularly (e.g. weekly) in conjunction with collecting basic weather data, for example using a maximum–minimum thermometer and a rainfall gauge. This provides data on the conditions experienced when they germinate and when they do not. It is thus very useful to show that seeds do not germinate in the cold winter weather but did so in the warmth of the spring. However, this does not distinguish whether they need cold stratification before germination (e.g. Baskin & Baskin 1988) or whether germination is simply inhibited by cold conditions (e.g. Baskin & Baskin 1981) and to distinguish such possibilities experiment using seeds evenly spaced on damp filter paper in Petri dishes at different test conditions as determined by the field data (for example, warm conditions compared with cold followed by warm). Some use a single Petri dish for each treatment but this technique suffers from pseudoreplication (see Section 5.16): a low germination rate may be because that dish was, say, infected with a fungus or atypically wet or dry. To reduce pseudoreplication problems there should be replicates, e.g. three dishes each with 50 seeds to confirm that the results are consistent across dishes.

Seeds should be sown evenly on moist compost with compost sprinkled on top and covered with glass. Planting seeds at high densities runs the risk of fungal infection. In general, seeds from the tropics should be kept at 19–24°C, those from subtropical and warm temperate at 13–18°C while those from cool temperate at 6–12°C. Keep away from sunlight. Once the seedlings are sufficiently robust to handle, transfer to individual pots. If planting out seed directly then a fine soil results in high contact between the seed and soil and so results in a more rapid water uptake, while compacting the soil after sowing further increases contact but over-compaction hinders seedling emergence (Iriondo & Pérez 1999).

Perennials and spreading root stock can be divided when dormant, but not in frosty or very wet conditions. Extract the plant using spade or fork causing as little damage to the root system as possible. If the roots are loose and fleshy, divide into clumps by hand and replant each. For those with fibrous roots or wood crowns that cannot be separated by hand, insert two gardening forks back to back and lever them apart. It may be necessary to cut through woody rhizomes but avoid growth buds and dust with fungicide before replanting.

Layering is a useful way of propagating some woody species. Wound the underside of a flexible stem and treat with rooting hormone. Peg under soil surface and tie the tip of the stem upright to a stake so it may form a new individual. Once roots have formed cut the old stem and transplant.

Growing new plants from cuttings is a very efficient means of propagating many perennials. The main methods are stem, root and leaf cuttings. If being transported back from the field then cuttings should be kept moist, such as in a plastic bag with wet cotton balls. Cuttings are best taken in the early morning when the plant is turgid. Dip in a rooting mixture, plant in sterilised soil and keep in a humid environment (such as a propagating case or within a supported plastic bag) until new roots develop.

11.5.1 Seed storage

It may sometimes be sensible to store seed, for example as a backup in case of extinction in the wild. Ideally collect including a wide range of genetic variability by selecting a wide range of individuals, for example at least one fruit from 50–100 plants in each population. Collect samples from a range of habitats, seasons, size classes and flower colours. Collect from a wide range of populations, but it is usual to keep each separate and labelled. Seed maturation can usually be detected by being full sized, the seed coat changing colour (usually from green to a darker hue), the stems are dry, the earliest formed seed is dropping, and the seed interior is too hard to squeeze between finger and thumbnail or easily bite through (Apfelbaum et al. 1997).

There is the possibility of harming the population by collecting seeds. Some suggest that no more than 10% or 20% of the seed should be removed over a year but the reality is more complex. The problem is just the same as exploitation (considered in Chapter 13) and it is the same as asking what fraction of eggs can be removed from a bird population. The response depends largely upon the degree of density-dependent seedling survival (if there is considerable competition within the species then removing some seeds is unlikely to markedly reduce the population) combined with the rate at which the population is capable of multiplying (see Fig. 13.1), for example the growth rate calculated when there is little competition from others of the same species. If seeds are either densely clumped as a result of being locally abundant or highly fecund, then competition between seedlings is likely to be severe so removing a moderate fraction of seeds is likely to have little effect on the number of seedlings that survive. If seeds are scattered and scarce, because the species is rare or has low fecundity, then removing even a small fraction may affect the population. Some perennials predominately spread vegetatively with seedling establishment being rare so seed removal is unlikely to greatly affect the population, although the occasional seedling is important for maintaining genetic diversity. Some species, especially annuals, have a large seed bank which will buffer the effects of collecting seeds in one year.

Seed can be collected by shaking the stems or fruit over a sheet of paper. Capsules can be cut off and allowed to dry: some open after drying by placing in a dry room for several months, while some fire tolerant species, such as *Banksia* spp. and *Hakea* spp., need to be burnt to encourage opening (Cropper 1993). Seed is usually cleaned by sieving as this saves space, is more hygienic and neater.

In general the drier the seeds are stored the better. Seed starts deteriorating as soon as it reaches physiological maturity, defined as the point of greatest weight, and starts deteriorating even while on the mother plant in the field (Roberts 1992). Moisture content is determined by weighing seeds and then drying at 135°C for 2 hours and allowing to cool in a desiccator before reweighing (Harrington 1970). Newly formed seed has a moisture content of about 50% (Roberts 1992). Deterioration occurs at the highest rate at about 30% for non-oily seeds and about 15–18% for oily seeds. Below about 12°C seed longevity doubles for each 1% reduction in moisture. Thus seeds should be picked at physiological maturity and dried as rapidly as possible, but without exceeding 40°C. Sun drying can be a useful temporary measure to bring the moisture to about 12%. Drying seeds to about 5% moisture content can be achieved by drying at 15°C and 15% relative humidity in a drying room, drying cabinet or in a sealed container with silica gel which has just been dried at about 200°C and cooled (this should equalise to a relative humidity of about 15%). Species adapted to arid conditions usually survive particularly well during desiccation. Dried seeds can be stored in clip top vials or screw top bottles with rubber seals but not in plastic bags as these allow water vapour through.

Most seeds should be stored cold. Under tropical conditions of 20–30°C, viability increases 1.5 to 2-fold for each 5°C drop in temperature. The norm for long-term storage is 5% moisture at −20°C (use a standard deep freeze). Most species, known as orthodox species, will remain viable for extended periods with this method (Hong *et al.* 1996) and probably for over 200 years (Alton 1998). The benefits of dry cold storage are enormous with cereal longevity estimated to increase about 11 000-fold at these conditions compared with the typical farm storage conditions of 13% moisture and 20°C (Roberts 1992).

Species are referred to as orthodox (or desiccation tolerant) if they are viable after drying to 5–10% moisture content or recalcitrant (or desiccation sensitive) if they become inviable once dried to this level (Chin & Roberts 1980). Recalcitrant species tend to be those with small seeds, fleshy seeds, very large seeds, those that germinate soon after falling or aquatic species. These have been stored at similar conditions to those at which they naturally occur. Thus a range of aquatic species have been stored in water at 3°C, temperate wood species have been stored at 5°C and high (85%) humidity and tropical species have been stored at room temperature and high humidity (Baskin & Baskin 1998). Long-term storage is usually likely to be difficult for such species and they are often conserved by sustaining a population of growing plants. Recalcitrant species cannot be stored below freezing due to the damage caused by ice crystals from the high water content. Ways of keeping them below freezing include treatment with cryoprotectant chemicals such as ethylene glycol followed by rapid freezing or storing the embryo axes in liquid nitrogen and growing using tissue culture (Iriondo & Pérez 1999).

If species are stored under dry conditions then when planted out they may experience imbition injury, damage due to rapid water uptake which may lead to death. This is prevented by keeping for a day suspended over water within a closed container. Drying may also harden the seed coat and thus prevent germination unless scarification occurs. Cold storage will often preserve or alter dormancy.

Thompson (1974) suggests testing seeds after one year, then at greater intervals. Experiments may have to be carried out to determine the temperature, light regime or scarification required for germination. Seed viability is usually checked by placing 100 or more seeds on damp filter paper in a Petri dish and storing in a warm location for a few weeks. These repeated germination tests will tell if the seed has deteriorated. Loss of viability is linked to chromosomal damage and heritable mutations (Rao *et al*. 1987) and to reduce this problem it is suggested seeds should be grown to produce fresh seeds once viability drops below 85% (Roberts 1992).

11.6 Re-establishments

The terminology is confusing and contradictory.

Re-establishment. The release and encouragement of a species in an area where it formerly occurred but is now extinct.

Introduction. An attempt to establish a species where it did not previously occur.

Re-introduction. An attempt to establish a species in an area where it had been introduced but the introduction has been unsuccessful. Confusingly, this term is also often used for the release and encouragement of a species in an area where it formerly occurred but is now extinct (as for re-establishment).

Re-inforcement. Attempting to increase population size by releasing additional individuals.

Translocation. The transfer of individuals from one site to another (for example to boost the population or save individuals which would otherwise be destroyed).

11.6.1 Determining feasibility and desirability of re-establishments

Re-establishments are fun and are a good source of publicity and funding. This can, however, lead to problems. They can detract effort from more important work, especially habitat management that benefits a wide range of species. They can also distort the priorities within a site such that the habitat management concentrates on assisting the re-establishment even though this might not be the most sensible overall management. Re-establishments also give the impression that species can be easily moved, for example to avoid a development, while in practice they often fail.

Re-establishments frequently entail transferring genotypes across geographic ranges and this has attracted criticism. The genetic consequences depend upon the importance of diluting local adaptations and breaking coadapted gene complexes balanced against possible advantages from increasing genetic diversity (Avice 1992, Ellstrand 1992, Bowles & Whelan 1994). If the species has marked geographical variation then it is rarely appropriate to move variants outside their usual range.

Every re-establishment should have a clear objective. Re-establishment is often the best way of understanding why species have disappeared from areas but this requires that it is genuinely experimental and properly monitored. Properly designed and monitored research may provide an exception to the rule that re-establishments should only occur if the reasons for the decline are known and reversed. Many re-establishments are called experiments when they are not designed in a way to produce useful information. As Soderquist (1994) cynically, but probably accurately, comments, the term experiment is often used to mean 'all the animals died and it is not my fault'.

Any re-establishment or re-inforcement programme should include the following criteria (Kleiman *et al*. 1994, IUCN 1998).

• Consult widely before deciding to attempt any re-establishment and speak to the relevant experts and government officials.
• There is a need to augment the wild population.
• The site is within the historic range of the species.
• Habitat is not already saturated with the species.
• The causes of decline are known and removed.
• There is sufficient protected habitat.
• Stock is available.

• There is no jeopardy to the wild donor or recipient population. Particular care should be taken if using root cuttings.

• The added individuals are taxonomically similar to those previously present and will normally be of the same subspecies or race and should be from sources with similar ecological characteristics.

• There is sufficient knowledge of the species biology and the species is compatible with the site conditions.

• The experience gained from previous attempts at re-establishments and re-inforcements for the same or similar species are used.

• Re-establishment technology is known or in development.

• Stock should be screened for infection and the possibility of infection during shipment excluded. Quarantining is often appropriate.

• Sufficient resources exist for the programme.

• There is no negative impact for local people.

• Community support exists.

• Government organisations and non-governmental organisations are supportive or involved.

• The re-establishments will be reported to the relevant recording bodies so that sightings do not confuse records.

• It conforms with laws and regulations. There is permission both from the receiving site and the source of material for re-establishment.

There is a long list of pathogens and parasites that have been released with wild or captive-bred animals which have hindered the release scheme, or even worse, affected wild individuals of that species or relatives (Woodford & Rossiter 1994). Quarantine prior to release for at least 30 days is then sensible, with screening for diseases and detailed clinical examinations by a veterinarian. While in quarantine it is important to ensure the animals cannot catch diseases from other individuals or the staff. Hygiene is also important in planting programs or translocations (see Section 5.4) to ensure diseases are not introduced, for example plants should be cut using clean secateurs that have been sterilised for 30 seconds in methylated spirits.

Kagu from Nouméa Zoo, New Caledonia, were released into a large natural pen but with artificial food provided. Those that did not make the transition to natural food were returned to the zoo. By then leaving the door open the others could make a gradual transition from the pen to the completely wild state (photo: William J. Sutherland).

11.6.2 Release protocol

The receiving site should be appropriately managed and of sufficient size to maintain a population. The numbers released should be large enough to secure re-establishment. Analysis of a range of release programmes showed the successful ones were often linked to community education programmes, which illustrates that local goodwill is often essential for success (Beck *et al.* 1994).

'Hard' release is simply releasing individuals into a new area while 'soft' release means that the individuals are gradually released. Soft release typically involves placing the animals into an enclosure within the release area with food and shelter provided so that they can become accustomed to their new surroundings. The doors are then opened but the food and shelter are retained so that individuals can gradually become more independent. Many bird and mammal re-establishments from captive populations now use soft release. If the adult animals are being transferred, i.e. caught from the wild and moved, then hard release is probably better to minimise the time in captivity, although it may still be sensible to provide food and shelter. With both techniques it may be appropriate to reduce the effects of predators and competitors.

Individuals can be monitored during soft release to confirm that they can cope. For example, Kagu *Rhynochetus jubatus* on New Caledonia were transported from Nouméa Zoo into a one-hectare pen in the release site and the diet of individuals was studied using faecal analysis (see Section 5.7). Those individuals that had not switched to a natural diet, which turned out to be those that had been raised in small cages with little access to natural food, were returned to the zoo.

Soft release for plants may involve removing competitors, watering or fencing to exclude herbivores. Introducing seeds into established vegetation is very rarely successful so it is necessary either to clear patches of vegetation or add established seedlings in plugs of soil. Tree planting is often most successful when trees are 1–2 years old ('whips'). In many cases it is best, if possible, to use soil from the re-establishment site to grow the plants as this reduces the likelihood of infection and reduces acclimatisation to unfamiliar conditions. Protection from grazing mammals is often beneficial in the short term; plants grown in well-watered, nutrient-rich conditions are likely to be particularly palatable! In general, adding nutrients will benefit competitors more than the re-established species and thus may be detrimental. Watering may be sensible in the short term until the roots are established but long-term watering is similarly likely to be counter-productive by encouraging competitors. Falk *et al.* (1996) and Akeroyd & Wyse Jackson (1995) describe strategies for the establishment of endangered plants.

One problem is to train species to recognise and respond to predators. It is likely that this will be most effective if it is realistic. Thus captive-bred Houbara Bustards *Chlamydotis undulata* were chased with Red Foxes on a lead before release which led to significantly higher survival (Heezik *et al.* 1998). Any such training has to be compatible with welfare considerations.

Some species, particularly social birds and mammals, may learn from watching others about predators, food, mates, migration routes or predator recognition and avoidance. The loss of such cultural knowledge can be a major problem in re-establishing captive-bred individuals as they may lack the most

basic survival skills (Sutherland 1998c). The technique used successfully for Golden Lion Tamarins *Leontopitecus rosalia* is to allow captive-bred individuals to breed in cages in the wild but allow the young free access with the result that young obtain sufficient skills. Training to recognise predators is one approach but it is very time consuming and has had limited success. The best policy is probably to ensure that highly social species do not go locally extinct, and thus avoid the considerable problems of re-establishing them.

11.6.3 Monitoring of re-establishments

There are very few accounts of successes and even fewer of failures, with the result that it is difficult to improve techniques. Release details should be meticulously reported and the success of re-establishment assessed and recorded. An example of a useful account of a failure is the description of three unsuccessful attempts to re-establish Wild Dogs *Lycam pictus* into Namibia (Scheepers & Venzke 1995). Captive-bred dogs were poor hunters, some were killed by Lions *Panthera leo* and four died from rabies. This frank and detailed description is clearly invaluable for anyone else considering such re-establishments.

Releases should be experimental. For example, the programme to transfer Hihi or Stitchbirds *Natiomystis cincta* from the species' only self-sustaining population on Little Barrier Island, New Zealand, to Kapiti Island involved a series of experiments (Castro *et al.* 1994). These showed that survival was higher if translocated birds were released immediately than kept caged for some time. Survival was the same whether released as a pair or group and there was a suggestion (though statistically non-significant) that birds had lower survival when released where Hihi were already resident. Carrying out and publishing such research is invaluable for other release programmes.

Re-establishment procedures are usually carried out on rare species so that the opportunities for experimentation are restricted. One technique is to use surrogate species. Neutered Siberian Ferrets *Mustela eversmanni* were used to test release

Brush-tailed Phascogale re-establishment: learning from experiments

The Brush-tailed Phascogale *Phascogale tapoatafa* is a rare Australian carnivorous marsupial. It became extinct in Gippsland (south-east Victoria) for unknown reasons and there were surplus juveniles from a captive breeding population so it was decided to introduce them to this area. Soderquist (1994) tested a series of hypotheses as to what determines the success of re-establishments. Individuals were released in two forest types which showed that the intake rate, body condition, growth rate and independence from supplementary food (a measure of hunger) was higher in box-ironbark forest than stringybark-silvertop forest.

Soderquist thought that sub-adult males would only disperse short distances but this was disproved in a release of five radiocollared males that moved an average of 6.5 ± 1.5 (SD) km in 2 days. He then hypothesised that the male dispersal depended upon the presence of females and their scent. He released two orphan males in an area occupied by females and neither dispersed over 2 km. He then released nine sub-adult captive bred males in areas where females had been released at least 2 weeks previously and these dispersed significantly shorter distances (mean 2.8 ± 1.1 km) than did those in the absence of females.

The third set of experiments examined the high predation (50%) in the week after release. A poison bait programme was successful in eliminating Foxes *Vulpes vulpes* (an exotic species in Australia) from the study area but predation by feral cats, monitor lizards *Varanus varius* and raptors still resulted in 39% mortality in the first week. To see if this problem could be removed by raising the Phascogales in the wild, three mothers and litters were transferred to separate nursery cages with portals so the young could explore but the mothers could not. Within 2 weeks the young were travelling hundreds of metres from the cage at night. After a month the portals were enlarged so the mother could also move. All young survived for 7 weeks and then experienced mortality at natural rates.

These experiments are not perfect. The sample sizes are small, as expected for rare species and the treatments are carried out at different times due to practical constraints, but this study shows the thoughtful approach which all such re-establishment projects should adopt. Most importantly the exact methods and results are described so that anyone else carrying out a re-establishment can learn from this experience.

procedures for Black-footed Ferrets (Miller *et al.* 1994). A common species, the Fluttering Shearwater *Puffinus gavia* was used to develop translocation methods that could then be used on rarer species (Bell 1994) while Valutis & Marzluff (1999) compared raising Common Ravens *Corvus corax* with a puppet parental model to test ways of reducing imprinting on caretakers for releases of endangered Hawaiian Crow *Corvus hawaiiensis* and Mariana Crow *Corvus kubaryi*.

12 Habitat management

12.1 Habitat management or wilderness creation?

If the objective is to maintain natural habitats then why not just minimise human influences? The increasingly accepted philosophy is to do just this by allowing rivers to take their natural course, coastal erosion to proceed and forests to burn naturally. The justifications for this are partly ecological, in restoring natural processes, and partly aesthetic, in creating a feeling of naturalness and wilderness.

The philosophy of restoring and encouraging natural processes is not without critics. Natural processes, such as fires or floods are often unacceptable to many, especially those who live alongside the wood or have built on the floodplain. For example, the current policy for managing Yellowstone National Park in North America, is to allow natural fires to persist, re-establish Wolves *Canis lupus* and no longer cull Elk *Cervus canadensis*. Many consider the fires unacceptable, even though the ecological evidence shows that this policy has been a great success (Lichtman 1998), the wolf re-establishment has been condemned by adjacent farmers and even the policy of no longer regulating the Elk population has been criticised as it resulted in heavy grazing, especially of willow *Salix* communities (Chase 1987, Kay 1997, but see Schullery 1997 and Huff & Varley 1999 for conflicting views). If encouraging natural processes in Yellowstone, which is one of the most famous conservation areas in the world, faces such a barrage of criticism then where will people consider it appropriate? Furthermore, Yellowstone is large and reasonably isolated and it is likely that the opposition to natural processes in smaller or more densely populated areas will often be overwhelming.

Even if it were socially acceptable to allow natural processes within small sites, it is likely that this would often fail to achieve the conservation objectives. The essence of the problem is that a remarkable number of species in temperate areas are dependent upon open habitats. The popular belief that the natural habitat is a continuous carpet of closed forest is clearly mistaken. Natural ecosystems often consist of a shifting mosaic of habitat patches with fires, tree falls, shifting rivers, landslips, droughts and floods clearing areas and setting back succession. Herbivores often also delay succession. A fragment of such an ecosystem cannot maintain such a natural mosaic and a fire or flood may destroy the entire remaining habitat. Furthermore, the site may be too

Early successional stages are created in natural habitats, such as this Latvian forest, by disasters such as tree falls, floods and fires. Reducing the impact of such disturbances can be deleterious for those dependent upon early successional stages (as well as those dependent on dead logs, flood and burnt habitat). An obsession with tidiness can be disastrous (photo: William J. Sutherland).

small to sustain a natural herbivore population. Conservationists protecting such fragments are thus left with a choice between allowing succession to proceed, and thus losing the early successional stages, or artificially maintaining the habitat at an early successional stage.

Conservationists visiting western Europe are often astonished to see nature reserves managed intensively by stock grazing, grass cutting or tree removal. Human activities, such as agriculture, burning, harvesting or grazing will create earlier successional stage habitats and in areas with a long history, such as western Europe, these largely semi-natural habitats may be important for early successional stage species. In practice, habitat management usually aims at mimicking many of the natural features such as habitat mosaics, rotation, complex mixes of habitats and interfaces between habitats. This has been very successful in retaining a subset of the natural biodiversity. If left to natural processes then, as succession proceeds, they will lose the species of interest but are unlikely to gain other species or communities of interest within the foreseeable future, particularly for isolated sites in which colonisation is likely to be slow. Although the idea of wilderness is attractive, many reserves in western Europe are isolated areas of early successional stages and would lose most of their conservation importance without human intervention.

Some argue that wilderness is a delusion. Chase (1987) points out that a management objective of Yellowstone National Park is to restore it as it was during colonisation by Europeans, but that the native Americans greatly reduced the herbivore populations and regularly burnt areas which kept them open. As Europeans colonised, the native American population collapsed, largely due to smallpox. The early documenters thus considered the areas unaffected by humans and saw the large population of herbivores flourishing in the open areas. Chase suggests that it is unrealistic to consider restoring the landscape without considering human influences. However, Whitney (1996) argues that the ecological effects of humans were localised with large areas being practically unaffected by native Americans.

There is increasing interest in the idea of restoring natural large-scale ecosystems, such as the Wildlands Project (Noss 1992). Some consider such wilderness proposals unattainable, but I believe four factors make the concept increasingly realistic. Firstly, ecologists are becoming more aware of the importance of interconnected habitats and there is increasing evidence that populations are unlikely to persist within isolated patches (Section 12.2). Secondly, the term wilderness originates from the medieval English *wildeor*, or wild beast, and is an ancient idea laden with religious meaning (Peepre 1998). At least in western Europe, the case for conservation has usually been argued largely on science but is increasingly being based upon cultural, spiritual, aesthetic, scenic or recreational values. We should not underestimate the thrill of knowing landscapes exist where you can walk or canoe without encountering roads or buildings and I believe people find this idea exciting even if they do not take advantage of it. Thirdly, we are at last appreciating the financial costs of destroying wild areas such as the costs of building on floodplains and then having to protect the buildings, building on eroding coasts and then having to maintain sea defences, rectifying pollution created on water catchments and facing local climatic and hydrological changes caused by habitat loss. Finally, there are considerable social and economic changes occurring within many agricultural and forestry communities which could provide exciting opportunities for conservation with sufficient vision. Rather than continuing to subsidise agriculture in areas where it is not economic, why not subsidise the restoration of natural communities as an option? If agriculture becomes increasingly efficient and there are greater opportunities for farmland coming out of production, then this could lead to the opportunity for wilderness creation.

Creating wilderness areas is likely to be more achievable if considered over a long time scale, for example by introducing tax concessions for those who promise to hand over their land after their death, or even after the death of their children, or permission for forestry or other activities being linked to eventual conversion to wilderness (Soulé

Wild nature in the Dutch Oostvaardersplassen

There has been a long tradition of intensive nature management in the Netherlands. However, there has been a recent movement towards more natural management, driven by the cost of intensive management and a change in philosophy.

The Oostvaardersplassen is the lowest part of a polder created in 1968, which was planned to become an industrial area. However, as a result of the huge number of birds it attracted during draining, increasing concern for nature conservation and a downturn in the Dutch economy, it was decided to create a reserve in 1974, extended to 5600 hectares in 1978.

The uniform straight ditches in the once agricultural part have been filled in or resculptured. Two hundred hectares of shallow pools are currently being dug for early successional species, because this habitat is rapidly disappearing in the Netherlands and because, as the 600 000 m^2 of clay can be sold for strengthening river and polder banks, the project will be almost free.

The underlying philosophy has been to keep human intervention to a minimum and see which species survive rather than managing to maintain a certain suite of species. Predators are not controlled and, as a result, ground-nesting birds have low breeding success.

Grazing was introduced to ensure variation in vegetation structure. As well as Roe Deer *Capreolus capreolus* already present, 150 Red Deer *Cervus elaphus* have been introduced. The natural European grazers have gone extinct so 200 Polish ponies, descendants of wild Tarpan (the original European wild horse), and 300 Heck, a breed of cow designed to recreate the extinct Auroch, have also been introduced. Tarpan and Heck can survive without human intervention.

Research suggested that allowing the water levels to fluctuate was highly beneficial, both in retarding succession in wet periods and in producing productive plant growth in dry years. During reflooding, the burst of productivity provided abundant seeds for waterfowl and highly fertile conditions for fish. Rather than artificially manipulating levels, it was decided to allow natural fluctuations within set minimum and maximum levels, only intervening if there are likely to be serious losses.

This wilderness management of the Oostvaarderplassen is an experiment that has only just started. The project has its critics. It is likely that some populations, such as waders, spoonbills and geese are lower than they would be if the area was managed intensively. Some doubt whether the natural grazing and flooding regime will be sufficient to keep large areas open. The area is, however, wonderfully rich with at present 60 Marsh Harrier *Circus aeruginosus* territories, 30 territorial Bitterns *Bolarus stellaris*, 60 000 moulting Greylag Geese *Anser anser*, 80 000 Siberian

Oostvaardersplassen with Black-tailed Godwits and Avocets (photo: Jan van de Kam).

(*Continued*)

(*Continued*)

White-fronted Geese *Anser albifrons albifrons*, 50 000 Black-tailed Godwits *Limosa limosa* and on migration up to 15 000 Avocets *Platelea leucorodia* (Whitbread & Jenman 1995).

My brief visit there was hugely enjoyable despite being on a cold, wet March morning. The absence of obvious man-made features, such as fences, gates or linear drainage ditches and the freely roaming herds of large herbivores helped to provide the wilderness experience so badly lacking elsewhere. It is clearly easier to have such a *laissez faire* management policy in a newly created or virgin habitat than in a nature reserve long famous for possessing certain species dependent on human intervention. While acknowledging that such wilderness management is not straightforward or necessarily the complete solution everywhere, I was convinced that this clearly shows the way forward.

1992). Obviously the more extensive the area is, such as the plan to reforest an entire catchment in Carrifan Wildwood project, South Scotland, the easier it will be to allow natural processes without upsetting others.

My personal perspective is that in heavily modified landscapes it would often be naive to abandon intensive management as this would usually lose species and communities of interest without an equivalent gain; we thus have to consider the history as a guide to current management. Living in England, I have spent far too little time in wildernesses, but my time spent in the Canadian Arctic, Sahara Desert, and various large tropical rainforests has left me excited, respectful and enthusiastic about the importance of restoring wild areas. Even in western Europe we should be seeking to establish large landscapes in which natural processes can occur so that management can then be less intensive or even unnecessary. At a smaller scale, people visit nature reserves in part for at least a hint of a wilderness experience yet often encounter a degraded landscape with ugly notices, fences and buildings; I think even on this scale conservationists should consider the aesthetics more seriously.

12.1.1 The need for research

As outlined in Section 6.2, reserve managers usually have a confidence in their understanding of the consequences of management that is probably often unfounded. There is a clear need for much

of management to be experimental and certainly to document the management and the consequences.

There will be variation within a reserve in the distribution and abundance of the plants and animals. An important early step is to understand the variation in the plant communities in relation to soil type, water status, altitude, aspect, herbivores and current or past management. It is also useful to understand the distribution and abundance of key plant and animal species in relation to the above factors, dispersal, predation, competition and exploitation. It is particularly useful to understand the role of those management aspects that can be altered, such as hydrology, visitors, grazing regime or exploitation.

It is often useful to consider the history of the site using a combination of documents and field evidence (Rackham 1998). The importance of understanding history for conservation is illustrated by the study of the Mediterranean island of Crete by Rackham & Moody (1997). They argue that the conventional view that the landscape is being ruined by overgrazing is simplistic as the historical evidence suggests that the vegetation is probably similar to that that the ancient Minoans would have seen. Furthermore many mammals have gone extinct, including seven species of deer, which must have had a considerable grazing impact. This research explains why most of the endemic plants are adapted to intensive browsing. The diaries of 19th-century explorers have been used to characterise aboriginal fire regimes in Australia which provides a basis for current management (Braithwaite 1991). Historical ecology is useful in helping understand why the area has the ecology it does, which is the first step towards deciding whether to maintain or change it. Historical ecology is also useful in showing what is likely to be possible. Rackham (1998) gives the example of foresters in southern Europe often planting arid orchid-rich hill-sides with pines to 'restore the forest cover' and who are dismayed when the trees inevitably die.

It is often useful to write a habitat management plan (see Section 7.4) before starting management. It is usually sensible to gain experience of the site prior to any management, for example by studying the area for a year.

12.2 Size, isolation and continuity

Many species can only persist in large sites. In small sites there is a high risk that in some years there will be insufficient suitable habitat, for example after a flood, drought or change in habitat management, while in larger sites it is more likely that sufficient suitable habitat survives. Small sites can often sustain only small populations and these are particularly prone to local extinction due to environmental stochasticity, demographic stochasticity, inbreeding depression and the Allee effect (see Section 5.14.1).

Heath Fritillary *Mellicta athalia* butterflies are poor dispersers yet also require recently cleared areas. To maintain such species it is necessary to ensure suitable habitat is always present and that it can be colonised from other patches. Without a continuous history of suitable and accessible habitat such invertebrates become locally extinct (photo: William J. Sutherland).

Small areas have a relatively large edge with a high proportion of individuals occurring near the edge, and this may lead to extinction. The climate may be different along a habitat edge (Laurance 1991) or there may be increased herbivory or predation (Paton 1994) or reduced seed germination (Bruna 1999). Interactions with edges may be complex, for example trees along prairie edges may delay snow melt and reduce the drying effect of sun and wind so that the edge does not burn and then acts as a reservoir for large populations of unwanted plant species (Kline & Howell 1987). Thomas & Hanski (1997) showed that the emigration of butterflies occurred simply due to individuals wandering off the habitat patch and then being unable to return. This emigration was greater in small sites and could then be too great for the population to sustain. Similarly, for various carnivore species, the mortality rate is considerably greater once they step outside the park (Woodruffe & Ginsberg 2000) so that population survival is dependent upon the protected area being sufficiently large. Across species, the extinction probabilities are correlated with the home range, so that nomadic species like Wild Dogs *Lycaon pictus* can survive only in huge reserves while the comparatively sedentary Jaguar *Panthera onca* often persists in protected areas under 100 km^2. Thus for both butterflies and carnivores, there is a size below which the species is unlikely to persist due to the loss of individuals along the edge. However, the scale differs enormously: Wild Dogs require minimum reserves of 3600 km^2 (Woodruffe & Ginsberg 2000) while Silver-spotted Skippers *Plebejus argus* require an area of only 0.0005 km^2 (Thomas *et al.* 2000).

Natural disturbances, such as fires or floods, often cannot occur within small blocks. Furthermore from a practical perspective small isolated reserves are usually relatively costly to manage. Against this some small isolated populations are very important and some minute reserves can be very effective, such as the 394 m^2 reserve at Badgeworth, Gloucestershire, England, set up in 1932 to protect Adderstongue Spearwort *Ranunculus ophioglossifolius* in its only British site (Marren 1999).

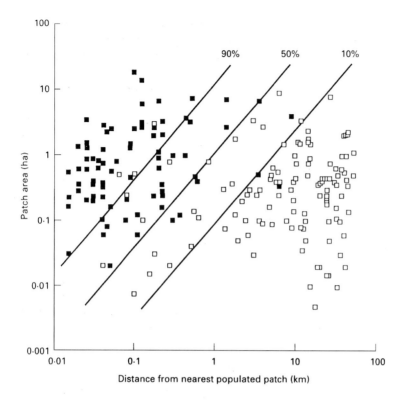

Fig. 12.1 Distribution of the Silver-spotted Skipper butterfly in relation to isolation and patch size. Points show vacant (open) and occupied (closed) patches. The lines show the combinations of isolation and area which give 90%, 50% and 10% probability of occupancy (from Thomas & Jones 1993).

Isolated sites are less likely to receive immigrants and are thus both more likely to experience extinctions and less likely to be recolonised. Immigration reduces the likelihood of extinctions simply by their capacity to restore populations which have happened to drop to low levels, a process known as the rescue effect (Brown & Kodric Brown 1977). As shown in Fig. 12.1, the effects of size and isolation act together: small isolated sites, with high extinction and low colonisation, are much less likely to hold a species than large sites adjacent to others. The pattern shown in the figure for the Silver-spotted Skipper butterfly is probably true for almost any species from bacteria to elephants, but with very different scales. This illustrates the importance of maintaining large sites, or sites close to others, and the difficulties of protecting small isolated sites. This concept leads to the suggestions of creating corridors to link areas, although their value has been debated (Simberloff & Cox 1987, Noss 1987, Simberloff et al. 1992). The empirical evidence on the success of corridors in maintaining populations is conflicting

and is likely to vary markedly between species and habitats (Hanski 1999). Even if they are beneficial, the question has to be whether it is better value to create corridors than either create new sites or extend existing sites. While narrow corridors between areas may be of restricted use, it is clear that maintaining connectivity is essential. An alternative is to improve the quality of the surrounding habitat matrix (Aplet & Keeton 1999), for example by providing education, grants or other economic incentives.

The realisation of the importance of local extinction and recolonisation has led to a flurry of interest in metapopulations, defined as populations of populations (Gilpin 1987, Hanski 1999). Thus colonisation of a patch is equivalent to an individual being born and patch extinction to a death. Although the original idea of identical patches with similar probabilities of extinction and colonisation seems to rarely resemble the processes seen in natural populations (Harrison 1994, Hanski & Gilpin 1997), it is clearly often important for conservationists to think about the interactions between the processes of colonisation and extinction. It is thus necessary to concentrate on connectivity, continuity, patch size and patch quality. As an example, the arrow bamboo, which is the main food of Giant Pandas *Ailuropoda melanoleuca,* undergoes mass flowering over large areas every 40–60 years and then takes another 15–20 years to regenerate sufficiently to sustain pandas. In the past, pandas found dispersing outside their habitat after mass flowerings were caught and kept in zoos. It is now realised that dispersal is an essential part of the strategy for surviving on bamboo and movements outside bamboo forests is a consequence of their habitat becoming isolated. The priority is thus to redesign the reserve network to connect different blocks of habitat to allow the population to persist (MacKinnon & De Wulf 1994). The processes of extinction and colonisation can act at a range of scales: at a smaller scale, larger logs take longer to rot and due to their steep sides are colonised more slowly by forest-floor bryophytes; rare, slowly dispersing bryophytes are thus largely restricted to large logs (Söderström 1988).

It is important to consider the continuity of habitat types. The manager of a site with a rare charismatic vertebrate would not dream of intentionally eliminating its habitat, even for a brief period. Many reserve managers do, however, regularly inadvertently eliminate all the habitat of less spectacular species, especially invertebrates. Many invertebrates have infuriatingly specialised requirements. A simple task such as clearing away dead wood, cutting a meadow or clearing a ditch can thus be devastating for a species. If a site is grazed or cut, so removing all of the seed heads of a plant on which an invertebrate is dependent, then the species will face local extinction. It is often not even sufficient to ensure that the required host species is present, for example, in general, the food plants of declining butterfly species are plentiful, even in sites where the species has gone extinct, but detailed studies show that the size, shade, whether adjacent to bare patches or nutrient content of the plants are of critical importance before females will use them for egg laying and this selectivity is, in turn, related to

the requirements of the caterpillars (Thomas 1991). Thus a short-term change in management may not change the abundance of the host plant but can drive the invertebrate extinct.

When considering invertebrates, reserve managers have an almost impossible task of managing for a suite of species whose habitat requirements are largely unknown and many of which the warden probably does not know are present. Furthermore, most invertebrates are short lived and the population cannot survive even temporary loss of breeding habitat, unlike plants with a seed bank, or longer lived species which can survive for many years without breeding. Invertebrates are thus particularly vulnerable to local extinction resulting from management decisions. Much of the solution is to ensure that tasks, such as ditch clearance, are done piecemeal with part done each year; the disadvantage is that it is usually more efficient to carry out the entire process at one time.

12.3 Disturbance

As discussed in Section 12.1, a common popular belief is that historically most areas were covered in permanent climax vegetation, particularly forest. However, a surprising number of species, such as many herbaceous plants and invertebrates, are dependent upon early successional stages. Furthermore, both Pedunculate Oak *Quercus robor* and Sessile Oak *Quercus petraea* do not regenerate in closed forests and Hazel *Corylus avellana* does not survive there, yet pollen analysis shows that these three species have been continually abundant in Central and Western Europe since the last ice age (Vera 1997). It is thus clear that the popular belief is mistaken and that there must have been a mosaic of habitats differing in successional stages, presumably caused by disturbance regularly setting back succession, combined with grazing retarding succession. Thus in the pristine Białowieza Forest, Poland, gaps are caused by fire and falling trees, which often pull down adjacent trees to create more extensive gaps. Deer and Wild Boar *Sus scrofa* then graze in these gaps which keep them open. Despite being a mature forest with little human interference, the Białowieza

Forest is famous for its annual plants and butterflies (Tomiałojc 1991).

Furthermore, following the metapopulation concept (Section 12.2), it is necessary to ensure that there is a continuous succession of suitable patches within dispersal distance of each other. Webb & Thomas (1994) studied a suite of species dependent upon disturbed ground on heathland (a short ericaceous community). For example, the ant *Myrmica sabuleti* requires a dry south-facing slope with a grass-dominated patch which has been disturbed in the previous 5 years. The number of such patches has declined from 42 in 1978 to six in 1987, which in itself is a serious problem but furthermore the mean distance from a patch in 1978 to the nearest other one in 1987 was 9.5 km. A landscape of isolated ephemeral patches is clearly insufficient to sustain this species and unsurprisingly, this and other species requiring these disturbed patches are declining.

In Finland, those bug (Hemiptera) species that are particularly prone to extinction are those either that require extremely stable habitat, for example the very old unburned forests found within peat bogs, or those that require early successional stages, such as those occupying shallow open water or burnt forest: those requiring intermediate disturbance are least prone to extinction (Petri Ahlroth personal communication). It seems likely that this pattern will prove to be generally true. Much of the challenge of nature conservation is to find ways of protecting old habitats (see Section 12.4) and maintaining early successional stages.

The ideal way of creating early successional habitats is to allow natural processes such as floods, tree falls and fires to occur and so create a patchwork of different successional stages. However this is often impractical (see Section 12.1) as areas are often too small or due to other constraints, such as public opposition to natural disturbances. Burrowing mammals can be important in creating a mosaic of successional stages on a small scale, herbivores keep areas open, wild pigs disturb the ground and beavers flood areas. While such species should be encouraged to play their normal role, on small reserves it is often impractical to retain such species at natural densities either because the site is too small

or due to intolerance from neighbours. If natural disturbances are unacceptable, yet it is necessary to retain early successional stages, then it is necessary to create them. However, the effort required in managing habitat is inversely proportional to its successional stage so that maintaining early stages can require considerable effort. The choice is then between retaining areas at a set stage (for example by grazing a meadow to keep it open or removing trees from a marsh) or alternatively creating a patchwork of disturbances and allowing succession to proceed, for example, by digging a succession of pools so that at any one time there is a range of successional stages.

12.4 Retaining old habitats

Many species are restricted to old habitats and many of the most important areas for conservation are old habitats. However these areas are often under considerable threat, especially old woodland due to timber extraction and old grasslands due to intensified agricultural management by rotovating, fertilising and reseeding.

There are two main reasons for the restriction of many species to old habitats. Old habitats often provide niches unavailable elsewhere, for example many species of conservation interest depend on old, dying or dead trees and thus are restricted to mature forests while other species require large tree holes to breed in and these are absent from young forests.

The second main reason for the restriction of species to old habitats is linked to dispersal. Some species are astonishingly poor at dispersing, for example many perennial plant species currently very rarely become established from seed and usually spread by vegetative means, such as bulbs, stolons, rhizomes or corms (they are more successful at establishing from seed if there is no competition, which presumably explains their spread after ice ages). Since such species currently disperse at an extremely slow rate, often at a maximum of a few centimetres a year, unsuitable conditions, such as a track or modified habitat, prevent growth and act as a barrier. Thus once the habitat has been removed, for example by the forest being felled or the grassland converted to agricultural land, it is very unlikely that such species would recolonise even if the habitat were restored. These poor-dispersing species are thus restricted to sites which have a long and continuous history. They are often referred to as indicator species and can be used to identify ancient habitats.

A 'virgin' habitat implies a lack of management over a very long time, 'primary' woodland has never been felled, while 'secondary' has at one time been cleared. In Europe 'ancient' has been used to refer to woods that have been in existence since at least 1600, and is a useful term as it is usually difficult to tell if a wood is primary.

12.5 Grazing

Grazing is often a key part of maintaining the required species. Many wish to create natural ecosystems as occurred before man had obvious impacts. However, it is also clear that the spread of man resulted in the extinction of numerous mammals (Diamond 1997). The choice is often now between recreating an incomplete ecosystem or artificially grazing, for example with cattle, sheep or goats, as substitutes for the missing species.

Grazing both reduces the sward height, so that shorter species may persist, and creates gaps, so allowing plant regeneration. Grazing is likely to favour those species that cannot grow through vegetative reproduction, such as stolons or rhizomes, and have to grow from seed. Very many invertebrates are largely dependent upon the bare areas caused by grazing. Dung is important for many species especially beetles (particularly scarabs) and flies, thus treating stock with anti-parasite drugs such as Ivermectin may affect both these invertebrates and the species that feed upon them (Madsen *et al.* 1990). Invertebrates are sensitive to the vegetation composition and structure, with many selecting particular parts, such as the seeds, while for ground-nesting birds the structure is likely to be of more importance than species composition (Ausden & Treweek 1995).

Monocotyledonous species, such as grasses, usually grow from basal meristems and thus grazing will have little impact on subsequent growth.

Dicotyledonous species, e.g. herbaceous plants, grow largely from apical meristems which will then be removed by grazing and hence severely impair growth. As a result grazing (or mowing) tends to produce a community of predominantly monocotyledous species, particularly grasses. The dicotyledonous species that survive will tend to be those that are unpalatable to herbivores or sufficiently short to avoid grazing or cutting.

There is a confusing literature on the consequences of grazing, as different studies show either increases or decreases in species diversity. However the underlying principle is simple: if the dominant species is palatable then grazing, by reducing the dominant species, will increase diversity. Conversely, if the dominant species is unpalatable then it will increase with grazing as the other species are reduced and so diversity will be reduced. The same principle explains the consequences of predators, pathogens and parasites. For example, the starfish *Pisaster ochraceous* selectively removes the mussels *Mytilus californianus*, which otherwise dominate the rocky shore and so exclude a wider range of other species (Paine 1974). Removing the *Pisaster* thus reduces the diversity as mussels increase.

Different grazing species produce differing sward heights, uniformity of sward and the amount of bare ground. Sward heights affect the microclimate and the capacity of plants to flower and seed. Bare ground is often beneficial to plants by allowing germination and for invertebrates and reptiles by providing suitable microclimate and foraging habitat. Patchiness is beneficial to the numerous species that need adjacent bare areas and tall vegetation. Thus different grazing species will produce very different communities.

The grazing species selected will depend upon the required results. For example, sheep nibble the vegetation to a uniform sward while their scattered faeces have little impact. By contrast, cattle pull clumps of vegetation, often break the sward with their hooves and their cow pats kill those plants that are smothered but enrich the nearby sward. As a result cattle produce a varied sward that can be beneficial for plants, invertebrates and nesting birds. Grazers such as rabbits that dig burrows further diversify the sward. Horses produce a patchwork of lightly grazed turf with latrine areas left ungrazed which are often rank; they tend to be active and so poach the ground markedly. Horses are particularly suitable for opening up rank vegetation. New breeds of livestock have been selected for high-input agricultural systems and are usually less suitable for grazing relatively infertile nature conservation sites than traditional more hardy breeds. Many site managers are thus using old breeds when they need to introduce grazing to semi-natural habitats. Mowing results in lower ecological diversity as it produces a fairly uniform sward with little disturbance. Mowing is usually used in small or urban sites where grazing is impractical. Wet areas that cannot be grazed have been mown when frozen in the Netherlands (Baines & Smart 1991).

Grazing can result in problems. Removal of stems, flowers and seed heads can have serious consequences for invertebrates. The trampling by stock may trample amphibians, bird nests or young mammals. High levels of trampling will encourage those plant species which are resistant to trampling. Intensive trampling can result in unwanted soil erosion when it is wet or very dry. The extent of trampling will depend upon the grazer, thus cattle and horses are heavier than sheep, horses are more active than cattle, young animals are more active than old.

Thus if grazing is used it is necessary to consider the method of grazing, whether rotational, continuous, seasonal or spasmodic, the type of grazing animal used and the intensity of grazing (Ausden & Treweek 1995). As always this depends upon the objectives for managing the site. A combination of these methods is sometimes used. The site often has to be divided into blocks.

Natural grazing. Using natural populations of wild species is usually preferred where possible and is often the only realistic method in very large sites. For smaller sites this will usually not work due to the difficulties of retaining wild populations within small blocks. It is also a problem if the grazers are hunted or objected to if they cross the reserve boundaries. Natural populations may be kept at low levels due to winter or dry-season mortality so that they may have less impact in the sum-

mer or wet season than required. High densities are often required if they are the main means of retaining early successional stages. When combined with other techniques, such as periodic disturbance, lower densities may be sufficient. A further problem is that natural populations may not graze when and where they are wanted.

Rotational grazing. Moving stock between areas has the considerable advantage that a mosaic of different habitat types can be created with some grazed intensively while others are allowed to flower and set seed. This can be particularly beneficial for invertebrates, especially as a method for protecting species whose requirements are unknown. Similarly, areas with ground-nesting birds can be avoided during sensitive periods. It is often sensible to ensure that each habitat type is divided into a number of blocks, often using fences or ditches, so that the site receives a range of management regimes. Temporary electric fencing can be very effective for dictating when each area is grazed. Adding fences can be expensive, unsightly, unpopular and, at least for some forest gamebirds, collisions are a serious problem (Moss *et al.* in press).

Continuous grazing. Keeping stock throughout the year, if carried out intensively, will result in a uniform short sward with few flowers, which will usually be of little interest for invertebrates, although for some species or communities a uniform short sward is desired. Each species of herbivore has a preferred sward height, for example sheep prefer short swards, so that if the stock density is low but the sward is high or fast growing, then sheep will graze some patches intensively and leave others ungrazed, which will result in a mixture of short and tall swards.

Seasonal grazing. Restricting the grazing to certain seasons is often very effective in setting back succession and creating open areas but without causing damage to particularly sensitive stages. Grazing during the winter will reduce the biomass and create gaps without preventing the plants from flowering and setting seed. Winter is usually considered the best time to graze important entomological sites. Grazing may be excluded from the site during the breeding season if trampling of bird nests and young is a particular problem. However, if the grazing is intense it will remove the seed heads and plant stems which may be important overwintering sites for various insects. If an objective is to manage for wintering grazing waterfowl then the best strategy is to graze heavily during the summer but leave the sward for a period to allow regrowth.

Spasmodic grazing. In some situations there are practical difficulties in grazing the land so that stock are only available for short periods. Although this may be beneficial in setting back succession, intensive bursts of heavy grazing and trampling can be very damaging to invertebrate or plant populations. Rapid defoliation can however also be beneficial, for example in creating feeding areas for wading birds and their broods (Benstead *et al.* 1997).

12.6 Burning

Burning often outrages members of the public and is a major conservation problem, especially in the tropics where intentional or accidental fires are destroying large areas of forest. Why then do conservationists sometimes burn areas or have a policy of not extinguishing natural fires? Habitat management often involves maintaining a mosaic of habitats of different successional stages (Section 12.1) and in some areas this mosaic has been maintained largely due to fire. In many habitats around the world fire is a natural and beneficial process. The need to retain or restore a mosaic combined with concern for restoring natural processes has led to conservationists tolerating or encouraging fires. Burning may be the cheapest option for creating early successional habitat, although if a large team is necessary for fire control then it can be more costly.

Early successional stages can alternatively be created by intensive grazing, cutting, felling or disturbing the ground and these are sometimes the most appropriate management. The main reason for burning is that a burnt landscape differs from a cut or grazed one. Fires can lead to markedly increased regeneration as a result of a wide range of processes: removal of the litter layer revealing bare soil and increased light; the release of canopy stored seeds;

breaking the dormancy in some species (often as a direct consequence of the heat or smoke) which leads to a flush of germination; while in some species the enhanced flowering resulting from the improved conditions after burning may also result in a flush of germination about a year after the fire. Burning the litter also releases large quantities of nutrients and makes acid soils more alkali, which also makes nutrients more available (Biswell 1989) and may temporarily reduce diseases (Wright & Heinselman 1973).

Other species may benefit from the pattern of regeneration after burning. Kirtland's Warbler *Dendroica kirtlandii* is dependent upon a mix of even-aged 5- or 6-year-old thickets of Jack Pines *Pinus banksiana* and patches of open grass. A programme of regular intensive fires maintains this habitat and the warbler. Herbivorous mammals and birds often benefit from the nutrient-rich regeneration. The reduced shading, litter and albedo all mean the bare ground heats up more rapidly (Adams & Anderson 1978) which may benefit some invertebrates and reptiles. However, the bare surface can also become too hot for some species during the day while at night is often colder than unburnt areas.

Burning may also be preferred over other methods as it can result in a complex variation in conditions, for example during the famous 1988 fire in Yellowstone National Park fire intensity differed markedly between areas as a result of variation in soil moisture, aspect and chance (Christensen *et al.* 1989). This led to a very different habitat from that which would have resulting from clearfelling of forest (Noss & Cooperrider 1994).

Some fires may be less damaging to soil inhabitants than expected. There may be a considerable difference in temperature between the soil surface and a couple of centimetres below. This is particularly true if some litter remains unburned (as it acts as a good insulator), if the fire passes over quickly and if the soil is wet (Hobbs & Gimingham 1984, Johnson 1992). Thus, although seeds on the surface may die those deeper in the soil may be unaffected. Plants with below-ground organs such as rhizomes may survive a quick hot fire but not a slow smouldering one that burns into the accumulated organic matter (Cropper 1993). Similarly, although it has been generally assumed that burning Common Reed *Phragmites australis* kills soil invertebrates, when we carried out controlled experiments, to our surprise, we could find no evidence at all for this. Our explanation was that the heat was insufficient to kill (Cowie *et al.* 1993, Ditlhago *et al.* 1993).

Mesozoic fossilised charcoal shows a pattern of regular and repeated forest fires so ecosystems have become adapted to fires since at least that period (Harris 1958). Some plants and a few invertebrates are largely dependent on regular burning. A wide range of plants show adaptations to surviving burning. For example, various species of Eucalyptus in Australia, Proteas in Africa and many conifers have their seeds protected in woody insulated fruits which release them after burning. The fire-tolerant species tend to thrive after a fire as they may grow faster or flower more, probably as a result of the lack of competition and increased nutrients. Many other species resprout from roots.

There is a long tradition of using prescribed low-intensity burns under suitable weather conditions to reduce the hazard of uncontrollable high-intensity fires (Whelan 1995). In 1887 Joaquin Miller wrote "In the spring . . . the old squaws began to look for the little dry spots of headland or sunny valley, and as fast as dry spots appeared they would be burned. In this way the fire was always the servant, never the master . . . . By this means, the Indians always kept their forests open, pure and fruitful, and conflagrations were unknown" (Biswell 1989). It is likely that this practice would greatly reduce the frequency of disastrous fires under hot dry conditions. Although prescribed burns tend to be effective in reducing fire risk for grasslands and possibly scrublands, fires in coniferous forests are largely independent of fuel quantity under extreme (hot, dry and windy) conditions.

While fires can be beneficial, there are habitats and locations where they are clearly inappropriate. Some species and communities are intolerant of fires and will decline in the presence of burning, for example the recent disastrous fires in rain forests where there is no history of burning. Primary forest disturbed by

logging or clearance is considerably more flammable than undisturbed forest (Uhl & Kaufman 1990) and as a result unwanted fires are likely to increase. Fires can be devastating to species without fire-tolerant adaptations, that cannot move away and escape or that are not protected from the heat of the fire, for example by being underground. The species affected will vary markedly between the seasons depending upon the life stage and behaviour at that time. Soil erosion in many habitats can increase markedly for 2–3 years after a fire. Although burning may promote regeneration, it usually kills any seedlings and saplings less than 10 years old (Savage & Swetman 1990, Bond & van Wilgen 1996).

It is useful to know the historical fire regime before deciding on the current management policy. The history includes not just the fire frequency but also the intensity and seasonality. The fire history will depend upon lightning, human activity, weather and type of fuel and it is likely that fire history of the past few decades will be different from the 'natural' fire frequency. Conifers are much more likely to have canopy fires than deciduous species as conifers have many more vertical layers of leaves, higher resin content, finer, drier leaves and finer branches (Johnson 1992). The best method to estimate the frequency of fires is to examine fire scars in the growth rings of long-lived trees (Whelan 1995) although it is usually difficult to assess the intensity or season. The time since last burnt can be determined using a range of methods if historical records are not available. If the burning kills part of the bark, then the fire date can be determined by comparing growth rings (see Section 5.8.1) in killed and surviving parts of the same tree. Young unscarred trees that have germinated since the fire can be cored to determine their age. Fymbos vegetation in South Africa should probably be burnt about every 15 years and the time since last burnt can be determined by counting the sympodial segments (see Section 5.8.1) of Proteas. The burning regime can sometimes be inferred from clearly delimited boundaries of even-aged blocks of vegetation (Johnson 1992).

When conserving populations of rare plants it can be useful to examine the morphology, phenology, longevity and age to maturity of the species when considering deducing the ability of species to withstand fires and thus an appropriate timing, area and intensity of burning. As there may be little information on the rare species it is sometimes useful to examine the associated species and the fire regimes they require.

Lighting a fire is a serious business that requires considerable thought, planning and consultation before being contemplated. Burning obviously has considerable safety and public relations consequences and there are many areas where it is inappropriate. When carrying out prescribed burns consider the following.

• Decide upon a pattern that is compatible with the history, priorities for the site, expected responses from species and communities, safety and the experience of the team.

• Check on legislative requirements.

• Burning should be carried out in an experimental way (Section 5.3) so that the experience gained can improve future management. Simple measurements of temperature, humidity and windspeed are useful. Experience can then be obtained of the conditions necessary to achieve the type of fire desired from the viewpoint of safety and ecological considerations.

• Create a firebreak. Where possible use existing features such as roads or lakes. Keep ground disturbance to a minimum. Wet the ground and if necessary keep it wet when the fire is burning. Do not use foam or chemical retardants as the ecological consequences are unknown (Cropper 1993).

• Avoid burning during pollution alerts or during temperature inversions, which reduces dispersal of smoke. Avoid burning when the wind would carry smoke towards sensitive areas such as schools or hospitals (Payne 1992).

• Ensure all traces of fire are extinguished before leaving the site.

• Fire crews can forget the location of sensitive sites in the excitement of controlling the fire and can cause considerable damage with fire trucks or bulldozers (Cropper 1993). Either mark areas that should not be entered with tape or allocate one person just to look after sensitive areas.

• Burn at the season, temperature, wind speed, wind direction, humidity and fuel moisture content at which the fire can be controlled.

• Hotter burns are generally better. Fire intensity can be partly regulated by restricting the amount of vegetation to be burnt at one time (Cropper 1993). In 'cold burns', narrow strips are burnt starting upwind of the firebreak, then once burnt out, lighting the next strip upwind. 'Hot burns' allow more vegetation to burn at one time but are less easy to control. Hottest fires are achieved by lighting around the perimeter, creating a convection column sucking a very hot fire into the centre.

• Fires are more likely to spread if there is abundant, dry, fine, accessible combustible material, and if it can progress uphill, as the fire can dry and heat the adjacent fuel (Johnson 1992).

• Surface fires are usually small scale and of relatively low temperature. Canopy fires tend to spread over a larger area, are usually of higher intensity and kill most of the canopy trees.

12.7 Hydrology

Managing wetland sites requires understanding of the hydrology. Altering a site without understanding the hydrology is likely to result in unexpected changes both within the site and elsewhere, for example blocking a channel may result in water accumulating in an unexpected location or simply exiting by another route. Errors in water management can be damaging to the conservation interest and can also cause economic damage elsewhere. Table 12.1 reviews the main hydrological measures.

12.7.1 Understanding hydrology

There are four sources of water:

Rainfall. Rain falling directly on waterbodies is usually insufficient to maintain water levels except for raised bogs which are entirely dependent upon surface rain and are thus restricted to areas of high rainfall. Rainwater is usually high quality but in areas of intensive agriculture may still contain levels of nutrients that can change nutrient-deficient habitats (Heil & Diemont 1983).

Table 12.1 A summary of hydrological measures and their ease of collection. Modified from Brooks & Stoneman (1997). The section in which each method is described or discussed is given.

Variable	Method	Section	Easy?	Approximate cost (US$ in 1999) of equipment
Rainfall	Collecting rain gauge	4.18.2	Yes	$20–250
	Recording rain gauge	4.18.2	No	$2500–4000
Water level	Site wetness mapping	4.18.3	Yes	None
	Dipwells	4.18.3	Yes	$100 per 20
	Water level range gauges	4.18.3	Yes	$80 each
	Water level logger	4.18.3	No	$1300
Pool water level	Stage board	4.18.3	Yes	$15 each
Evapotranspiration	Lysimeter	4.18.5	No	$2200–3000
Stream flow	V-notch weir	4.18.4	No	$1500–5000
Seepage	Piezometers	4.18.3	No	From £200 per set

Surface water. Surface water, flowing from adjacent land, contributes to most water bodies and is often the major water source for water bodies on clay soils. Pollutants can accumulate rapidly due to low throughflow rates so that water quality then depends critically on the adjacent land use. Buffer zones of semi-natural vegetation can be used to intercept pollutants and these should be wider where the problem is greatest.

Ground water. If the water body is below the water table then ground water can be important. Those on sand, gravel and some peats are usually dependent on ground water and usually have relatively high-quality water due to filtration and dilution.

Inflows. These derive from springs, streams or rivers. Streams and rivers usually carry sediment and if the catchment is agricultural, industrial or urban then these are likely to be polluted. Springs are usually of higher quality.

The first sensible stage in considering water management of a site is to prepare a hydrological map (Brooks & Stoneman 1997), often in conjunction with preparing a topographic map. It is thus sensible to determine the heights of the main water features within the site. It is often also useful to map the site either by measuring the height of a grid of points or by paying for aerial photogrammetric mapping, although it is expensive, may not be available and as it measures the top of the vegetation it may not provide the information required.

For a hydrological map, take a map of the site, ideally with topographic features already marked, and plot:

• Route and direction of all flowing water. This can be determined by accurate levelling or by simply observing the flow. It can be more easily accomplished after heavy rain although the routes may then be atypical.

• Areas of permanent water.

• Areas of temporary water if the survey is repeated at different seasons. Observing the sequence of flooding can be useful and provide the required topographic information.

• Anthropogenic features that influence the hydrology, such as ditches, mine shafts, pipes or drains.

• Main water catchments.

• Inflows and outflows.

• Vegetation types that indicate wet conditions or dry conditions. The distribution of certain species can also be useful, for example rabbit burrows are only found in dry areas.

• Diffuse flows. These may be inferred from the presence of sloping wet land but on fairly flat land can sometimes only be detected by comparing piezometer (see Section 4.18.3) readings.

The water budget is a fundamental concept in hydrology. The underlying concept is obvious: arriving water is either stored or lost to the system, but thinking in terms of a water budget is invaluable for understanding why a wetland exists and what the consequences of change might be. In reality, it is rarely possible to have accurate values for each term but it is still useful to think about how the water balance might operate.

The basic water balance equation is:

$$P - E - U - G - \Delta W = 0$$

where P is the net precipitation, the amount of water that hits the ground; E is the actual evaporation resulting from both transpiration by plants and evaporation from the soil and open water; U is the lateral seepage; G is the vertical seepage and ΔW is the change in storage in the soil (Gilman 1994).

A water budget, as shown in Table 12.2, is invaluable for answering a range of questions such as: by how much is the water likely to drop over the summer? How many cubic metres of water have to be stored during the wet season to maintain a particular water level in the dry season? How much water can be taken from the stream leading into the site before there are problems? What will be the consequences of blocking off a water source because the water is of low quality? What will the consequences be of a period of unusually wet or dry weather?

A water budget is easiest to create in an isolated area in which lateral and vertical seepage are negligible. Average rainfall can be determined by using data from a nearby meteorology station or by using methods described in Section 4.18.2. Evapotranspiration rates can be obtained from tables in some areas while Section 4.18.5 gives some means of direct estimation. Subtract the evapotranspiration from the rainfall to give the water balance – or water deficit if the evapotranspiration exceeds rainfall. Multiply these figures by 10 times the area in hectares to give the site demand per month in m³. The water demand must be made up from water input through streams, from the soil or from a falling water table. Thus, although the figures in Table 12.1 show a surplus over the year, there is a deficit expected for five months. Either the site will have declining water levels or additional water will have to be found from elsewhere. The extent of water can be estimated by multiplying area by depth. Thus flooding 1 km × 150 m to a depth of 20 cm requires $1000 \times 150 \times 0.2 = 30\,000$ m³.

Soils are permeable, although to greatly varying extents, resulting in movement from high to low water levels. The rate at which water moves through the soil depends considerably upon the gradient of the water table and soil type. Water typically moves over a million times faster through gravel than through clay (Shaw 1994). This has considerable consequences for management. Thus a high water level in a ditch will wetten a peat soil over a great

Table 12.2 A water balance for a 100 ha wet grassland site in central England (RSPB *et al.* 1997).

	Average rainfall (mm)	Wet grassland transpiration	Water balance (mm)	Water deficit (mm)	Site water demand (m³/month)	Site water demand (m²/day)
January	62	1	61	0	0	0.00
February	47	10	37	0	0	0.00
March	42	32	10	0	0	0.00
April	43	56	−13	13	13 000	433
May	50	81	−31	31	31 000	1033
June	46	93	−47	47	47 000	1566
July	58	93	−35	35	35 000	1166
August	64	76	−12	12	12 000	400
September	60	47	13	0	0	0.00
October	64	22	42	0	0	0.00
November	76	5	71	0	0	0.00
December	54	0	65	0	0	0.00
Year total	677	516	161	138	138 000	

distance but will have an extremely localised effect for a clay soil. To wetten the clay soil it is necessary to ensure the water floods onto the field.

12.7.2 Water management

A useful approach is to divide the site into compartments using dams in order to increase the control over the hydrology. These dams are usually small scale. Common designs include a vertical series of wooden planks (drop plates) across a ditch which may be added or removed to adjust waterlevels or an earth bank through which a pipe leads and if the pipe is flexible then adjusting the height at the outlet alters the water level at which water flows out. Pumping can also be used to move water between compartments but is clearly more expensive, often slow and can be troublesome.

It is sensible to use the method described in the previous section to consider how much water will be needed when. Consider the possibility of higher than expected evaporation or reduced rainfall. Manipulating water levels can obviously have implications for others, for example by using water that others desire, or by impeding the drainage of others; such wider consequences need to be carefully considered.

12.8 Water quality

Water quality is often an important conservation issue. For dramatic pollution incidents it is usually moderately straightforward to identify the problem and source if environmental and biological samples (Section 5.12) are taken immediately. It can, however, be difficult to locate the origin of short pulses of pollution especially in flowing water and where observations are infrequent. A rapid response to an incident makes tracing considerably easier and establishing procedures for detecting and responding to incidents can be invaluable. This can be as simple as ensuring that those likely to notice incidents appreciate the importance of immediate reporting and have the means to do so, even outside office hours. Problems resulting from gradually accumulating pollution can be particularly difficult to identify and usually require a mixture of chemical and biological monitoring (Chapter 4) and diagnosis (Section 6.3).

There is an extremely slow natural progression from a geologically young oligotrophic lake with few nutrients and high oxygen levels, to a mesotrophic lake with some nutrients, to a eutrophic lake with thick sediment, high nutrient levels and possibly low oxygen; pollution can, however, enormously accelerate this process, resulting in cultural eutrophication (Henderson-Sellers & Martland 1987). Although plants need about twenty chemicals for growth, the levels of nitrogen and phosphate are critical as their natural availability is low compared with the amount needed (Moss et al. 1996). Thus eutrophication results from increased levels of both phosphorus, particularly phosphate from sewage or detergents, and nitrogen, particularly nitrate from agricultural runoff.

Phosphorus is relatively insoluble, and thus often the chemical least available in relation to plant requirements and its abundance often determines the extent of eutrophication.

Eutrophicated lakes may have lush plant growth leading to high detritus levels and thus anoxic conditions, while dense phytoplankton populations can shade out the submerged rooted plants. The problems of eutrophication include considerable loss of conservation interest, lower quality drinking water, health hazards for swimmers and reduction in fishing due to reduced fish stocks. Although polluted deep-water bodies are usually of little conservation interest, polluted shallow bodies can still maintain emergent plant communities.

In a waterweed-dominated community the weeds absorb the nutrients while zooplankton, and especially Cladocerans (such as *Daphia*) are very effective at removing phytoplankton, often resulting in reasonably clear water. Fish that are capable of eating zooplankton may be abundant but the zooplankton may also remain common as plants act as refuges (Schriver et al. 1995). When nutrients have been experimentally added to such communities they are simply absorbed by the growing plants (Balls *et al.* 1989, Irvine *et al.* 1989). In the alternative state of no weeds but abundant phytoplankton, the phytoplankton shade out the plants while, as the zooplankton have nowhere to hide, the fish can easily deplete them. Thus this state may persist even if the nutrients are reduced as the lack of weeds benefit phytoplankton which in turn result in a lack of weeds. Although the waterweed-dominated state is linked to low nutrients and the phytoplankton-dominated state to high nutrients, simply adjusting nutrients may thus not always be sufficient to result in a change of state (Moss *et al.* 1996) and other processes may be necessary to bring about the switch.

Even if nutrients have been gradually increasing, the switch from a waterweed-dominated community to a phytoplankton-dominated one may take place suddenly and dramatically. It may be more likely if there is plant removal by: cutting; intentional or accidental addition of herbicides; herbivory by fish, birds or mammals; atypically hot calm weather or extreme fluctuations in water level that may expose or destroy plants (Andrews 1995, Moss *et al.* 1996). Extreme fluctuations are a particular problem in reservoirs. Factors that remove zooplankton such as toxins, a saline intrusion or introduced zooplankton-eating fish may also favour a switch: murky green fish ponds are a testament to this process.

The modern approach to restoring eutrophicated waterbodies is to first reduce nutrient inputs, for example by isolating from polluted sources, by encouraging phosphate stripping in upstream sewage farms, by reducing agricultural run-off or, more rarely, by flushing with nutrient-poor water (Oglesby 1969, Welch & Patmont 1980). One option to improve the water quality is to extract water from rivers only when water levels are high as any pollutants then tend to be more diluted. Reducing inputs is often sufficient in deep lakes but may be insufficient for shallow lakes if there are large stores of phosphate in the sediment. It may then be necessary to pump the mud out, but this is a substantial undertaking and presents the problem of disposing of large quantities of mud. If, as outlined previously, each of the alternative states is stable then it follows that it may also be sensible to force the switch from the phytoplankton-dominated state to a waterweed-dominated one. The techniques include introducing plants and protecting them from grazing, removing zooplankton-eating fish, and encouraging naturally occurring fish predators, such as Pike *Esox lucius*. The latter can be achieved by flooding in the spring to enhance spawning, providing suitable spawning sites, or adding further individuals, although the technique is only likely to be effective if high predator densities can be achieved (Søndergaard *et al.* 1997). Adding non-native predatory fish is sometimes recommended but is obviously unacceptable (Section 11.3). Once the waterweed-dominated community is restored, a natural community of zooplankton-eating fish should no longer have a significant effect.

12.9 Habitat creation, restoration and translocation

Habitat creation usually refers to the creation of a new habitat on a site, restoration refers to modification of existing semi-natural habitat and translocation refers to moving a habitat, usually because it would otherwise be lost to development. In America the term restoration includes creation (Gilbert & Anderson 1998). In this section, I just consider establishing new habitats by creation or translocation.

Habitat creation clearly has exciting potential for making new sites and linking remaining patches: why then are many conservationists so dismissive? One reason is that habitat creation has a mixed track record with the successes overshadowed by more cases where enthusiastic promises turn out to be poorly planned schemes using inappropriate methods, species and varieties. A second criticism is that habitat creation is regularly used by developers as a justification for destroying existing habitat. Not only are newly created patches usually a very poor substitute as they are rarely completely successful and often fail, but such creation also misses the point that the history and cultural importance are often critical to a site. Simply creating the same species combinations and claiming it as a substitute is like building a fountain in a mound as a substitute for the Old Faithful geyser, in Yellowstone, America, or placing stones on top of each other as a substitute for Stonehenge, England: even if the replica was so good as to fool experts it would still not be a substitute. My opinion is that habitat creation should not be considered as an acceptable alternative for destroying habitat. Creation is still essential if we are to restore our fragmented landscapes and created habitats may be very welcome even if clearly inferior to the real thing.

Any habitat creation project needs a plan which considers both the short-term and long-term objectives. These could be to establish certain species, establish specific communities, provide educational opportunities or create amenity areas. At the start of a project it is important to be realistic about the long-term management and whether there is really likely to be any long-term funding and commitment. Some habitat creation projects are devised on the assumption of long-term management even when this is clearly unlikely: it is then sensible to plan a scheme requiring negligible management.

Any plan should be placed in context. Is the proposed habitat appropriate for that area? Does it help link existing habitats? Does it contribute towards wider conservation objectives such as maintaining key species or communities? Is it detrimental to adjacent areas, for example by changing the microclimate or encouraging predators?

The objective should be to leave something better. This seems trivially obvious but many important wet meadows are converted into fairly dull ponds and, as conservation is often considered synonymous with tree planting, many excellent grasslands have been forested, frequently with conservation grants.

It is usual to start with a site survey (Gilbert & Anderson 1998) identifying the following:
• General topography and aspect including vehicle access.
• Hydrology and likely seasonal variation.
• Local climate including those areas which will differ from the surroundings such as exposed ridges, frost hollows and sheltered areas.
• Main vegetation types.

• Species of plants and animals present and their current conservation importance.
• Water and soil characteristics such as pH, nutrients, water hardness, salinity, heavy metals and soil compaction.
• Adjacent communities and their conservation importance.
• Any constraints such as underground services or overhead powerlines.

Habitat creation can be moderately straightforward providing there is the correct chemical composition, water regime, soil structure, subsequent management and that the desired species can either naturally colonise or be added (see Table 12.3). One useful approach is to examine the conditions in existing habitats that are being mimicked and assess the soil structure and chemical composition especially pH, nitrates and phosphate.

Fertility is a key determinant. High fertility is essential for lush growth but often results in a community of low diversity. Many species are either dependent upon disturbance or on some form of stress such as water shortage, low nutrients, high salinity, acidity, alkalinity or shade, which prevents the competitors from flourishing. In many sites the secret of increasing diversity is to reduce the fertility and especially the nitrogen and phosphorus (Section 12.8; Gough & Marrs 1990). Reducing fertility is, however, often difficult. Stripping away the soil surface works if the lower layer is less fertile but, unless the topsoil can be sold or used elsewhere, it may be difficult or costly to dispose of. Natural leaching is faster on sandy soils than clay soils (Gough & Marrs 1990) but is often too slow a process. Mixing the topsoil with a fine subsoil can reduce the fertility. Furthermore fine grades of crushed concrete can provide a relatively infertile calcareous soil while colliery shale can produce an acid soil (Ash *et al.* 1994).

Derelict land, such as that used for industry, may often have the opposite problem of being too infertile. Growth may be considerably enhanced by adding nitrogen and continuing to do so for a number of years (Bradshaw 1983). Establishing nitrogen-fixing species such as legumes is an effective long-term solution. Table 12.3 outlines the suggested

Table 12.3 Problems with restoring derelict land and suggested treatments (after Bradshaw 1983).

Category	Problem	Immediate treatment	Long-term treatment
Physical			
Structure	Too compact	Rip or scarify	Vegetation
	Too open	Compact or cover with fine material	Vegetation
Stability	Unstable	Stabiliser/mulch	Regrade or vegetation
Moisture	Too wet	Drain	Drain
	Too dry	Organic mulch	Vegetation
Nutrition			
Macronutrients	Nitrogen	Fertiliser	Legume
	Others	Fertiliser + lime	Fertiliser + lime
Micronutrients		Fertiliser	
Toxicity			
pH	Too high	Pyritic waste or organic matter	Weathering
	Too low	Lime or leaching	Lime or weathering
Heavy metals	Too high	Organic mulch + /or metal-tolerant cultivars	Inert covering or metal-tolerant cultivars
Salinity	Too high	Weathering or irrigation	Tolerant species or cultivars
Plants and animals			
Wild plants	Absent or slow colonisation	Collect seed and sow or spread soil containing propogules or plants	Ensure appropriate conditions
Animals	Slow colonisation	Introduce	Ensure appropriate habitat

treatments for restoring derelict land suffering from a range of physical or chemical problems.

Much species diversity in natural habitats arises from the variation in topography and soil depth. If creating a habitat on an area that has been smoothed for agriculture, forestry or industry then is it usually sensible to restore some structure and variability at an appropriate scale. If restructuring a site then it is often sensible to incorporate appropriate variation in soil depth, water regime and substrate size.

An important decision is whether to allow natural colonisation or introduce the desired species. For small areas surrounded by suitable habitat then natural colonisation may provide a desired range of successional stages. However for isolated or large sites, natural colonisation is often too slow, especially as there is often pressure to produce quick results. Furthermore, as sources of suitable colonists become rarer, it becomes increasingly important to aid conservation. One compromise is to plant some areas while leaving others to be colonised and thus producing a more natural age structure. The first individuals to arrive may greatly influence the subsequent community so allowing natural colonisation can be risky. This is a particular problem if the surrounding landscape contains many undesirable species, such as non-native species. It is sometimes sensible to plant in a sequence, for example by adding the fast-growing or aggressive species later once the slower growing or poor competitors have become established.

Much of the skill lies in combining the actual or expected environmental conditions with selected species and communities to produce realistically sustainable habitats. It is usually much easier to select the objectives and thus the species on the basis of the existing environmental conditions rather than try to change the environment to fit the species. Thus try to use local variation positively rather than always considering it as a problem. An early stage should be to map out the site including pH, slope, water content and shade and use this to decide upon the species which will be introduced. Identifying the species already present and finding their requirements is an invaluable step.

While it seems obvious to use native species, there is a disgraceful history of incorrect species and varieties being added. As a result of errors or ignorance, many native planting schemes have included species alien to the country. Gilbert & Anderson (1998) describe how a 'Nature Conservation Blend' seed mix for sale in Britain in the 1970s probably did not contain a single native seed! Many consider that if a species occurs in a country then it is appropriate to plant anywhere, regardless of local distribution. Habitat creation should mimic the communities in nearby equivalent ecosystems. A more subtle problem is not using local genotypes. In Britain many sellers of wildflower seeds are selling central European stock (Akeroyd 1994). Non-native genotypes may flower and fruit at different times, which may affect associated frugivores and nectivores, and they may have slower growth and survival (Worrell 1992, Jones & Evans 1984).

The species planted should usually be the commoner members of the community. Part of the 'meaning' of a species is its distribution, habitat and abundance so planting a rare or restricted species, such as an ancient woodland specialist alongside a new road, is usually inappropriate, even if it occurs within that region. It can be useful to visit examples of the habitat that is being mimicked to see the distribution and arrangement of the components. For example, does a particular species grow in clumps or as isolated plants? Is it rare or common? Does it grow in the shade or open?

It is sometimes sensible to plant in single-species groups, where this resembles natural communities, as this reduces competition between species, so increasing the diversity of species surviving in the medium term. Faster spreading species should usually be planted at lower densities than slower spreading ones.

Seedlings or transplants are easily destroyed by grazing, competition or soil erosion so it is necessary to ensure that there is sufficient protection. This may entail fencing to keep out grazers, removing competitive species or providing a mulch.

Turfs can be cut and transplanted as a source of plants. This can be successful but on a large scale usually causes unacceptable damage to the donor site and is expensive. Turfs may fail if there is a drought or animal damage and the community usually changes from the initial composition (Section 12.9.5).

Employing contractors is often essential for many large-scale habitat creations. However, contractors have a poor record including often failing to use the specified species or varieties, failing to prepare the ground properly, or not adapting the procedures to account for local conditions. A common problem arises from the fact that contractors are usually judged on their ability to produce neat formal habitats, while for aesthetic and habitat diversity reasons irregularity is required. Asking contractors for random spacing usually produces regular spacing. Similarly, leaving bare ground is often important, yet most contractors consider this a measure of failure. In creating waterbodies it can be difficult to persuade contractors to produce a natural irregular pattern with varying depth as they are traditionally assessed by their straight edges and uniform slopes. It is necessary to either specify exact requirements in detail and supervise regularly or carry out extensive training. Once these peculiar requirements are fully explained some contractors can be enthusiastic, skilled and committed.

The tendency is to treat an entire area in one way and say the method either works, fails or is partially successful. It is not, however, possible to say which of the steps were necessary or which of the omitted steps should have been incorporated. By experimenting (Section 5.3) the methodology can be improved.

12.9.1 Waterbodies

Creating waterbodies is one of the easiest ways to quickly improve the conservation interest of a site. It may be sensible to create wetlands in two phases: first create the main waterbodies and then adjust the profiling once the water levels are clear. The following are guidelines for creating ponds and wetlands (Baines & Smart 1991, Williams *et al.* 1997) which obviously have to be considered in relation to the objectives.

• Ensure that the habitat flooded is not more important than the waterbody to be created. Interesting marshes and grasslands have often been destroyed in the name of conservation to create relatively uninteresting ponds.

• Locate new ponds so as to minimise risk of pollution from inflows.

• If possible, link new ponds to established wetland areas such as streams, fens and ditches.

• Base design on natural wetlands. It is often preferable to create pond mosaics and wetland complexes rather than single isolated waterbodies.

• Vary the main factors such as depth, permanence and size to create a range of communities, including some that are very small such as 0.5 m across.

• Create irregular margins.

• Ditches should zigzag or be curved to improve cover and impede wind and wave erosion (Payne 1992).

• Include very shallow pools near water level. Even those 5 cm deep are valuable.

• Ensure that some shallow and deep pools are only linked at very high water levels.

• Create shallow areas which will be exposed as the water recedes (drawdown zones). If necessary dig steeply to highest seasonal water table along some edges and then flatten out but leaving irregular hummocks and hollows. Most new waterbodies have a narrow drawdown zone unlike most natural waterbodies. Creating an irregular extensive drawdown zone is likely to enhance the diversity of marginal, shallow water and semi-terrestrial plants.

• Relate topography to subsequent management. If drawdown zones are managed by grazing or cutting then slight topographical differences result in varying communities. If the drawdown zone is unmanaged then tall emergents of one or two species are likely to dominate and if this is undesired then a greater topographical variation will increase species diversity.

• New ponds and lakes are easiest to construct at the natural water level by either excavating down to the water table or damming a water course. Unless the dam is small its construction will require guidance from a qualified structural engineer.

• Make positive use of the extracted spoil, for example as a screen to reduce disturbance from visitors.

• Temporary pools, or those that dry out irregularly, often provide refuge for species that are intolerant to predation by fish and predatory invertebrates dependent on permanent water (Andrews 1995). It is tempting to deepen such ephemeral pools to prevent drying out but, by allowing predators to persist, this may be counter-productive.

• Shelter by trees reduces erosive wave action and results in a warmer water and air microclimate which may benefit flying insects. However, shade from trees can restrict growth of water plants and should thus be restricted to areas where they will cause little shade. If spray drift from adjacent agricultural land is likely to be a problem then screen with scrub or trees.

• Islands are particularly beneficial for nesting birds as they are relatively inaccessible to predators. One design for waterfowl is to have a basking beach facing the sun and dense vegetation with easy access to the water for nesting on the shaded side.

• Bare mud for invertebrates, plants or feeding birds is best created by grazing or keeping flooded early in the growing season.

Planting of waterplants is easiest soon after the site is created. Most species only occur at a given water depth range. Marginal species can be planted directly into the soft mud; in shallow water the roots can be pushed in by hand and kept in place by a stone. Protection from erosion and waterfowl grazing may initially be useful. For deeper water, some species can be introduced by weighing down with a stone and dropping in or placing in a sack with some soil so that the bag sinks and new shoots grow through the bag and colonise the substrate (Baines & Smart 1991).

12.9.2 Trees and shrubs

Contrary to popular belief, at least in many temperate countries, trees can actually grow on their own! As long as the climate is suitable and there is not excessive soil disturbance or grazing then most areas will revert to woodland. Much of the planting that has taken place in the name of conservation is thus completely unnecessary. Planting is sensible if it is necessary to speed up the process, maintain control over the distribution or modify the composition, for example by establishing species which are otherwise unlikely to colonise or plant before unwanted species do colonise. Natural colonisation is the best method if adjacent to suitable woods. One solution is to plant some areas and then allow natural recolonisation to create a more realistic age structure.

Forestry advice, in which the objective is to produce the greatest volume of straight timber, is often followed yet it is often inappropriate for conservation, in which the objective is to produce a range of habitats. Plant at a range of densities, usually at a much lower general density than suggested by foresters, retain a variety of open areas and allow natural recolonisation. Plant shrubs in clearings. Create a scalloped edge rather than a uniform straight edge. Consider the wood in relation to surrounding areas, for example it may be beneficial to extend adjacent woods but detrimental to create a tall forest adjacent to an important marsh or meadow.

Tree and shrub seeds can be added directly but the major problem is to prevent competition from weeds and damage by herbivores. The proportion becoming established from seed is often likely to be very low unless competitors are removed by herbicides, disturbance or fire and grazing is restricted.

Planting small (0.4–0.8 m tall) bare-rooted trees is cheaper and usually results in better survival and growth than planting older specimens. These are usually germinated in a seed bed and after a year transferred to another bed for a year before planting out. Plant trees in a hole and press the soil firmly with your feet. After planting add plenty of water (e.g. 5 litres of water per tree). Bare-rooted trees are easily killed if allowed to dry out before or during planting and should be kept covered and damp until planted. If there is a delay before planting then cut a notch in the ground with a spade, insert the trees along the notch, cover with soil and ideally add water.

Trees can be protected from grazing by fencing or individual tree shelters. Polypropylene shelters supported by a stake result in increased survival and growth. Reducing competition greatly increases

survival and growth even at 2 m radius (Hibberd 1989). One common approach is herbicide spraying. Mulches or matting are very successful but too labour intensive for large projects. Cutting the grass around trees can increase the grass growth, thus increasing the water soil moisture deficit and thus increasing tree mortality (Gilbert & Anderson 1998).

12.9.3 Grass and herbaceous communities

Although natural colonisation will usually quickly occur, restorers often wish to speed up the progress of succession. Native grasslands or herbaceous communities can be cheaper to produce than amenity grasslands as the best results are often on nutrient-poor soil or even subsoil, so saving the need to buy topsoil or fertilisers.

Many commercial seed mixes are inappropriate. Even for an agreed mix, suppliers sometimes make inappropriate substitutions so make it clear that substitutions are unacceptable without consultation.

Lower application rates allow more opportunities for dicotyledonous species (= herbs) to grow and for natural colonisation to occur. The aim should not usually be for complete cover in the first year. Thus although it is common to add seed at $30-40\,kg\,ha^{-1}$, chalk grassland can be formed even with $1\,kg\,ha^{-1}$ (Stevensen et al. 1995) which may even result in improved growth of many species of interest. Furthermore, at low densities the reduced cost means a lower proportion of grass can be used, further increasing the diversity. The rush to establish cover may result in too high a grass density which may reduce the natural colonisation. However, if there is a great risk of colonisation by unwanted species then it may be sensible to reduce the amount of bare ground. Creating cover quickly may sometimes be necessary to reduce soil erosion or satisfy a public demand (Gilbert 1995).

Various ratios of grass: dicotyledonous species are often used including 80:20 and 50:50 but a higher proportion of dicotyledonous species may lead to a more attractive and diverse community although it is more expensive (Gilbert & Anderson 1998).

Legumes can markedly increase the nitrogen level which can be either beneficial or detrimental and should only be introduced if beneficial. Adding hay from the desired community can be very successful (Jones *et al.* 1995b). The damage to the invertebrate community and plants of the donor site should be considered, for example only partial harvesting of a site may be appropriate. Seeds may be drilled directly into bare ground or, less successfully, broadcast on the surface. Seeds are difficult to spread evenly, so allocate a given amount to each area and mix with sand or sawdust. A thick mulch layer may hinder establishment.

A common error is to assume annual plants can be maintained in an unstressed undisturbed habitat. Either the conditions have to be too severe to allow competitors to establish (e.g. saline, heavy metals, drought) or there has to be regular disturbance to prevent domination by perennial species. The spectacular display of annual plants in the first summer after seeding is thus lost unless the site is disturbed regularly, for example by rotovation.

Seeds rarely grow on established habitats as the competition is so severe but the following techniques can be used to alter the species composition:
• Create open areas by slot seeding – spraying a band of herbicide and drilling the seeds into a cut slit within the sprayed band (Squires *et al.* 1979) which can be successful for many species (Wells 1989).
• Strip seeding by cultivating and sowing in a strip.
• Insert seeds to natural or artificially created bare patches.
• Use a bulb planter to insert young plants; this technique is very successful but labour intensive (Wells 1989).

In nutrient-rich habitats, it is a common practice in the first growing season to cut and remove the cuttings to encourage slow-growing species. In subsequent years the pattern of cutting, grazing or burning should follow the methods described in Sections 12.5–12.6.

12.9.4 Reefs

Natural coral reefs are frequently destroyed as a result of pollution, by extraction for the construction industry or by dynamiting to capture fish. There has been increased interest in creating artificial reefs, particularly as a means of attracting fish. If planned and managed effectively, they may increase fish productivity, improve ecological diversity and be popular diving locations, but without planning or long-term management they may contribute little and can be considered as little more than pollutants (Chou 1997).

It is well known that sunken ships are excellent fishing and diving sites and many ships have been sunk for this reason. Artificial reefs have also been made out of a range of materials including rubber tyres (Branden *et al.* 1994), concrete blocks (Danna *et al.* 1994), PVC pipes (Omar *et al.* 1994), quarry blocks (Palmerzwahlen & Aseltine 1994), limestone boulders (Cummings 1994), blocks of pulverised fuel ash and gypsum (Jensen *et al.* 1994), suspended horizontal and vertical PVC plates (Oren & Benayahu 1997), redundant oil and gas platforms (Wright *et al.* 1998) or derelict train carriages (Clavijo & Donaldson 1994). Successful reefs tend to be those that are structurally complex but even these have fish communities that differ markedly from equivalent natural sites (Clark & Edwards 1994). Juvenile corals may be established in the laboratory and successfully transferred into the field (Oren & Benayahu 1997).

Important issues to consider when designing artificial reefs are the safety for the marine environments, durability and stability. A structure that disintegrates is likely to be of little benefit, may damage nearby communities or cause other problems. Cars, oil platforms and ships may contain considerable quantities of pollutants. Tyres should be submerged in concrete but may leach some toxic chemicals. As a result of the problems of disintegration and pollution, recycled artificial reefs, such as those made from vehicles and tyres, are now banned in many countries. PVC may become brittle over time. Concrete is excellent reef material. If taken from a demolition site it should be clean and any reinforcing rods that could entangle divers should be removed. Fresh concrete tends to have a high pH due to the calcium hydroxide, which may delay settlement of certain species, but this problem is reduced if it is allowed to weather in the rain first.

Reefs for fish are usually placed at such a depth and location that they do not interfere with shipping. Although some corals grow in water over 40 m deep, significant reefs are only found in water less than 20 m deep. Reefs in deeper water developed when the water was shallow and continued growing as the land fell relative to the sea level (Darwin 1842).

12.9.5 Translocation

Translocation, the movement of entire habitats, is very rarely sensible. It is often perceived as the obvious solution where development projects are planned on important conservation sites. Although some may express confidence in translocating habitats, our understanding of the underlying processes is usually woefully insufficient. In practice translocations are often unsuccessful (Parker 1995, Gault 1997) especially as:

• Vegetational changes often occur.
• Only a small proportion of associated animals are likely to be translocated.
• Not all animal species will be able to recolonise.
• If part of a habitat area is removed both donor and recipient sites may be too small to support viable populations so reduced diversity and increased edge effects are likely.
• Landscape-scale dynamics of species are likely to be disrupted.
• The soil structure may be irretrievably altered.
• The community may be dependent upon a particular hydrological regime which may be hard to mimic.
• The community may be dependent upon the surrounding habitat.

The problems of translocation are illustrated by the long-term monitoring of an expensive project to move large areas of sedge-cottongrass, meadow and fen that was in the way of the extension to Zurich airport runway (Klötzli 1987, Loucks 1994). Large (90 × 130 cm) blocks, 30–50 cm deep were carefully placed on pallets, moved on by trucks, and placed in position in the original pattern using forklift trucks and a team of 8–10 men per pallet. Despite all the

care and expense, the translocated habitat showed a substantial range of changes with considerable declines in many of the characteristic species, largely as a result of the disturbance caused by the move affecting the availability of water and nutrients. Cavities in the turf led to mineralisation of humus, fissures led to a mixing of humus, subsoil and the surrounding habitat resulting in an influx of nutrients.

Ironically, the process of moving habitats usually makes good publicity and so the move is often perceived as a success while the gradual deterioration of the translocated habitat is not newsworthy. Although many developers are prepared to pay for the actual translocation they are usually unwilling to guarantee long-term management or monitoring.

Thus translocation should not be accepted as a solution to a development problem, but if the habitat is going to be destroyed then in some cases it might be worthwhile to try to move it elsewhere while accepting that the new site is likely to be a poorer substitute. If translocation does occur, then:

• There must be considerable research to ensure that the hydrology, climate, future management and surrounding habitat are compatible.
• The soil must be removed to a considerable depth to buffer the vegetation from the new substrate.
• A large area will have to be taken to ensure that sufficient representatives of the species assemblage are present.
• The new site will have to be sufficiently large to maintain viable populations.
• The material must be handled very carefully as disturbing soil structure can cause considerable problems.
• There must be a commitment to perpetual management.
• The translocation should not cause damage by destroying an important recipient site.

12.10 Managing access

Access should be encouraged both for its role in education and from the principle of not making unnecessary restrictions, however access can lead to problems as a result of disturbance or direct physical damage (Vickery 1996). The aim is to find appropriate visitor levels for a site and ways of minimising impact. Access can be controlled by means of fences, signs and instructions. Although such controls undoubtedly have a role to play, where possible access is usually best managed in more subtle ways, for example by careful planning of access points, paths, roads, car parks, by making some routes look enticing while hiding others and by creating natural scrub or water barriers.

Although crime control in New York may seem to have little relevance to nature reserves, Bratton's (1998) concept of 'zero tolerance' probably applies widely, for example failing to repair vandalism or to prevent obvious infringements may create an atmosphere where such rule breaking is tolerated, encourage further more serious problems and create a mentality where infringements are not prevented or reported by others (Kelling & Catherine 1998). However, this needs to be balanced against the risk of overreacting and creating enemies. Creating a sense of identity and pride by encouraging volunteers to help, giving talks, providing guided walks, visiting local schools and groups and welcoming visitors is likely to reduce problems.

12.10.1 Zoning

Much of the success of controlling access is based on carefully planning the distribution of visitors in relation to their requirements and the damage they could cause. The first stage is to consider which areas are sensitive and to what. The main concerns are usually disturbance of birds or mammals and trampling of the vegetation. The next stage is to consider what visitors are likely to do and decide if there is any real conflict. A solution is to simply ban disruptive activities but this may be unrealistic or undesirable. A common solution is to zone the site so that the disturbing activities are concentrated and avoid the sensitive areas. The zoning may be gradual, for example from a recreation area, to an area where wildlife can be watched, to a sanctuary area with no visitor access. The zoning needs to be planned bearing in mind the psychology of the visitors. One obvious solution is to ensure that there are good facilities in the recreation area for the required activities but that it is difficult for them to carry out these activities elsewhere. If there is a species of considerable popular interest then it can often be worth enabling visitors to see it in one area in a controlled manner to reduce pressure elsewhere.

12.10.2 Car parks and footpaths

Most visitors stay remarkably close to the car parks so their positioning is critical. Parking places for wheelchair users should be at least 3 m wide, and preferably 3.5 m, with level access to paths.

Clay and peat soils are most vulnerable to trampling, then sandy soils, loams and silts, while substrates with rocks and stones are least sensitive. Wet paths are much more easily damaged and steep slopes are more sensitive to erosion, especially if people walk downhill, thus provide a circular walk up the steep sections. Straight paths are more visually intrusive but people will tend to cut across corners if they can see an unobstructed route to the path ahead, thus either ensure the track cannot be seen, provide obstacles such as water or vegetation or psychological barriers such as ditches, lines of logs or stones which makes it clear that people are off the path. The feeling of solitude is increased if everyone walks the same way round circular walks. This can be engineered by making one way the obvious route at the start with the return route obscured.

12.10.3 Visitor centres and hides

Visitors centres provide the opportunity to explain the importance of the site, describe the conservation issues, improve the visitor satisfaction and sell membership and other materials. They will often be positioned so as to regulate the access. Visitors centres should obviously not be situated so that the environmental damage exceeds the educational benefits. One sensible solution is to have the visitor centre at the entrance to the reserve but outside the sensitive habitat.

Observation hides (or blinds) can greatly improve views while minimising disturbance. They should be unobtrusive both to the wildlife but also to the landscape by being made of natural materials, set back from the viewing area and placed below the skyline. Wheelchair access requires a slope not exceeding 1 in 12, sufficient space inside and out to manoeuvre and accessible windows either by excluding some seats or making some removable.

13 Exploitation

13.1 Why manage exploitation?

Overexploitation is a major threat to many species as a result of increasing human population size and increased exploitation efficiency due to guns, mechanised transport and chain saws, along with increased access to markets. It is useful to distinguish between harvesting, the removal of individuals from a planted crop or stocked fish pond and exploitation, the removal of individuals, or parts of individuals, from a wild population.

There is often an ethical aspect to the regulation of exploitation. For example, I would like the hunting of great whales and apes to end regardless of whether they can be exploited sustainably. It is important, however, not to muddle such ethical considerations with direct conservation arguments.

It is easy to be outmanoeuvred in attempting to control exploitation, so it is unfortunately important to be careful and cynical as the following examples illustrate. In creating Lundy Island Marine Reserve, England, catching fish with traps was not prohibited

Allowing legal but sustainable levels of hunting for species in Zimbabwe has helped change attitudes of landowners and communities towards wildlife from being pests towards being assets (photo: William J. Sutherland).

as it was rarely done. Although intended as a sanctuary and agreed with local fishers, trapping increased markedly after the reserve was established (Gibson & Warren 1995). In the Waddensee, Netherlands, Mussel *Mytilus edulis* fishing by underwater dredging was banned due to the destruction of Mussel beds. The last remaining Mussel beds were destroyed by the fishermen dredging at low tide from boats resting on the mud which, not being underwater, counted as hand fishing! A frequent problem of setting up protected areas and managing them for wildfowl is that the surrounding land may be bought by hunting organisations which may greatly reduce the usefulness of the protected area.

13.1.1 Benefits of exploitation

As well as the considerable economic and cultural benefits, exploitation may have conservation benefits. Many areas are protected and cared for in order that certain species may be exploited. This may be due to individuals protecting their own land, governmental or private measures to protect game or even measures by private companies profiting from hunting. For example, in 1980 a corporation, International Paper, started charging fees for hunting deer, turkeys and wildfowl on its million hectares of forest in America and this now produces 25–35% of its local profits. In order to improve the game populations they now leave buffer strips around the edge of forests and have 70% smaller clear cuts (O'Toole 1991) and these measures are likely to benefit a wide range of species. Similarly, in Latvia the spring hunting of Capercaillie *Tetrao urogallus* is valuable with German hunters paying high rates which leads to the protection of old forest. During the period when hunting was not allowed, many lekking sites were cleared and the population declined (Maris Strazds personal communication).

Profits from the sale of hunting permits may enhance local attitudes towards conservation. For example, participants in the CAMPFIRE project in Zimbabwe gain meat and profit from hunting permits with the result that they sometimes dig wells for African Elephants *Loxodonta africana* and provide fodder for Hippopotamus *Hippopotamus amphibius*, although traditionally they disliked these species as they damaged crops (Pye-Smith & Feyerabend 1994).

The positive aspects of exploitation for conservation are thus improved habitat management, a larger pool of people with interests in the environment and a greater respect for species that are being exploited. The negative aspects are the potential reduction in the exploited species, intentional or accidental killing of other species and at least in some countries, a restriction in access to the countryside and the killing of large carnivores and birds of prey (Bibby & Etheridge 1993).

13.1.2 Why does overexploitation occur?

Despite intense theorising, data collection, analysis and management, most fish populations are overexploited and many fisheries have collapsed (Ludwig *et al.* 1993). This pattern of overexploitation also applies to innumerable other species of animals and plants. There are four main reasons

Species, such as Green Turtle *Chelonia mydas*, of high economic value but with a low population growth rate, are particularly susceptible to overexploitation (photo: William J. Sutherland).

for overexploitation: insufficient knowledge, the tragedy of the commons, discounting and multi-species exploitation.

Insufficient knowledge. Many fisheries have collapsed because it was realised too late that they were being overexploited and by then it was politically too difficult to reduce the level of exploitation to the level that would have stopped the population from crashing. It is difficult to detect declines due to overexploitation from other population fluctuations or sampling error unless the declines are considerable and persistent.

The tragedy of the commons. If the population can be exploited by a number of individuals, then it is in each individual's interest to continue exploiting, even if the long-term yield is greatly reduced or even if the population becomes extinct. The reason for this is that restraint by one individual is unlikely to result in an increased yield to that individual in the long term, but just gives the opportunity for the other individuals to take a larger share. This process, referred to as the 'Tragedy of the Commons' (Hardin 1968), only applies if there is competition. With sole exploitation rights, an individual showing short-term restraint will gain from the larger yields in the long term. As an example, each family of the Topnaar people in the Namib Desert, Namibia, had a traditional area in which they held exclusive rights to collect !Nara melons. Following social upheaval there is now free access and instead of waiting until the melons are ripe and gently tapping them off with a stick, the melons are often hooked off while still unripe. This not only results in poorer fruit, but the physical damage to the !Nara bushes is considered by the elders to have markedly reduced the yield (A. Shilombaleni, J. Henschel personal communication). The same process occurs on a much larger scale when companies or governments are fishing communal stocks or competing for logging rights. The tragedy of the commons is probably largely responsible for the overexploitation of most important shared resources on the planet. The logical extension is that granting property rights to resources to individuals or communities can provide the incentive for more sustainable exploitation. There are, however, a few situations where resolving ownership can lead to overexploitation. For example, in Vanuatu, areas of remaining forest are often those with disputed ownership and resolving such disputes may result in the owners selling the logging rights.

Discounting. Discounting is the preference for money now rather than in the future, which explains why banks pay us if we leave money with them, but charge us if we spend money before earning it. It also explains the preference for overexploiting now and having the profit immediately, rather than accepting the continuous profits from sustainable exploitation (Clarke 1973, Lande *et al.* 1994). This is a particular problem with slow-growing species, such as trees, whales or large fish, where the benefits of showing restraint are gained over a long period. An individual's discount rate will depend on social factors, for example those suffering from poverty may have high discount rates as the short term is of particular concern. Similarly politicians who are unlikely to retain power for long may greatly discount the benefits of leaving resources to exploit in distant future, when they are unlikely to accrue the benefits.

Multispecies exploitation. If a species is exploited in isolation then, once it becomes rare, it may not be worth exploiting, which then protects it from further decline. However, if other species are present it may be worth continuing to seek these while still taking any of the rare species encountered. Thus Blue Whales *Balaenoptera musculus* continued to be exploited because it was profitable to hunt Fin Whales *Balaenoptera physalus* (Clarke 1990) and Black Rhinoceros *Diceros bicornis* hunting persisted in areas where they were too rare for it to be profitable due to hunting for African Elephants *Loxodonta africana* in the same area (Milner-Gulland & Leader-Williams 1992). As exploitation persists even when the species is rare, this process is probably an important ingredient in the local and even global eradication of many species.

With multispecies exploitation, those species that are easy to catch or whose populations are especially sensitive to exploitation are particularly likely to decline. Large species are generally disadvantaged on both grounds. They are more likely to be caught, as they are more vulnerable to nets or more prone to

detection during hunting and generally more likely to be pursued once detected. Furthermore, larger species are usually more sensitive to any additional mortality as they typically mature at a later age and have a lower fecundity (Jennings *et al.* 1998).

13.2 Determining sustainable yields

The logic of sustainable exploitation is simple: it is to remove individuals at the rate at which they would otherwise increase (Caughley & Sinclair 1994). Thus if a population is increasing at 5% a year, then this 5% can be removed each year while keeping the population at a stable level. However, for an unexploited population at equilibrium the birth rate will balance the death rate so the population does not grow and there are none to remove. This annual population increase is usually induced by reducing the population so that, as a result of density dependence (see Section 5.13.1), there is an increase in average survival or breeding output (see Fig. 13.1). Thus if once the population is halved it then increases by 8% per year, this 8% can be removed continuously while keeping the population constant.

Extraordinarily, some papers and books describing sustainable exploitation consider density dependence only as a hypothetical factor which might exist and may play some role, while in reality it is the

Regular monitoring of the population and exploitation levels has enabled these South African Fur Seals *Arctocephalus pusillus* to be one of Namibia's main tourist attractions while still being exploited for meat and fur (photo: William J. Sutherland).

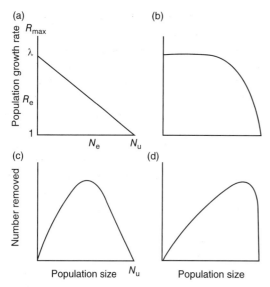

Fig. 13.1 The basic model of sustainable exploitation. (a,b) Relationships between the population growth rate, R, and population size. Population growth rate declines with population size as a result of density dependence. (c,d) Relationships between the numbers of individuals that can be removed without reducing the population and population size. Note that an unexploited population N_u will not increase ($R = 1$). For the linear relationship in (a) the highest yield occurs at half the unexploited population size (see (c)). If density dependence acts most strongly at high densities, as in (b) then the highest yield will be above half the unexploited population size (see (d)). A major source of confusion arises from the fact that R can be measured at different levels. For an unexploited population at equilibrium N_u then $R = 1$, the highest rate of increase at low densities is the finite rate of increase λ, at intermediate population levels N_e where exploitation may occur there will be an intermediate level R_e. The highest possible rate of increase for animals in ideal conditions is R_{max}.

obligatory essential core of sustainability. Without density dependence exploitation would be analogous to mining – remove a hundred individuals a year and the population would decline by a hundred per annum until extinct. The simple fact that species persist despite being exploited confirms the ubiquity of density dependence.

There is a popular, but mistaken belief in conservation circles that a population decline is evidence for overexploitation. It follows from Fig. 13.1 that sustainable exploitation is usually only possible if the population is reduced. As a rough approximation, populations will be reduced to about 50–80% of the initial population size when the population is sustainably exploited (see Section 13.2.6). If the population is unexploited then there is no population increase that can be sustainably exploited.

If a unexploited stock is exploited there may be a short-term bonanza in which the population can be reduced to the level at which it will produce a high sustainable yield. For long-lived species there is often a striking contrast between the considerable short-term profits from mining a stock of large old individuals and the much smaller profits obtainable from removing a sustainable share of the annual growth and recruitment. There is thus often considerable political pressure to allow the bonanza to continue, despite the long-term consequences.

Exploitation is also likely to change the age structure, resulting in fewer old individuals, as shown in comparisons of the proportions of juvenile Agouti *Dasyprocta sp*, Paca *Agouti paca* and Collared Peccary *Tayassu tajacu* across sites differing in hunting intensity (Robinson & Redford 1994). A shift in age structure occurs if exploitation affects all ages equally but is particularly pronounced if the exploitation concentrates upon larger, older individuals. However, if exploitation is restricted to the youngest stages, such as eggs or seeds, then there will either be no change in age structure or the reduced density may even increase survival and thus result in more old individuals.

The age structure can thus be used to indicate the extent of exploitation, with a younger age structure usually indicating greater exploitation. This is not straightforward as other factors may influence the age structure, such as past conditions affecting mortality and recruitment. A further complication is that it is usually difficult to age individuals in the population and thus size structure is often used instead. However, individuals may grow more slowly at high population densities and thus a paucity of large individuals may either be due to heavy exploitation (which may lead to a recommendation to reduce exploitation) or due to dense populations retarding growth (for which an increase in exploitation, particularly of young individuals, may be sensible). It is clearly essential to distinguish between these contrasting alternatives, for example by determining the growth rates of individuals.

There are many reasons to exploit at a lower rate than might initially seem sustainable. If the yield is set too high, the population may crash, as has been observed repeatedly in the world's fisheries. This can result from an undetected decrease in breeding output or survival (Beddington & May 1977, Bayliss 1989) or from undocumented illegal exploitation. Furthermore, there may be errors in the estimate of the sustainable yield, so that the greater the uncertainty, the lower the exploitation should be. In practice, for a variety of reasons, the main methods, as described in this chapter, tend to overestimate the sustainable yield. Unfortunately, once overexploitation has been identified, either due to a history of overexploitation or due to recent poor recruitment or survival, it is usually politically difficult and unpopular to reduce the catch or effort sufficiently to allow recovery. This inability to respond sufficiently can lead to a downward spiral in abundance with the necessary reduction in exploitation becoming increasingly severe yet never being achieved, so the population inevitably crashes. Globally important examples of this process include the collapses of Peruvian Anchovetta *Engraulis ringens* (Clarke 1981) and the Cod *Gadus morhua* of the Canadian Grand Banks (Walters & Maguire 1996) but exactly the same process occurs on local scales.

The estimate of the sustainable yield should thus be considered as an overestimate and the actual exploitation should always be less. Caughley & Sinclair (1994) suggest a safety margin of 25%, and higher if the population is variable, the data are poor or if monitoring is irregular. This margin should also be particularly high if it is likely to be difficult to reduce the exploitation if overexploitation occurs. Unfortunately, politicians and policy makers often consider estimates of sustainable yield as a lower starting point for negotiation.

It is not always sensible to interfere with existing exploitation practices: there may not be the legal authority to do anything, it may be impractical to regulate, the loss of goodwill may cause greater conservation problems than the benefits from reducing exploitation or there may be conservation problems of higher priority.

With the exception of some game and commercially exploited species, a major problem is the lack of data (Johannes 1998). A crucial first step is usually to initiate a programme of population censuses in each region, exploitation effort in each region, an estimate of the number caught and ideally the age and sex composition of those removed. If this data is inaccurate then the exploitation should be particularly conservative. The data should be systematic, standardised and consistent between years and stored in a way that can be analysed later. The consequences of any modification in data collection should be quantified, for example by collecting the data in both the old and new manner for some time period. It is extremely useful to have data on the unexploited or barely exploited population in order to assess the natural mortality rate in the absence of exploitation. If it is possible to collect data on subdivisions of the total area then this can be invaluable, especially if the areas are harvested at different intensities. It can also be useful to see how the exploitation patterns have changed across areas, for example the composite data may show little decline in catch per unit effort with time while the detailed data may show that the catch per unit effort in the nearest site has plummeted with everyone now travelling to the distant site. Such data is invaluable for determining future policies.

It is usual for exploiters to distrust and dismiss both scientists and those regulating the exploitation. Scientists and policy makers may not respect the judgement and motives of the exploiters. An increasingly used method is for exploiters to be fully involved in analysing the data and agreeing regulations. Not only is this likely to improve trust and relationships but the science may improve as the exploiters can provide additional information.

The basic models of exploitation consider a single population experiencing uniform exploitation. In reality there may however be different populations using a site, for example of a migratory species, and it may then be easy to overexploit one population, such as the one that arrives earliest or stays longest. A second problem is if the population is subdivided and unevenly exploited, for example if there were different populations within a lake and the fishing pattern resulted in one being much more heavily exploited.

There is a huge literature describing methods of data analysis and modelling of fisheries (e.g. Hilborn & Walters 1992, Quinn & Deriso 1999, Jennings *et al.* 2000) and sophisticated commercially available computer packages and I assume that anyone with substantial data sets will already have access to this information. In this section I review the main methods (see Table 13.1) but particularly have in mind someone faced with managing the exploitation of a species for which there is little data and ecological information.

13.2.1 Surplus yield models

The underlying concept of surplus yield models, which are also known as production models, is that if the population is reduced then as a result of the reduced competition the birth rate will exceed the

Table 13.1 A summary of the main methods used for managing exploitation.

Method	Section	Data needed	Uses
Surplus yield models	13.2.1	Annual catch and total effort or catch per unit effort of a subsample. More sophisticated models require further data	Straightforward if long run of data, exploitation methods have stayed constant and there has been some heavy exploitation. The most simple models are likely to be inaccurate. The more accurate ones need considerable data
Yield per recruit models	13.2.2	Mortality and economic value of each age class. Relationship between population size and recruitment	Requires considerable data. The main method for those involved in commercial fisheries and forestry
Robinson and Redford models	13.2.3	Either direct estimate of rate of population increase (e.g. when recovering from catastrophe) or from age-specific birth and death rates	Can give a rough measure of sustainable yield but often misused
Relating yield to recruitment and mortality	13.2.4	Recruitment and mortality rates	Of limited use and often misused
Adjusting in relation to population changes	13.2.5	Population size. Information on annual variation in recruitment is useful	Widely and successfully used where it is reasonably straightforward to adjust exploitation effort
Lotka–Volterra model	13.2.6	Unexploited or expected unexploited population size. Actual population size	Very useful if only data is on abundance and it is possible to estimate unexploited population size
Full population model	13.2.7	All demographic parameters and how they change with density	Excellent but requires considerable data and understanding. Usually impossible
Adaptive management	13.2.8	This method uses experiments and data collection to improve analysis	All exploitation methods should include some adaptive management

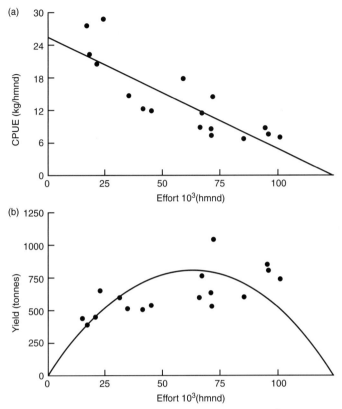

Fig. 13.2 An example of the simplest form of the surplus yield model applied to a Barrumundi *Lates calcarifer* fishery in Australia (King 1995). (a) The relationship between the catch per unit effort (CPUE) (in kg per 100 m of net per day) plotted against effort. The regression equation is CPUE $= a + bE$ where a is the intercept, b is the slope and E is the effort. (b) The Schaefer curve (yield $= aE + bE^2$) based on the above regression. This gives the fishing effort that results in the highest yield. As described in the text, this basic method is considered highly inaccurate and has been replaced by more complex analyses but which still use the same underlying principle.

death rate so producing a 'surplus' which can be removed each year while still keeping the population at that level (Fig. 13.1). If the population is very low then the 'surplus' population will obviously also be low. If the population is not reduced then there is no 'surplus'.

The extent to which the population is reduced will depend on the intensity of exploitation effort. If there is very little effort then the population will be large but few are caught. Surplus yield models simply determine the intermediate level of effort that produces the largest sustainable yield. In this section I will describe the simplest method (the Schaefer method) to explain the general principles. In practice

more sophisticated means of analysis are usually used.

1 Collate information on the numbers or biomass exploited each year in a given area or the catch per individual for representative exploiters.

2 Collate information on the annual amount of effort, such as the number of days spent hunting. It is necessary, although often difficult, to express the effort in constant units allowing for changes in efficiency due to changes in behaviour or technology. Thus if the same number of days are spent hunting, but a change in transport, equipment or access results in hunting being twice as efficient, then the effort doubles. In practice there are often substantial

changes in methods but the consequences are unknown; surplus yield models are then of little use.

3 Plot catch per unit effort (CPUE) against effort E (as in Fig. 13.2(a)). Calculate the regression as $CPUE = a + bE$ where a is the intercept and b is the slope of the regression against effort.

4 Calculate the yield, Y, for different levels of effort using the equation $Y = aE + bE^2$ (Schaefer 1954), as shown in Fig. 13.2(b).

5 Determine the level of effort which results in maximum yield.

6 Be conservative. One problem with the analysis of catch per unit effort is that the population does not have time to reach an equilibrium for each level of effort. For example, a sudden increase in effort one year may have only a moderate short-term effect on the population and thus result in a high catch per unit effort, yet if this effort were to be maintained over a longer period the population may collapse. This is thus most suitable when there are a number of periods in which exploitation effort is reasonably consistent and so the population can approximate to an equilibrium.

Another problem is that it is often difficult to assess where the maximum sustainable yield occurs (Hilborn & Walters 1992) with the relationship between catch per unit effort and effort often only becoming clear once the population has been considerably overexploited, whereupon it is usually difficult to reduce the exploitation level.

In reality the Schaefer model is no longer used although current, more complex, methods use the same principles. More sophisticated methods allow for a range of relationships between catch per unit effort and yield, instead of assuming a linear relationship, use different methods for finding the best fit of the relationship to the data and allow for the fact that the population may not have reached an equilibrium (Quinn & Deriso 1999).

The great advantage of surplus yield models is that they can be applied without having to collect extensive ecological data, as they only require data on exploitation effort and catch. Although surplus yield models were initially invented for fisheries, they can be used for other groups, for example Courtois & Jolicoeur (1993) applied this approach to Moose *Alces alces* hunting in Quebec using data on population density, exploitation rates and hunting effort.

Catch per unit effort data can also be useful on its own. If the catch per unit effort remains constant over time, as for most of the birds and mammals caught by the Siona-Secoya people in north eastern Ecuador over a decade (Vickers 1991), then this suggests the populations are not declining. Of course, it is important to ensure the hunting efficiency has not increased, as this would disguise any population decline.

13.2.2 Yield per recruit models

Older individuals are often more valuable: older trees produce more timber and hunters often prefer to kill mature individuals. However, the longer exploitation is delayed the fewer individuals will have survived natural mortality. Yield per recruit models (Beverton & Holt 1957) can be used to determine the optimal compromise and with sophisticated modifications have become the major technique for those fisheries which have the capacity to collect and analyse the extensive data required. It is easiest to consider a single cohort but the conclusions are usually the same as for the realistic case of a number of cohorts being exploited simultaneously.

1 For each age determine both the natural mortality (see Section 5.11.2) and the value of an individual of that age, e.g. the weight of meat, the timber volume of a tree or the price hunters will pay to hunt.

2 Tabulate the natural mortality of each age in a column in a computer spreadsheet. If males and females are considered separately then use a separate column for each. Introduce a recruitment of, say, 1000 individuals. Multiply by the natural mortality to give the number alive at the end of that year. Repeat, multiplying by the mortality of each age to give the number of each age surviving.

3 Rerun the analysis in stage 2 but with an additional fraction exploited, say 10% of all individuals each year above the age of 5. An extra column in the spreadsheet (0.1 for each year above 5) describes this strategy. Rerun as in stage 2 but incorporate both natural and exploited mortalities to give the number surviving. A further column multiplying the numbers of each age by the exploitation mortality (i.e. 10% of each age over 5) gives the actual numbers of that age exploited.

4 Tabulate the value of individuals of each age. This column is then multiplied by numbers exploited (from stage 3) to give the value of the exploited individuals of that age and again stored. Adding these gives the total value of all exploited individuals. This process can be repeated with different fractions exploited at different ages to determine the strategy that produces the most valuable result.

5 In reality, it is usually impossible to predetermine the exploitation level of each age class and it is usually necessary to use approximate measures, such as regulating the minimum length that can be removed, the youngest animal that can be caught or the mesh size that can be used. Furthermore, instructions not to remove individuals below a certain length may not always be strictly adhered to or some small fish may be retained in a mesh through which they are theoretically capable of escaping. It is very useful to quantify this selectivity expressed as the relative probability of capture. Thus if no one-year-olds, half the two-year-olds and all three-year-olds and older are retained in a net of a given size, the values in this column would be 0, 0.5, 1, 1, 1, 1,...1. For a larger mesh the values might be 0, 0, 0.1, 0.6, 0.9, 1,...1.

6 Run stage 3 incorporating selectivity and explore the consequence of realistic combinations of effort and the selectivity expected from various methods, such as different minimum length regulations or different mesh sizes. The best approach is to plot yield per recruit against combinations of mortality rate and methods (such as mesh size) to determine how yield per recruit varies with these parameters. Determine a combination of effort and method which produces a sustainable high-value yield yet is practical to implement.

7 Be conservative. The basis of yield per recruit models is to avoid growth overexploitation, the removal of small individuals which could be more profitable if left to grow. The other major problem is recruitment overexploitation, which is removing so many adults that the recruitment is seriously reduced.

The yield per recruit models presented here assumes that the recruitment is constant. However, there must be a population size below which recruitment drops. It is thus necessary to determine the relationship between stock size and recruitment and incorporate it into the model. One problem is that, for many fisheries, recruitment is reasonably constant over a wide range of stock sizes but plummets once the population is below a certain size; this threshold can usually only be determined empirically.

13.2.3 Robinson and Redford models

As shown in Fig. 13.1, the aim of sustainable exploitation is to remove individuals at the rate they can be replaced. It thus follows that the fraction that can be exploited can be determined by calculating the population growth rate at the population size which gives the highest yield. Robinson & Redford (1991) outline an approach for measuring this that has been widely used.

A major source of confusion is that the population growth rate R can be calculated for a range of population densities and conditions (see Fig. 13.1) but each means of calculation derives a different measure. One measure is the maximum possible population growth rate R_{max} often calculated using data from captive animals in the absence of competition. A second measure is the finite (or intrinsic) rate of growth λ, the growth rate in natural conditions but in the absence of competition. Thirdly, if population growth rate is measured in unexploited stable populations, each individual must, on average, leave one descendent so that $R = 1$, but with errors in estimating parameter values it is easy to produce values exceeding 1 suggesting population growth. Finally, for calculating sustainable yields the required measure is the growth rate at the actual density at which exploitation will take place. This final Figure is rarely available and one of the other measures is usually substituted.

An extremely useful indication of the potential population growth rate can be obtained from populations reduced by severe weather, disease or exploitation or those introduced to a new area. For example, the Mantled Howler Monkey *Alouatta palliata* population on Barro Colorado Island, Panama, had an annual population growth rate of 16.7% following a yellow fever epidemic (Froehlich *et al.* 1981), while Muskox *Ovibos moschatus* colonising unoccupied areas can increase at 25–30% per year (Gunn 1999). The growth rate at low densities can be considered as an estimate of the finite rate of increase λ, thus for the Howler Monkeys, $\lambda = 1.17$ while for Musk Ox, $\lambda = 1.25$–1.3. Of course if measured at too high a density the value of λ will be an underestimate.

The most widely used approach is to estimate the maximum population growth rate R_{max} using demographic data on age of first breeding and breeding output in captivity (Robinson & Redford 1991). This method is:

1 Determine a, age at first reproduction in years; b, annual birth rate of female offspring and w, age at last reproduction in years.

2 Solve R_{max}, the maximum rate of population increase, iteratively from Cole's equation $I = R_{max}^{-1} + bR_{max}^{-a} - bR_{max}^{(w+1)}$ by using a range of values until the equation balances.

3 This method assumes all individuals start breeding when first reproduction can occur and all survive until the age of last reproduction. This is likely to greatly overestimate the population growth rate. A solution, proposed by Robinson & Redford (1991), is to assign a proportion of the annual production that can be exploited. They suggest that 20% of the production can be removed in long-lived species (age of last reproduction is 10 years or greater), 40% for short-lived species (age of last reproduction between 5 and 10 years) and 60% for very short-lived species (age of last reproduction less than 5 years).

Alternatively, if there is information on the age-specific survival and breeding output then the population growth rate R can be calculated from

$$1 = \sum_{x=a}^{w} l_x m_x R^{-x}$$

where l_x is the probability of surviving at age x years and m_x is the number of female offspring produced

per year by a female of age x. Again the value of R is determined by using a range of values until the equation balances.

If it is reasonable to assume that survival rate and breeding output are independent of age, then R can again be determined by iteration from

$$1 = pR^{-1} + l_a bR^{-a} - l_a bp^{(w-a+1)}R^{-(w+1)}$$

where p is the yearly probability of survival and l_a is the probability of surviving up to prereproductive age (Slade *et al.* 1998). In practice this gives more conservative results that does the Robinson & Redford (1991) method (Slade *et al.* 1998).

4 The population growth rate R can be used to calculate the proportion of the population that can be exploited without reducing the population as $(R - 1)$. Thus if, at a particular density, the population increases at 8% a year ($R = 1.08$) then up to 8% can be removed without lowering population size, subject of course to the usual cautions of being conservative (see stage 7).

5 It is essential to consider the population density at which exploitation will take place. If exploitation is already occurring then one approach is to use the current population density with the estimate of population growth rate R_c for this density. Thus for a population of current density D_c the following yield can be obtained without reducing the population further

$$\text{yield} = D_c(R_c - 1).$$

An alternative approach, and that suggested by Robinson & Redford (1991), is to use the Lotka–Volterra method (Section 13.2.6) and reduce the population to a proportion p, such as 0.6, of the unexploited density D_u. It is, however, necessary to determine the likely growth rate R_p for this density. A common technique is to use the maximum growth rate R_{max} which must be a considerable overestimate as it measures growth under ideal conditions without competition. Such high values are thus likely to lead to overexploitation. If it is possible to estimate λ, population growth rate at very low densities, a very rough approach could be to assume that the density-dependent relationship is linear (as in Fig. 13.1(a)) with a growth rate of λ at very low densities and 1 at the unexploited population size. In this case the growth rate R_p for a population at a proportion p of the unexploited population density is:

$$R_p = (\lambda - 1)(1 - p)$$

With an estimate of growth rate it is possible to calculate an estimate of sustainable yield

$$sustainable\ yield = pD_u(R_p - 1).$$

The value of p which provides the highest sustainable yields will depend upon the shape of the density-dependent response, as shown in Fig. 13.1, but a value of 0.6 is frequently used.

6 If it is possible to estimate the likely unexploited density for a site then it is possible to use this equation to calculate the number that can be sustainably removed. A species with a value of R of 1.15 at the target density of 60% of the unexploited density of 1000 individuals can thus sustain an annual exploitation of $0.6 \times 1000\ (1.15 - 1) = 90$. Some people measure the actual number exploited and compare this with the sustainable exploitation from this equation. This can be useful for estimating if the number exploited is unsustainable (i.e. over 90 in the previous example) but not for determining if it is sustainable. If only 80 were left then removing all 80 would count as being sustainable! The population size thus needs to be considered alongside the number exploited.

7 Be conservative. There are numerous problems with this method as Robinson & Redford (1991) readily acknowledge in their original paper. This method depends critically on the age of first breeding and breeding output but this data is often collected from captive animals which may be very different from the mean value in the wild. Population data from one area or habitat may not apply elsewhere. The method calculates the maximum potential rate of increase, yet what is needed is the rate of increase at the actual population density, which will usually be much lower. If the simple objective is to determine whether a site is

overexploited I would usually suggest comparing densities if practical (Section 13.2.6).

Robinson and Redford models have become popular for assessing sustainability of terrestrial vertebrates and especially forest mammals (e.g. Fa *et al.* 1995, Fitzgibbon *et al.* 1995, Wilkie & Carpenter 1999a, Peres 2000). Despite the considerable problems, and my concerns that this is often misapplied by using λ, R_{max} or R_u rather than R_p, this approach is often the only one available for estimating yields and thus is still useful.

13.2.4 Relating yield to recruitment and mortality

Consider a population with 20% recruitment into the adult population and 5% annual adult mortality. It is clear that 15% can be removed each year without reducing the population (see Fig. 13.1). By measuring recruitment and mortality it should then be possible to determine the sustainable exploitation. However, for unexploited populations at equilibrium the recruitment and mortality will balance, so there will be no discrepancy. This method can thus only be used for populations already exploited and or reduced by natural processes to about the target population size (say about 60% population size, see Section 13.2.6). This method is thus rarely applicable. It is sometimes used mistakenly for trying to estimate yields from unexploited populations.

13.2.5 Adjusting in relation to population changes

Adjusting in relation to population changes simply entails relaxing or tightening the exploitation in relation to evidence of population increases or decreases. This method is improved by adjusting the exploitation each year in relation to observed or predicted changes in breeding output or mortality. Although this is a simple method requiring little data it probably has a better track record of conserving populations, and providing sustainable yields, than any other. It has the considerable advantage that it concentrates upon the population size which makes it easier to detect overexploitation. It is largely used for those populations of birds and mammals which are straightforward to count.

1 Collect data in a systematic and repeatable fashion on the population size and, if possible, age structure. It is also very useful to have systematic data on the pattern of exploitation (how many people exploit, how long for, at what seasons, where does the exploitation take place, what size and sex of individuals are caught) and the total numbers exploited.

2 Collect data to see whether exploitation seems to be having an effect on the target population. The three methods are: to examine any historical data to see whether the exploited population has declined, to compare the abundances in exploited and unexploited sites, or to examine the age (or size) structure to see if there are indications that older or larger individuals are becoming scarcer.

3 Consider if the current exploitation rate is too high or whether it could be increased. The objectives may be complex as the aim may be to maintain high populations for conservation reasons or maintain low populations, for example to reduce agricultural damage. If the objective is to obtain a high yield and maintain the population at a proportion of the unexploited population size (say 60%) then this approach is identical to the Lotka–Volterra model (Section 13.2.6).

4 Adjust the regulations such as hunting seasons (see Section 13.4) if necessary. Consider whether changes should be severe (if the population is declining fast) or slight (if only slowly declining).

5 Evaluate the response of the population to the change in regulations. In reality there will usually be a number of simultaneous changes, for example in the regulations, habitat, weather or equipment used by the exploiters, which make such evaluations difficult. Ideally introduce experimental management with different regulations in different areas and examine the consequences.

6 Continue annual monitoring of population trends. If available review data on breeding success and mortality. An alternative is to monitor environmental conditions, such as the severity of the winter, which have known effects on breeding success or mortality. For example, in North America, biologists have determined the correlation between the area of breeding lakes, which varies considerably with the annual rainfall, and the productivity of waterfowl. Adjust regulations as a result of population changes or exceptional events. Thus if due to poor recruitment one year the population is expected to decline by 10% then if effort is reduced so that the yield is also likely to decline by 10% then the mortality due to exploitation will stay constant and not exasperate the problem. It is often sensible to reduce the effort further, so allowing some population recovery.

7 Be conservative. There is often considerable resistance to changing the rules making it difficult to make fine adjustments. Be particularly conservative if the population estimates are poor or infrequent or if reducing exploitation is likely to be difficult.

13.2.6 Lotka–Volterra model

If exploitation is incorporated into the classic ecological models of population biology (such as the Lotka–Volterra equation) it leads to the prediction that the greatest yield occurs when the population is halved (Schaefer 1954), as in Fig. 13.1. A possible, but exceedingly crude, method is thus to assume that the highest sustainable yield is at approximately half the natural density and adjust the exploitation to maintain the population at that level.

This method is absurdly simplistic which is both its weakness and strength. The highest yield is in theory at 50% of the unexploited density due to the assumption that population growth rate declines linearly with population size. Thus the line in Fig. 13.1(c) is symmetrical and centred around 50%. As shown in Fig. 13.1(b,d) increasing density dependence at high populations will result in an asymmetric line with an highest yield at levels above 50%. There are numerous reasons including food depletion, interference and territoriality for expecting density dependence to increase disproportionately with population size (Sutherland 1996c). In general highest sustainable exploitation levels are likely to be above 50% of the unexploited population size. A figure of 60% is widely used.

Despite being naive and simplistic this method has a number of advantages. It requires little data

Methods of Moose management in Finland

The Moose *Alces alces* population in Finland has fluctuated considerably. It became completely protected in 1923 following an intense decline during widespread hunting as a result of food shortages in the First World War (Nygrén 1987). By 1933 the population had increased sufficiently for hunting to be reintroduced. The large harvest of adults in the 1960s resulted in a rapidly declining population and total protection was again introduced in many areas during 1969–71 (Nygrén & Pasonen 1993). The decline seems to have resulted from the large number of licences issued and the small proportion of calves shot, which by maintaining a young age structure resulted in a low reproduction rate. Following some recovery, a moderate harvest was introduced but this now included many calves. As expected this increased the mean age and so resulted in increased calf production from 29 calves born per 100 adults in 1961–62 to 50 in 1973–83.

However, as the population increased the damage to forests and car accidents increased and so the current management involves setting maximum density goals for different regions, taking into consideration the interests of hunters, forestry, agriculture and road safety. In 1998 this was 4 per km^2 in the south, 3 per km^2 in the middle and 2 per km^2 in the north. The present strategy is to reduce the currently male-biased sex ratio and maintain a population of under 80 000 Moose with an annual harvest of 30 000–43 000 animals.

The main source of data is questionnaires completed for each hunting day on the numbers seen of each age and sex. At the end of the season, or once all the licences have been used, the hunters also estimate the number of moose still alive in their hunting area (these estimates seem to be accurate except in the vast hunting areas in the north) and the size of their area. With nearly 100 000 hunters from about 4000 hunting clubs, this provides considerable data. This data can be used to estimate for each region the number of cows per bull, the percentage of cows with calves, the percentage of twins, the number of calves per 100 adults and the number of moose per 1000 ha of dry land after the hunt.

Each spring a meeting is held between the Ministry of Agriculture and Forestry, the Hunters' Organisation and the Finnish Game and Fisheries Research to decide the strategy for the next season. The Finnish Game and Fisheries Research biologists report on the status, give the results of their population model on the consequences of different strategies and give recommendations for quotas on the number of licences and the proportion of calves and bulls in different regions after taking into consideration the goals of different local or national authorities. After consultation with others, such as traffic and agricultural organisations, it is decided how many may be shot within each region and the age and sex composition.

Exploitation strategies are often not simply designed to produce the highest possible harvest. The objective may be to maintain a high population for conservation reasons, or maintain a low population to reduce grazing or accidents. The driver survived with minor injuries, the moose died and the vehicle was written off (photo: Alaska Department of Public Safety).

CASE STUDY

Goose management in North America

In the late 1970s the populations of White-fronted Goose *Anser albifrons*, Cackling Canada Goose *Branta canadensis* and Pacific Black Brent *Branta bernicla*, had all severely declined. These populations largely breed around the Yukon–Kuskokwin Delta in Alaska, where they were illegally hunted during the spring and autumn by the native Tup`ik people, and winter in California where they are hunted for sport. As a result of the declines some Californian sport hunting groups threatened legal action against the United States Fish & Wildlife Services for failure to enforce the Migratory Bird Treaty Act (Hunt 1997).

As attempts at enforcement had failed before, the Fish & Wildlife Service decided to organise a series of discussions involving all parties within both California and Alaska. The solutions agreed at these meetings were that Cackling Canada Geese would be closed to both subsistence and sport hunting, White-fronted Geese and Pacific Black Brent could be taken by the Tup`ik in the spring until they nest, while the numbers of White-fronted Geese and Pacific Black Brent that the sport hunters could take would be halved. It was agreed that no eggs of any of these species could be taken nor could they be captured during moult.

Since this agreement the populations of all three species have increased markedly, almost certainly as a result of this plan. Cackling Canada Geese increased from 26 200 in 1983 to 161 400 in 1995, and their protected status was then revoked. The White-fronted Geese increased from 112 900 to 343 000 while the Pacific Brent Geese increased more slowly from 109 300 to 129 900.

The goose management seems to have been successful as the scientific evidence made it very clear that there had been severe declines, the reasons for the declines were clear, groups were demanding change and threatening legal action, and the authorities were prepared to bring the sides together for difficult negotiations.

This collaboration of the Tup`ik people, federal agencies, state agencies and sport hunting groups has been such a success that similar plans have been written to conserve the populations of Caribou *Rangifer tarancus*, Moose *Alces alces*, Brown Bear *Ursus arctos* and various fish.

The successful management of migratory species requires close collaboration between authorities along the migration route. As a result of successful measures the populations of almost all geese are increasing in Europe and North America. Greater Snow Goose on Bylot Island, Canada (photo: William J. Sutherland).

and does not require a knowledge of population ecology, exploitation effort or success. The only essential data are unexploited population size and current population size. Although there may be dispute about the level to which the population should be reduced, whether the population is reduced to 70% or 60% is probably a small issue compared with the possibilities for overexploitation using other methods. Many of the other methods do not focus on population size and thus population crashes can occur undetected, for example because catches remain high as the remaining individuals are aggregated or because the catching efficiency has increased.

1 Estimate the population size that would occur in the absence of exploitation. This is obviously easiest if exploitation has yet to start. This could also be obtained by using old population estimates prior to exploitation or, most commonly, from population estimates from comparative studies in similar habitat. The greatest errors are likely to arise from this stage, especially if it is necessary to extrapolate from very different sites.

2 Decide upon the target population size. For example this could be at 60% or 70% of the unexploited population size and will depend upon conservation objectives as well as exploitation objectives.

3 Determine the current population size and assess whether it is less than the target.

4 Determine the current effort (ideally number taken) and method used.

5 If overexploitation appears to be taking place then reduce the exploitation level, using the methods in Section 13.3, until the current population is at the target population level.

6 Be conservative. Use a higher target population size if the population estimates are poor or if it is likely to be difficult to tighten regulations in response to population declines. Alter regulations if there are marked changes in recruitment, survival or in the number of exploiters.

13.2.7 Full population model

For a few species there is sufficiently detailed information that it is possible to create a reasonably complete model of the population ecology. It is necessary to have a good understanding of how the components of the birth rates and death rates change with population density. It is usual to examine components of mortality, such as predation, disease, starvation and exploitation and the components of breeding output such as the ability to find a mate, fecundity and survival of young.

1 Determine the nature of all density-dependent processes and determine their strengths.

2 Create a population model incorporating all density-dependent and density-independent processes including the current level of mortality due to exploitation.

3 Explore the consequences of changing the exploitation rate for the annual yield in both the short and long term.

4 Be conservative.

An example of this is the extensive studies of the Grey Partridge *Perdix perdix* (Potts 1986, Potts & Aebischer 1995) in which it was discovered that the main density-dependent stage was nest predation, as at high densities predators learn how to find nests but are much less likely to do so when nests are scarce. One interesting discovery was that most populations were underexploited so that more birds could be shot without reducing further yields. This model was also used to consider the consequences for the yield of improving the chick feeding by leaving a narrow margin unsprayed with herbicides and insecticides on the yield (Potts 1986). This approach is excellent but requires an enormously detailed study which would need a team of biologists working for many years. Even then, most species are too difficult to study for this approach to be practical.

13.2.8 Adaptive management

The essence of adaptive management (Walters 1986) is to create models, as in the previous methods, but then manipulate the system in a systematic manner to improve the assessment. Thus the traditional approach has been to treat assessment and management as separate. The approach adopted here is to use management to improve the assessment which then feeds back to the management.

1 Collate the biological information required for one of the previous approaches.

2 Produce models based on the ecological information. Be conservative when suggesting exploitation levels.

3 Consider where there are doubts in the information. Use a range of models with different

assumptions to explore the consequences of this uncertainty. Determine the areas where greater certainty would lead to improved management.

4 Design, run and analyse management experiments that are both acceptable to those exploiting the population and will improve the knowledge in those aspects considered important. A particularly powerful method consists of manipulating the management differently in different areas.

5 Return to stage 1, incorporating the improved information and continue forever improving knowledge and management.

This approach is clearly sensible and the way forward. My only concern is that it is astonishingly difficult to get people to accept the concepts of experiments and uncertainty and so I fear this is often impractical. I hope I am wrong.

13.3 Controlling exploitation

There are a number of ways of controlling hunting levels as listed below. Which technique is most appropriate depends largely on social and practical considerations. Those hunting purely for fun often object to rules that stop them hunting but may accept limits on the number or size of individuals they can kill or the equipment they can use. Closed seasons or closed areas are relatively easy to monitor while total catch or the size taken is sometimes difficult to control.

Total bans can be sensible in extreme cases and as they can be easy to enforce it is then not necessary to distinguish illegal and legal exploitation. They may be necessary to restore populations to a level at which sustainable exploitation can be carried out. Total bans are the quickest means of restoring populations but are often unacceptable.

Closed areas in which no exploitation is allowed can be excellent both for conservation and in creating healthy populations which can restock surrounding areas. Closed areas have the enormous advantage of being relatively easy to monitor. Closed areas can also benefit other species if the exploitation technique is damaging. Each reserve needs to be large enough to retain a viable population with larger grids for more mobile species (McCollough 1996).

As an example, marine reserves have been set up, for example in New Zealand, and local fishers believe these increase recruitment in the adjacent areas. A meta-analysis of the published studies suggests that marine reserves provide conservation benefits for fish populations (Iago Mosquera *et al.* in press). Large target species were three to five times more abundant in marine reserves than surrounding fishing areas. However, some non-target species were also more abundant in marine reserves, which suggests that there may be a bias to locate reserves in richer habitat, which may confound any analysis.

A fixed quota per season is in theory very sensible as this is the variable that should be regulated. It is necessary to decide how to allocate the quota. This total may be divided between groups or individuals so they can exploit when they wish to. If the quota is a group target it encourages individuals to rush to exploit to gain a disproportionate share followed by irritation when exploitation is stopped. However, a quota per individual per season can be difficult to monitor and even a daily quota is easily abused. If the quota is set too high, due to an initial overestimate or a subsequent undetected decline in natural survival or breeding output, then fixed quotas rapidly lead to overexploitation as each season the fixed quota becomes a larger fraction of the continually declining population. This is a particular problem if the population is not carefully monitored, so the decline is not detected until there has been a dramatic decline. Despite such problems quotas can sometimes work well. For example, Canadian Beaver *Castor canadensis* trappers are given sole hunting rights to areas, monitor the number of dams annually and are then given a quota. It is obviously not in their long-term interest to overestimate the population.

Regulating effort is much better than regulating quota, as if the population declines the number caught for a given effort is also likely to fall. This is not foolproof. If the population halves, the exploiters may work harder or have to spend less time dealing with the catch so the number caught need not also halve. Effort can be regulated by restricting the numbers who participate or the days when exploitation may occur.

Regulating seasons is a frequently used method. It is often used to avoid sensitive periods, such as the breeding season. Exploitation should ideally take place before the main period of density dependence. For example, for many birds in temperate regions there is severe competition for food in the winter and reducing the population before the winter reduces the competition so that more of the remainder can survive (Murton *et al.* 1974).

Restricting the techniques that can be used is often important. Efficient methods may be banned if likely to be damaging with current levels of effort. Some techniques are considered so damaging to the ecosystem by professional exploiters (such as poisoning or dynamite fishing), or so 'unsporting' by those hunting for enjoyment (such as using live decoys) that they are easy to forbid. The further environmental consequences of the different techniques, such as the damage caused by lead shot if swallowed by waterfowl or snares which capture other species, are often important in determining which are acceptable.

Compulsory training can sometimes be highly beneficial to introduce, for example with a written and field test to ensure that participants can identify legal and protected species, know and understand the rules and laws, practise in a safe manner for themselves and other participants and practise in a manner that does not cause unnecessary damage or suffering. Such training may also be beneficial in restricting the exploitation to those with a serious interest.

Sex-biased regulations are sensible. For polygamous species, such as most mammals and fish and many birds, in which the males do not care for the young, removing a proportion of males will not reduce recruitment. Furthermore, by reducing competition for food it may even increase productivity (Ginsburg & Milner-Gulland 1993). It is thus common for there to be rules which bias the exploitation towards males.

Size limits are widely used by dictating mesh sizes or stating which sexes cannot be removed. There is a common belief that it is a mistake to take young individuals, as it is these that produce future generations of breeding individuals. Curiously there is

less concern about taking adults which have succeeded in reaching the age at which they can now reproduce or in taking those large individuals which in species with continuous growth will have particularly high fecundity. The considerable advantage of setting a minimum size which allows some reproduction to take place before removals is that exploitation cannot eliminate the population and this measure could even be the sole means of regulating exploitation. However, with a maximum size it would be possible to prevent any from breeding so this may have to be accompanied by other restrictions.

13.4 Discouraging illegal persecution

Species may be protected by law and yet are still persecuted. A common objective is to create the impression that illegal persecution is no longer worthwhile and that crime enforcement agencies have infiltrated everywhere. Most sucessful programmes reduce illegal persecution using a combination of approaches from the following:

• Encouraging the government or authorities to strengthen the law or increase the consequences of it being broken.

• Increasing public awareness of the conservation problems to make the market unacceptable.

• Encouraging the government or authorities to strengthen or enforce existing laws on buying or selling products of illegal persecution.

• Identifying the weakest point where action is most effective. For example in India, there are large numbers of poachers and many small town traders, smugglers and retailers. The very few city-based wholesalers buy from the town traders and sell to smugglers or retailers. If these are shut down the conservation benefits are considerable (Kumar & Wright 1999).

• Running education programmes to explain the law if there is a lack of knowledge as to what is illegal.

• Increasing public awareness to make persecution unacceptable.

• Creating and marketing alternatives, such as jade knife handles to reduce the market for handles made

CASE STUDY

Anti-poaching strategy to protect the Amur Tiger

In 1993 the Amur Tiger *Panthera tigris altaica* was considered to be in danger of imminent extinction with a world population of 200–300 and an estimated 50–60 being lost a year to commercial poachers. But by 1997, as a result of Operation Amba, the poaching had been dramatically reduced and there was evidence of population recovery (Galster & Eliot 1999).

The fragmentation of the Soviet Union resulted in escalating crime across the region including poaching and trade in Brown Bear *Ursus arctos* gall bladders, Musk Deer *Moschus sp* musk glands, Wild Ginseng *Panus quiquefolia* roots and tiger bones, pelts and organs. The general increase in crime stretched law-enforcement officers while a reduction in funding reduced the number of wildlife rangers and their resources.

In December 1993, Russian authorities and foreign non-governmental organisations met and agreed on a plan. The agreed objective was to reverse the situation so that the tiger population was increasing by the year 2000. It was clear that the most urgent need was to reduce the poaching, while the longer term aims were to secure sound habitat and ensure sufficient deer and wild pig prey.

Interviews with enforcement officers showed the following patterns:

- Poaching was either carried out by opportunistic poachers or members of organised groups. Both would sell the parts or sometimes entire tigers to middlemen in cities, especially Khabarovsk, Nakhodka, Plastun, Ussuriysk and Vladivostok.
- Most tiger parts were smuggled to The People's Republic of China, South Korea or Japan. The middlemen were usually Russian, Russian-Korean or Chinese.
- Smuggling to China usually took place by road or train from either Pogranichniye or Poltovka. Smuggling to Korea and Japan took place by boat from Nakhodka, Plastun or Vladivostok or by air from Khabarovsk or Vladivostok.

- Poachers and smugglers were often linked to high-level organised crime groups, 'Mafia', who would often lend firearms, vehicles or provide protection for poachers. Sometimes the Mafia would simply buy the parts from a poacher who was unconnected to the group. In some cases poachers openly advertised to sell remains from a tiger they had killed.
- Tigers were usually killed in winter when it is easiest to track animals in the snow.

In exchange for external funding for the operation, the government agreed to create a new 'Tiger Department' within the Ministry of Environment branch in Primorski Krai. This was code-named Operation Amba after the local name for the tiger.

The operation only had 15 rangers and a huge area to cover. They decided to use psychological operations, which are a key element of insurgency and counter-insurgency measures world-wide. The objective was to make Amba appear large, powerful and effective so that it seemed as if the risk of being caught had increased dramatically. The uniforms and new vehicles gave a professional and effective image. Good vehicles and radios allowed for effective co-ordination so regularly allowing them to converge on poaching gangs. Groups of rangers moved between areas to randomly check vehicles emerging from the forest, hunters in the forest and hunters' shacks. During their patrols, rangers told local people and authorities about their role and invited them to provide information on poachers and traders. Gradually a network of informants was established which was the key to the success of the operation. Press coverage was used to publicise the project and any successes.

In the first winter Amba rangers collated information so that they could increase their understanding of the underground wildlife trade. Amba then organised underground investigations to expand their knowledge. The following three early cases were typical of

(Continued)

(Continued)

many that followed. The first concerned professional hunters killing tigers and selling them to Chinese citizens in the city of Khabarovsk either whole for US$5000 each or US$300 per kilogram of bone. These were smuggled to companies in Heilongjiang and Jilin Provinces in north east China. Investigators visited these firms and discovered the tiger bone was mixed with other bones and converted into wines, pills and plasters and then sold to companies locally, in South Korea or in Singapore. The second case involved corrupt officials in a government department in Khabarovski Krai employing poachers and using a government truck to transport tiger carcasses. The officials sold the tigers to members of a Russian–Korean community in Khabarovsk, who then sold them to Chinese or South Korean merchants. The third operation involved a collaboration between a Russian government employee and a driver for a joint Russian–Korean logging company who hunted from his vehicle. To export them they used two contacts: an employee of a shipping port who would sell the remains to Korean workers or Russian sailors heading by ship to South Korea or Japan and a customs officer who would organise the secret export by ship.

In investigating these operations there were initially no attempts to make arrests but just understand the system. It became clear that traders were well organised and well connected. Amba thus set out to improve relationships both with the police and the FSB (formerly KGB) in Ussuriysk and Luchegorski and future operations were jointly organised with them. One priority was the traders in Ussuriysk, which informants made clear was the centre for tiger smuggling. By using information from an informant (an ex-hunter) they arrested two traders, who each gave details of the other traders and poachers before being indicted. Although the effort was concentrated on dealers, they also used information to capture poaching gangs in operation. Video was used when visiting dealers, both to provide evidence for use in court and also for publicity afterwards.

Such work is not without risks. After uncovering details of a Mafia group that paid hunters to kill tigers and smuggled tigers, arms and drugs, the Amba rangers gave their information to police and prosecutors in Vladivostock District. The police said that they were already aware of the problem so Amba should not interfere. The lead ranger involved was subsequently beaten up outside his home and a police investigation confirmed that this was linked to the discovery of the smugglers.

The Global Survival Network found a means of approaching the Russian Prime Minister who in 1995 issued a decree 'On Saving the Amur Tiger and Other Endangered Fauna and Flora of the Russian Far East'. This high-level support resulted in the Russian courts taking the issue more seriously. Seven people were indicted for poaching or selling tiger products the next year compared with only two in the previous 3 years.

Amur Tigers are now increasing and this project regularly encounters trade in other protected species as well as other illegal activities such as drug and arms trafficking.

from rhino horn. This usually requires a deep understanding of the real needs of the market.

• Increasing the efficiency of prosecution, for example by educating the police or pressurising to increase the effort and efficiency spent in enforcing the law. In many cases the main requirements are to ensure that there is the necessary equipment and staff, motivate the staff already present, and ensure a sensible amount of time is spent in the field preventing persecution.

• Signalling that protection is taken seriously, such as maintaining signs and fences and having a conspicuous wardening presence or by ensuring that prosecutions are well publicised. Planes can be particularly effective at reducing illegal exploitation in large open areas.

• Rewarding a lack of persecution, for example by giving payments if successful breeding occurs or an annual payment if a species is present.

• Providing a reward system for individuals who provide information or take action in uncovering illegal activities.

• In extreme cases it may be necessary to provide direct protection. In some populations Black Rhinos

Diceros bicornis have been under such a threat from persecution that each individual has been allocated its own armed guard. Similarly, the one remaining Lady's Slipper Orchid *Cypripedium calceolus* in Britain has a warden camping next to it each summer. Visitors are discouraged as the site is sensitive and the area is private. Anyone visiting is only allowed to see the orchid if they give their name and address and how they learnt of the location. They are asked not to return or to tell anyone else.

Galster (1999) suggests a ten-point plan for reducing tiger poaching (see Case Study) which probably has wide application.

1 Thoroughly and comprehensively identify the threat. Review the research and discuss widely, for example with rangers, biologists, hunters, farmers, foresters and police officers. It is especially useful to talk to poachers and smugglers either covertly or by discussing with those who are no longer operating but are willing to talk.

2 Bring all the information on threats together in a species action plan (Section 7.3, Box 7.1) which, like all plans, is only likely to work if devised in consultation with those that will carry it out and if the personnel, equipment and money required is realistically obtainable. The action plan may either consider all problems faced by the species or just those relevant to poaching. The plan should consider all stages from the poaching to the final use and all approaches to the solution including education and political support. The plan should be used to agree upon priorities.

3 Gain high-level political support by presenting the action plan to sympathetic high-level government officials and involving them in the action planning process.

4 Establish good relationships with the police and other enforcement agencies. Understanding how they operate, when they are likely to act, what they need and their strengths and weaknesses, is likely to ease the working relationship.

5 If the poachers and smugglers are likely to be armed then the rangers and investigators need the capacity to defend themselves with reliable equipment.

6 It is often important to be able to move quickly and ideally much faster than the poachers. Having reliable fast transport with sufficient funds to buy fuel makes a considerable difference.

7 Effective communication, both within members of a team and between teams and the base, greatly increases the capacity to respond.

8 Each team should have someone trained in investigations (see Section 13.5) to identify the geographical areas where poaching and trading is most frequent, identify poachers and smugglers and through surprise arrests project a magnified perception of the anti-poaching force. Interrogations following an arrest can lead to much useful information.

9 Secure local support and co-operation. The leader of the anti-poaching team can meet with representatives of the local community, to understand their concerns and determine shared ground between their goals and those of the anti-poaching programme which can then be built into the programme. Options for promoting the project include education programmes, brochures, posters and media publicity.

10 Ensure effective management to maintain the capability and motivation of the anti-poaching team. This includes sensible budgeting, reliable cash flow and equipment maintenance.

13.5 Criminal detection

The principles described here apply to identifying both illegal exploitation and other illegal activities such as vandalism or theft. With increasingly sophisticated technology, monitoring persecution is becoming increasingly easy. Techniques used include cameras set up to monitor protected species, and various means of identifying individuals such as adding PIT tags (Section 5.10), marking with ultraviolet sensitive marks and DNA analysis of hairs or blood (see Section 5.15).

There are often risks associated with detecting and enforcing laws. Even people checking fishing licences have been shot dead and in one year 3.6% of Canadian Wildlife and Fisheries Officers were assaulted; over half of the offenders were under the influence of drink or controlled substances (Scarlet *et al.* 1996). Evaluating risks is clearly essential before deciding how to behave and it is also important to understand the relevant laws and procedures.

Successful identification and prosecution usually involves some of the following stages.

1 Upon discovering an incident, such as finding a felled tree, hearing gunshots or discovering a nest has been robbed, start by considering whether the culprits could still be in the area and could be identified or safely apprehended.

2 Minimise the disturbance in the area surrounding the incident, both so that evidence is not destroyed and so that evidence can be properly collected. Search widely and systematically beyond the incident, including possible entry and exit routes and around likely vehicle parking places.

3 For each incident, document in as much detail as possible any information on the location, date, time, entry point, route taken, exit point, exact method, number of people involved, descriptions of individuals if seen and transport used. Tyre marks, footprints and the damage caused by equipment, such as for tree climbing, are often highly individualistic. They should be photographed or plaster casts made. The ideal is to find trademarks such as characteristic techniques, litter or damage that is unique to that individual. There are computer software packages available that can be used to collate and analyse the information.

4 Question nearby witnesses and those that may have witnessed the entry or exit. Revisit the site, especially at the same time and day of the week when the event occurred, to produce further witnesses. Negative information can be useful in determining the timing and route. Document the precise information, e.g. 'X said "I was harvesting fruit from 2 o'clock until nightfall, which must have been after 6, and no-one walked along the river" ' rather than a vague 'did not leave along the river'. The reason being that it might turn out that X can only see the near bank of the river, the offender might have waited until dark before leaving or X may even have been involved. With precise information it is possible to return to it for reinterpretation.

5 If prosecution is intended, then ensure that the collected information is suitable for evidence. Thus witnesses should produce written and signed statements as soon as possible after the event. The statements should include date, time, distance of observation, lighting conditions, visibility, period of observation and what drew their attention to the incident. If possible culprits are encountered then describe fully, including clothing, behaviour, items carried, any conversation, and whether the person involved was known to yourself and whether seen before and if so when.

It is necessary to be able to convince the court that any physical evidence presented could not have been intentionally or unintentionally modified or exchanged between removal from the incident and the courtroom. This requires marking evidence before removal from the incident, for example with location and date, or by making a file mark upon a cartridge, marking by inserting a PIT tag (see Section 5.10), marking birds by removing a couple of primary feathers or marking fish by clipping a fin. It is also important to ensure a continuity of possession, for example by sealing critical evidence in a bag with the seal signed by a witness, packaging it to prevent damage and then keeping it in a locked cupboard before it is taken to the court. The marking and possession should be by someone who the court would consider reputable. Photograph and describe all evidence before touching. Keep evidence in its original state and do not remove mud, rust, dirt or fingerprints or add fingerprints. Sketches and photographs of the scene are much more convincing evidence than written accounts.

6 If likely to become a witness then it is very useful to know the law. Each offence leads to a series of 'points to prove' and it is useful to have some knowledge of the common responses that are made in defence so that they may be covered as fully as possible by the evidence. Thus if it is illegal to 'intentionally kill a wild otter' the points to prove are: (1) It was intentional. The defence may be that it was mistaken for a squirrel as it was far away/ against the sun/obscured by vegetation/up a tree so that evidence on the distance, direction, view or habitat may be critical. Alternatively, the defence may be that it was found already shot, in which case the critical evidence may be what was seen or heard of the person's behaviour. (2) It was killed. Evidence may include a corpse, with evidence of continuity of possession, to show it was the same one removed from the incident, possibly including a post mortem to show that it was shot and the ammunition was compatible with that used by the culprit. (3) It was wild. The defence could be that it was an escaped tame otter, in which case any evidence on the otter's behaviour may be important. (4) It was an otter. The defence could be that it really was a rat and these are not protected. If a corpse is not retained the description of the animal may be important.

7 The main standard investigative technique (Weston & Wells 1997) is to follow leads such as descriptions of individuals, routes taken, vehicles used, determining those who have the opportunity to carry out the incident, determining those who have the necessary information, implements used, fingerprints, *modus operandi* (see stage 8) and sale of illegal items (see stage 9) with the objective of identifying and narrowing the group of possible culprits until those responsible can be identified.

8 Police investigation often uses the fact that each criminal is usually fairly consistent in the method, or *modus operandi*, used (English & English 1996). The *modus operandi* may be pieced together by documenting each case in detail and then collating the accounts. The *modus operandi* can be used to determine if all incidents follow a similar pattern or if they suggest that more than one individual or party is involved. For example, it might be that all the incidents can be attributable to either a single individual who enters alongside the river from the south by bicycle accompanied by a dog or to a small group who enter by car at weekends and at least one of whom smokes. As the picture becomes more detailed from documenting and collating more cases it is easier to intercept or identify the participants. It is useful to be able to recognise the *modus operandi* of known culprits, and evaluate whether these are compatible with the observed incident. A further use of the *modus operandi* is to provide insights into the perpetrators' behaviour and lifestyle, which may

lead to identifying possible culprits (or ways of intercepting or dissuading them), for example are the incidents during usual working hours or not? Are they outside school hours? Is transport involved? Is the person alone or in a group?

9 While illegal exploitation can often be carried out discreetly and at night, the sale of removed items often has to be more public. Tracing illegally obtained products is often highly productive. Posing as a buyer or someone who wishes to participate in an illegal activity can be highly effective but can be dangerous.

10 Collect information from anyone with relevant information who is prepared to talk to determine participants, entry points, methods used, areas where exploitation occurs and markets. Question past and current persecutors. A network of informants who are paid for reliable information can be invaluable. In Nepal a reward equivalent to US$10 000 is given for information leading to the arrest of poachers of Tigers *Pantheris tigris*. At the Aberdare National Park, the Kenya Wildlife Service have two full-time employees who visit villages and collect information. Informants can also be unreliable and may have motives for accusing others, so the information should be corroborated. In Canada, there are rewards for providing information that leads to the arrest of criminals and this is being promoted as a means of arresting poachers. To keep the process anonymous, the informants phone and are given an identification number. They phone again after an agreed delay and if their information has lead to an arrest they are eligible for a reward, paid in cash, delivered to a pick-up location which they can collect by quoting their identification number.

11 Consider if it is possible to intercept the persecutors or identify them by using surveillance such as observing or using cameras. This can be most effective by first looking for patterns to determine where, when and how most persecution occurs and use this to consider where and when the persecutors are most vulnerable. This may be when in the field, when returning, at home or when selling the produce.

14 Integrating development and conservation

14.1 Why combine development and conservation?

Conservation and development have often been considered separately by different organisations, sometimes with conflicting consequences, such as the numerous examples of conservation organisations trying to protect a forest while development organisations provide free sawmills. It is now widely accepted that development projects need to consider their environmental consequences and conservation projects are most likely to be effective if considering the needs of local people. In recent years there has thus been an enormous increase in the number of projects that integrate conservation and development.

Conservation problems largely result from increasing human populations, new technologies and our increasing expectations. The traditional approach to conservation has been for authorities to impose restrictions, such as preventing farming, grazing, hunting, timber extraction or access, often with little consideration of the social or economic consequences. Many such restrictions are often ignored as a result of local antipathy, or even antagonism, combined with an inability or lack of motivation of the authorities to enforce the regulations. In some countries the imposed structure is a remnant from colonial times which may reinforce the resentment. Some radical rethinking has resulted from the patchy success of imposed regulations, combined with changed social attitudes and increased awareness that it is neither realistic nor ethical to exclude people from their traditional livelihoods unless providing an alternative. Over the last two decades, conservation organisations have increasingly tried integrating conservation and development with the hope of producing the dream package that both preserves biodiversity and reduces poverty, so generating co-operation instead of conflict.

Eradicating poverty worldwide is obviously exceptionally important, but is not what this chapter is about. I thus only consider aspects of development which aid conservation and the role of conservation in reducing poverty.

There is a need for clearly thinking precisely how the proposed development will aid conservation. One common belief is that raising living standards will automatically reduce pressure on natural resources, but this does not necessarily follow and increased wealth can even lead to greater capacity to exploit resources or encourage local immigration. Another common approach is to provide economic incentives to support conservation and sustainable management. However, in reality, it is usually very unlikely that the incentives can be changed sufficiently that some individuals would not profit from encroaching, exploiting or polluting, even if this is detrimental to the community as a whole. To be successful, such programmes usually have to simultaneously consider social or legal controls. Such problems can often be prevented by continually concentrating on the conservation objectives; many projects become deflected from these.

An example of how projects intended to benefit conservation may be severely detrimental is Okomu Forest Reserve, Nigeria (Oates 1995). Since the 1960s, and especially in the 1980s, there was very heavy settlement by farmers into the reserve from elsewhere in Nigeria because of the adoption of the *taungya* system in which farmers are allocated an area of forest which they can clear and farm but have to plant trees such as Teak *Tectona grandis* which eventually replace the food crops. This was so popular that the poorly funded forestry departments were unable to provide sufficient trees or supervision and, unsurprisingly, the farmers were more concerned with their crops than tending trees. It was realised that the *taungya* system was destroying the forest and was abandoned. In 1990 a conservation and development plan was introduced, involving a team of consultants, a socioeconomic survey and a development officer. The consultant sociologist stated that 'conservation should foremost have the well-being of the resident inhabitants as a priority' even though most had only recently settled in the forest. Following the report's recommendations, livestock-rearing projects were established, a facility to process cassava was built and road access improved. The increased standard of living of the migrant workers is likely to encourage them to stay and others to settle. Following this programme, logging and poaching have increased, which was not helped by the fact that although funding was available for development, the state forestry department, with the job of protecting $5600 \, km^2$ of widely scattered forest, did not even have a functioning vehicle.

It is clear that reserves and parks are likely to be more effective if they involve people living near them, but these should be compatible activities, such as sustainable exploitation, reserve protection, tourism or research, rather than incompatible agricultural activities (Oates 1995).

A review of integrated conservation development projects in Indonesia (Wells *et al*. 1999) concludes that even if local villages do cut timber, hunt wildlife or plant crops illegally, they are usually not the major threats in most Indonesian protected areas. Most threats arise from illegal activities organised and financed by outsiders, such as logging, mining and wildlife poaching, road creation or urban expansion. These threats are best dealt with by law and policy enforcement and influencing regional planning and development rather than local community development. Altering regional planning requires influencing national or regional politicians and policy makers (see Chapter 10).

14.2 Approaches for combining development and conservation

There are a range of ways in which conservation and development can be combined. Many projects will include a number of the following five approaches.

14.2.1 Regulations to restrict access or use

It is regularly said that the traditional approach of regulations constraining use has failed to conserve species and habitats. However, there are many areas throughout the world where such regulations have provided good protection, or at least better than no protection. Contrary to popular belief, in such cases simply introducing a development project is unlikely to reduce the pressures on species and habitats and may just destabilise satisfactory arrangements. Enforcement for protected areas or species is perfectly acceptable within the framework of integrated conservation development projects, if accompanied by local development and genuine attempts to improve relationships between parks and local people. As well as regulations established to control use by individuals within the community, for integrated conservation development projects to work it is often necessary to introduce regulations to prevent exploitation by those outside the community. Thus almost all projects will include some element of introducing legislation or regulations or in enforcing existing regulations.

Imposed regulations are also appropriate when considering the consequences of development over large scales. Development affecting habitats, such as coral reefs, mangroves, forests or wetlands may have implications for coastal defence, fish recruitment, soil protection or flood defence over a much wider area and it seems appropriate to allow some higher level decision about their future rather than delegating to a local level. Thus, the current project to isolate and drain Yala Swamp, Kenya, by diverting the Yala River so that it goes straight into Lake Victoria is understandably popular locally, as the creation of new agricultural land is likely to benefit the local community and the costs are paid for by an external development agency. The loss of Papyrus swamp is likely to greatly reduce the conservation interest of this important site (Oliver Nasirwa personal communication) and the loss of the cleansing capacity of the swamp combined with creating farmland adjacent to the lake is likely to further the eutrophication problems in Lake Victoria. Eutrophication is affecting the fish stocks resulting in declining catches.

This project is thus not only damaging to the conservation interest of the swamp and lake but may not even reduce overall poverty. In such cases it is entirely reasonable to introduce legislation that protects habitats or species which produce wider benefits to society.

Many communities have traditional regulations, such as sacred lands and taboos, which have sometimes been used as a form of conservation management. For example, on Mount Elgon, Uganda, killing of young or pregnant animals is forbidden, as is cutting more bamboo or firewood than can be carried as waste angers the ancestors (Scott 1998). New taboos may be created, for example Peter Fidelio, a local tourism entrepreneur, was concerned about overfishing on the coral reef at Wala Island off Malakula, Vanuatu, and established a marine reserve by persuading the local chief to set up the south side of the island as a taboo area, which, like all taboo areas, was then marked with palm leaves. Through his brother-in-law, Peter Fidelio also arranged for the chief on the neighbouring island to designate an adjacent section to be a taboo area. Similarly, in Ghana, concepts of taboo are still recognised and widely understood so can be given new powers by the support of modern legislation and enforcement (Coulthard 1996). However, with a loss of traditional customs and decline in the authority of traditional leaders the acceptance of taboos has often become eroded.

In some cases communities lack the power to enforce regulations on the use of their resources which are thus overexploited. Caldecott (1996) describes how providing power boats to people in the Cyclops and Cenderawasih areas of Irian Jaya and the authority to limit illegal exploitation led to enthusiastic protection.

14.2.2 Increasing the value of natural resources

Most conservation problems arise because individuals or organisations gain more from destroying the resources than from maintaining them. Increasing the value of the natural resources can thus help their protection. It can also be helpful to raise the profits gained from products. One approach is to add value,

for example by selling worked products rather than the raw materials. Another approach is to increase the value by marketing, for example by emphasising that the product is sustainably exploited or part of a conservation project. A third approach is to create new markets for traditional products. Of course, there can be situations where a high value may lead to over exploitation for the reasons outlined in Section 13.1.2.

Butterfly farming has been established at the Sokoke Forest, Kenya, by Birdlife International/ Nature Kenya (the 'Kipepeo project') in which local species of butterfly are raised on plants collected from the forest and then sold abroad. As a result of the profits made, the pressure to clear the forest has been reduced.

One approach is to allow the exploitation of large mammals with the profits going to local communities. The best-known example is the Communal Areas Management Programme for Indigenous Resources (CAMPFIRE) in Zimbabwe (Metcalfe 1995, but see also Murombedzi 1999) in which private landowners and communal landholders were given the right to manage wildlife and administer the exploitation. The underlying principles are that effective management of wildlife is best achieved by giving it clear value for those who live with it and that there must be a positive correlation between the quality of the management and the magnitude of benefit. This programme required close collaboration between communities, local authorities and the Ministry of Environment. It has resulted in a change of attitude, with game often considered as beneficial and encouraged; the area devoted to wildlife has expanded with numerous private nature reserves set up. The financial gains from safari hunting can be substantial as the following annual figures in US$ indicate. Tanzania: gross revenues almost 14 $million; trophy fees 3.6 $million. Zimbabwe: trophy fees almost 4 $million. Namibia: revenues over 6 $million, trophy fees 2.8 $million (Wilkie & Carpenter 1999b). Hunters are prepared to travel further than typical tourists and visit a much wider range of sites. Hunting can thus provide alternative incomes to a considerably larger number of communities than can tourism.

Abruzzo National Park in the Apennine Mountains, Italy, has been developed in a manner to benefit wildlife and the local communities. Despite considerable pressures, the park management managed to exclude high-impact activities and kept intact the landscape and wildlife (including 70–100 Brown Bears *Ursus arctos* and 40–50 Wolves *Canis lupus*). A strict zoning system was established with a protected area of 440 km^2 and a buffer zone of 600 km^2. The 2 million visitors a year have ensured real economic benefits for local communities. In 1989, Rochetta and three other neighbouring villages voted to be included in the park, so extending the park's area by 10% (Adams 1996).

In each of these three examples – butterfly breeding in the Sokoke Forest, game management in Zimbabwe and wildlife tourism in Abruzzo National Park – the success is due to a programme linking financial benefits to local people to conservation and thus changing attitudes. In combination with providing the financial benefits it is usually also important to provide legislation and a framework for preventing infringements.

Many natural resources have sufficient value in terms of flood defence, water storage, coastal defence, soil protection, sustainable exploitation or tourism to fully justify preservation. The requirement is then to ensure that the economic consequences of destruction are fully understood and that there is the ability to create regulations or incentives to protect them.

14.2.3 Alternatives to damaging exploitation

Sustainable exploitation (Chapter 13) is often central to modern conservation thinking and often encouraged. Where current levels of exploitation are incompatible with the conservation objectives, one approach is for conservationists to encourage the use of alternatives. It is sometimes sensible to ban or restrict exploitation and this is often ethically easier if an alternative is provided. Providing alternatives will only reduce exploitation if they are really used as an alternative rather than a supplement. Although some conservationists dislike exploitation, it can often provide an incentive for habitat

protection (Section 14.2.2), so if the populations are sustainably exploited then providing alternatives is not necessarily beneficial.

Honey gatherers in the Aberdare National Park, Kenya, light fires to smoke out the bees prior to collecting the honey. Although honey collection in itself seems perfectly sustainable, it sometimes results in accidental fires which burn large areas of forest. To overcome this problem, the Kenya Wildlife Service provide bee hives and courses on bee keeping. The honey provides an alternative income and this programme has greatly reduced the fires. The marketing and labelling emphasises the fact that it is part of a conservation project which seems to increase sales.

Turkey has numerous species of beautiful bulb plants, many of which are endemic. However, the demand for bulbs by gardeners in western Europe led to severe population declines. The initial response by conservationists was a successful programme to encourage gardeners to boycott wild-grown bulbs and only buy those propagated in nurseries. This then led to a Fauna and Flora International programme to encourage Turkish bulb collectors to grow their own bulbs in fields, so providing an alternative income.

Other uses of this technique include providing kerosene stoves as an alternative to firewood, encouraging the use of jade-handled knives instead of rhino horn and encouraging chicken keeping rather than hunting primates. Despite these examples, in practice there has usually been little success in discovering alternatives to resource-use practices that threaten protected areas (Wells & Brandon 1993).

14.2.4 Development as part of a package

A blinkered interest by conservationists towards biodiversity alone can be considered offensive, particularly where there are obvious social or economic problems. To overcome this, many reserve managers provide some assistance that helps maintain good relationships with neighbours and some projects carry out work without direct conservation benefits. The Himalayan Jungle Project in Pakistan had the initial primary objective of protecting the Western Tragopan *Tragopan melanocephalus* and it started by identifying the best area for the species (the Palas Valley) and determining that logging was the major threat, but that illegal hunting of the species also occurred (Duke 1994). Using the techniques of participatory rural appraisal (Section 14.5.2), the local people drew diagrams, maps and charts to describe their social and economic practices, identify problems and constraints and formulate for themselves conservation and development opportunities.

The initial plans were disrupted in 1992 by heavy rains and severe flooding which caused considerable damage. The project team then co-ordinated relief work as it was familiar with the region and had links with the local community, donor organisations and the government; collectively they airlifted 50 tonnes of food to 10 000 stranded villages. Engineers from the project, in conjunction with the community, identified the priorities for alleviating the flood damage, while the project team also provided advice and appealed to the government on their behalf. At the jirga (tribal council; Fig. 14.1) of the Bar (interior) Palas Valley (1000 km^2) in 1993 an agreement was signed between the tribespeople and the project which, in exchange for rehabilitation assistance from the project, formally recognised the concept of Palas becoming a special area for conservation,

Fig. 14.1 The jirga (tribal council) in the interior Palas Valley, Pakistan, which resulted in the signing of a conservation and development agreement. In exchange for assistance with rehabilitation following floods, it was agreed that the area become a special area for conservation, all hunting was instantly banned and formal dialogue would be initiated to improve the forest management (photo: G. Duke, Birdlife).

agreed to formal dialogue on improved forest management and an instant ban on all hunting. Although there was considerable flood damage in the Palas Valley, the damage was much worse in other more heavily deforested valleys, leading to a government moratorium on timber extraction. The conservation success of this project has thus been dependent upon the trust and goodwill between the conservation project and the local people built through development assistance and the emergency aid package.

A strategy for managing protected areas may be to combine strong protection with assisting local development for those whose livelihoods are affected, with the aim of improving their conditions and reducing conflict. It is essential that the links between the development and conservation are made clear.

14.2.5 Benefit sharing

In some cases individuals have become displaced by parks or have had their rights curtailed. If the park makes a profit from entrance fees, then it only seems fair and moral that the local community gets a share, both as compensation for their losses and to encourage a positive attitude towards the park. Solutions include distributing a proportion of the park fees to the local community or investing locally, such as in schools and clinics.

Benefit sharing is not, however, as straightforward and inevitably beneficial as it seems. One problem is that benefit sharing or development may lead to increased expectations and immigration to take advantage of the improved conditions. This may increase pressure on the park, especially if the project falters due to funding, personal or political problems. Another common problem is the lack of an explicit link between the benefit sharing and the conservation objectives so that there may still be resentment and infringements. Simply allocating profits to the local people may not lead to the desired alleviation of poverty if they are not widely distributed and in practice benefits are often distributed very unevenly, usually to individuals that pose negligible threats to the park (Wells & Brandon 1993). The poorest individuals and the landless are those that are particularly likely to gain from exploiting or degrading the area yet are least likely to gain a share of the benefits. It is likely that effective benefit sharing leading to conservation gains will require strong local institutions, an appropriate legal framework and an understanding of incentives and the cause of degradation (Alpert 1996).

14.3 General principles for integrated conservation development projects

It is important to have a clear understanding of how the development may benefit conservation. It is even possible that there may be conflicts between the objectives of participatory development and conservation. For example, it is often likely to be profitable, particularly in the short term, to exploit those resources that conservationists seek to protect. One problem is that most people and organisations have experience in either conservation and development and thus projects tend to be either conservation or development projects.

After reviewing World Wildlife Fund's experiences with integrated conservation development projects, Larson et al. (1997) made the following ten recommendations:

- Ensure that integrated conservation development projects focus on biodiversity objectives.
- View integrated conservation development projects as one tool within the region's conservation strategy.
- Seek consensus on conservation agendas among key interest groups.
- Address external factors.
- Support integrated conservation development projects over the long term.
- Plan, monitor, learn and adapt.
- Build on what exists.
- Clarify who controls what.
- Work in strategic partnerships and act as a facilitator.
- Generate economic benefits for local people.

Although the emphasis is usually on local development, it is important to ensure political commitment at higher levels, such as the regional government, political figures and authorities.

Existing authority structures or government bodies may limit empowerment, especially when it may reduce the authority of local leaders or administrators. For example, it is often necessary to ensure that there is an administrative structure for community decisions to be acted upon and there may need to be legal changes to allow local control. Another common problem is that the authority of managers of protected areas is restricted to within the area boundaries, making it difficult for them to be involved in development projects involving those affected by the park but living in adjacent communities. There may also be conflict between park authorities and other government agencies (McNeely 1995), which can be manifested in disagreements over money or policy. Constructive dialogue and public support for the project can help reduce such differences. High-level political support may give the authority and capacity to resolve local issues and can lead to more effective enforcement of legislation by local government (Wells et al. 1999).

Small-scale projects are more likely to be successful (Chambers 1993). Ambitious, complex projects provide too many opportunities for failure, not least due to the high demands on local institutions and leaders. It is at least sensible to start with small groups and small problems, advancing to larger problems once confidence and knowledge has increased.

Smaller, cheaper projects also have the advantage of attracting those committed to change. Rich, large projects with vehicles, high salaries, offices, equipment and excessive funding often attract people for whom conservation and development are not priorities (Bunch 1982). Many development projects have the intention of creating a self-reliant project but, by providing excessive money, just increase dependence (Burkey 1993). The CAMPFIRE project of Zimbabwe (see Section 14.2.2) has largely been successful in spreading the profits of hunting and tourism and in re-establishing a spirit of local independence with responsibilities. However, Bird (1995) suggests that the provision of considerable external aid to parts of this project was counterproductive by resulting in a loss of local control and squabbles over spending.

Combining development and conservation in Kilum-Ijim Forest, Cameroon

Kilum-Ijim Mountain Forest project is a Birdlife International project started in 1987 with the objective of protecting one of the largest blocks of montane forest in Western Africa. The forest is the only remaining location with viable populations of two birds: Bannerman's Turaco *Tauraco bannermanni* and Banded Wattle-eye *Platysteira laticincta* (Coulthard 1996) and is the only known location for several plants including the orchid *Disperis nitida* and the toad *Xenopus amietti*. The forest had no legal protection, was in a densely occupied area with 200 000 people within a day's walking range (Thomas in press) and was being rapidly destroyed.

The forest is important for the local community. It is a source of medicines, food, and other produce and is important for water conservation. The forest plays a major role in many aspects of the culture of the local communities. Primary feathers of Bannerman's Turaco are collected from the forest during moult and are worn by Chindohs (traditional council members). When someone dies birdsong is mimicked on the Njang (a form of xylophone) for three days.

The approach used by the team from Birdlife International, in collaboration with the Ministry of Environment and Forests, was to discuss the issues with the local people and find mutually compatible solutions. The major reason for the loss of forest was the need for new farmland due to the deterioration of current farmland from soil erosion during the rainy season. A soil conservation programme was established by introducing new techniques to construct ridges along contours and then stabilising these with trees and perennial crops. Fallow crops were planted to improve fertility.

A second major problem was that fires are used to promote fresh growth and clear areas but accidentally spread. Fires were a major source of conflict between the farmers and the graziers, who are different cultural groups, with each blaming the other. The project facilitated co-operation between these two groups, first through a series of meetings with each group and then by encouraging joint meetings (Gardner *et al.* 1999). The eventual result was a group of three farmers and four graziers who patrolled for fires and created fire breaks. The two groups have subsequently collaborated on other issues.

Honey production was encouraged as it both provides income and is dependent upon the existence of the forest. To reduce pressures on the forest there has been an introduction of new crop varieties, afforestation, and new techniques in livestock health and husbandry.

The project brought groups of community, local authority and project representatives together to agree on a definitive boundary to the project. Although the final decision was unpopular with the few individuals that had created farms within the forest, the community agreed that the boundary was for the common good. The forest has now been formerly designated as Cameroon's first 'community managed forest'

Birds are an important part of traditional culture. This juju dance team are all wearing masks incorporating model Bannerman's Turacos (photo: Mark Edwards, Birdlife).

(Continued)

(*Continued*)

---

under their 1994 Forestry law, in collaboration with the Ministry of Environment and Forests. The original project was planned to cover both Kilum and Ijim forest areas but the respective occupants, the Kom and Oku people, had a major land dispute so a project based in one could not be extended to the other. The project is now organised out of two bases.

The project includes systematic monitoring of various measures of forest condition and wildlife populations. Each year a week is dedicated to reviewing the project and considering future plans.

The success of this project seems to be because the local culture is strong and traditional leaders are respected, and because the conservation organisations found development projects that have clear conservation and development gains.

Many projects make changes almost immediately after arriving in the area but, as the project develops, the realities and problems become clearer and it is either necessary to make major adjustments, which can cause disillusionment, or persist with a clearly unsatisfactory approach. It is usually better to research for some time (e.g. 6–18 months) and introduce ideas gradually so that they can be tested and modified using a repeated cycle of analysis, action and reflection (see Section 14.4). To establish the capacity for communities to participate in their own development, a commitment to long-term involvement (at least a decade) is likely to be needed. Against this there is often a requirement from funding agencies for early results, there is a sense of urgency within the project to deal with urgent threats and there is the need for the project to gain respect by tangible results. These may result in bypassing participatory processes with the risk that in the rush for short-term results the opportunity for long-term participatory development is lost. One solution is to introduce a small scheme soon after starting to gain interest, support and respect while continuing research towards the main long-term project. Once the project is running there is a need to encourage feedback on the effectiveness of the programmes by frank comments from participants.

14.3.1 Participatory development

The two overwhelmingly important causes of poverty are a lack of access to earning opportunities and an inability to take advantage of these opportunities (World Bank 1990). In theory, removing poverty is thus straightforward by creating the opportunities for households to put their labour to good use and ensuring that households have members who are healthy, educated and skilled. However, it is now widely realised that simply providing potentially profitable schemes, a clinic, a school and training by no means ensures that poverty is removed.

It might seem that the most effective way to remove poverty is for a team of experts to collect data, go away, analyse it, draw conclusions and then suggest to the government or aid agency a solution that should be imposed. However, such practices tend to fail. The dissatisfaction with such imposed approaches to development and their high failure rate has led to a radical shift in methods used. It is now realised that development is only likely to be a success through the full support of the community. A common approach now is for communities to analyse their own problems, decide priorities and draw their own conclusions with the aim of empowering local people to decide their future and produce changes that they consider necessary and desirable (Burkey 1993). There has also been increased appreciation of the importance of local knowledge, cultural norms and social institutions, both for ethical reasons and to increase the likelihood of success (Adams 1998).

The emphasis is now on enabling people to do things for themselves rather than doing things for people. Burkey (1993) suggests that if the poor are to manage and control their own development then they need to gain self-confidence, learn to be assertive, develop faith in their abilities and trust in their comrades. However, many aid programmes achieve exactly the opposite by convincing the poor that they are ignorant, backward and helpless as they require foreigners and strangers to help them progress. Unless development workers respect the abilities of those they are working with they are unlikely to succeed. Self-confidence is particularly likely to be broken by outsiders expecting rapid progress.

Even experienced development workers find it hard to establish participation. A project using participatory techniques in India ended in conflict and two deaths (Shah & Shah 1995), while in Cameroon the community representatives were treated as scapegoats for problems, hounded out of the community and imprisoned (Brocklesby *et al.* 1999). Conservationists without development training are likely to encounter difficulties. Local participation can occur at a number of stages: information gathering, consultation, decision making, initiating action and evaluation. Although virtually all of the 23 major integrated conservation–development projects reviewed by Wells & Brandon (1993) stated a commitment to local participation, in most cases there was very little genuine local participation and where there was participation it was usually in just one or two of the above stages. Thus participation needs to be thought through carefully, ideally with guidance from experienced development workers.

Those involved in implementing an integrated conservation development project must be involved in its creation. Planning is more likely to be successful if it fully involves those affected. An early stage must be to identify the affected groups (often called 'stakeholders') and incorporate them in the process. Groups need to be sufficiently cohesive and committed before starting an activity and are more likely to be successful if motivated, voluntary and having a common interest. Marginal groups such as the poor, the landless or newly arrived migrants often have the greatest impact on natural resources but are easily excluded. Women need to be sufficiently represented.

Participatory development usually requires direct involvement by those setting up the project, such as staying in the community in order to be accepted and

be seen to care. Short-term visits are less likely to result in genuine participatory development.

Many suggest that development should primarily consist of providing the opportunity for groups and communities to make decisions and carry them through (Burkey 1993). While accepting that this may often be the best approach for a pure development project, conservationists are, however, only likely to invest in a project having established that there is a conservation problem and that it may be alleviated by an integrated conservation–development plan and thus usually enter a project with proposals and suggestions.

14.4 The project cycle

The approach almost universally used by development agencies (e.g. Overseas Development Administration 1995) is project cycle management (see Fig. 14.2), which simply describes the sequence of identifying a project, designing it, monitoring and evaluating it so that the lessons learnt can be applied to improve future projects. Thus the Lak Project in New Ireland, Papau New Guinea, attempted to gain customary landowner support for a sustainable forestry project while competing with an already running commercial logging company. It failed, but the frank evaluation (McCallum & Sekhran 1997), which

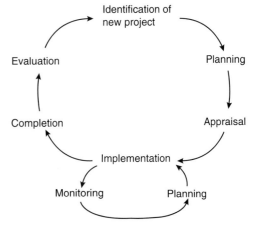

Fig. 14.2 A project cycle for running integrated conservation development projects. Note that planning is not restricted to the start of the project but also occurs in response to the results of monitoring.

CASE STUDY

Coral reef and fisheries management in the Philippines

The coastal areas of San Salvador Island, Philippines, faced familiar problems: overfishing, sedimentation from deforestation, and dying coral due to fishing with explosives and cyanide. Cyanide stuns the fish so that they are easier to catch but also kills the coral, other fish and invertebrates and is a health hazard. Over two thirds of the coral in the Philippines is damaged by repeated use of cyanide. Although the problems on San Salvador were obvious, the communities were poorly organised and the local authorities showed little interest, so conditions continued to deteriorate.

The Marine Conservation Project for San Salvador was started in 1988 with the objective of setting up a community-based marine park to prevent coral reef destruction and improve the fishery (Buhat 1994). A peace corps volunteer spent the first year living in the community, learning the language, determining their needs and becoming accepted and trusted. He surveyed coral damage by snorkelling, presented the coral damage data at various council meetings, made contracts with related organisations and made a successful funding application for a community-based project with the Haribon Foundation (a Philippine environmental non-governmental organisation) being the implementing agency. A formal ceremony to mark the beginning of the project in December 1988 was attended by guests from supporting organisations and 150 residents.

A community organiser and two field officers were employed on the project and together with the peace corps volunteer made socio-economic, land use, resource use and fish yield surveys. All the data was presented to the community. The project encouraged community development through the creation and support of core groups, who were responsible for managing the marine resources and generating alternative income. The core groups constructed both a guest house and a community centre which increased support for the project.

The success of the project was dependent upon the involvement of various local organisations: the Department of Agriculture, the Haribon Foundation and the municipal government. Part of the project involved providing training and support to these organisations to improve their capacity to assist.

Education was an essential component of the project. As well as monthly slide shows, role playing and lectures, there were children's outings and an environmental art competition. The project paid for seven San Salvador fishermen to visit a similar community sanctuary project on Apo Island, Negros. As a consequence of their experiences, the fishermen set up their own environment management committee, the Lupong Tagapangasiwa ng Kapaligiran (LTK) which encouraged others to participate in the project.

Due to a series of community assembly meetings, a resolution was drafted to create a 127-hectare marine sanctuary with no fishing at all while dynamiting, sodium cyanide and beach seines would be banned from around the rest of the island due to their impact on coral. In July 1989 the LTK and field workers presented the resolution to the island council who passed it. The regulations were sent to all villages for all 255 households to sign which practically all did.

The only exceptions to signing the resolution were 31 households in the village of Cabangun. This is inhabited by aquarium fish gatherers from the southern Philippines who had not fully integrated into the resident Sambal community. They usually use sodium cyanide to catch their fish and previous disputes over this method had further increased their isolation. They were in favour of the sanctuary but wished to continue to use sodium cyanide elsewhere. As it

(Continued)

(Continued)

became clear that the alienation of this group was becoming a problem there were particular efforts to fully incorporate them within the LTK and the island council by running several education and training programmes, and efforts were made to ensure that their fishing rights were respected. As a consequence the Cabanguns became supporters of the proposals.

Training courses were provided by the Haribon Foundation for those using cyanide to learn how to use barrier nets, horseshoe-shaped fine nets placed around a group of coral heads, into which the fish are chased with those required being caught with hand-held nets while those not required can escape. A consequence of this training was the establishment of the Cabangun Aquarium Fish Gatherers Association to ensure that nets were used and improve the profitability of fish gathering. As well as being more environmentally acceptable, captured fish have higher survival if caught by nets than by cyanide so the Association succeeded in agreeing on higher prices for net-caught aquarium fish, which further encouraged the spread of this technique. On a larger scale, the Philippine Government, assisted by the International Marine Alliance, has introduced a cyanide testing laboratory at the main airports. The aim is to introduce a certification scheme to create a demand for fish caught in an environmentally responsible manner. Barrier nets are also much cheaper, about US$25 for a net compared with over US$500 a year for sodium cyanide.

The sanctuary and reserve was marked with buoys in October 1989 and a guardhouse was built. Some illegal fishing has taken place but this has largely been by outsiders unaware of the ordinance. A few were fined but most were just given a warning. The level of dynamite fishing dropped from an average of 3.2 incidents per day during the main season before the project to just one recorded incident since the reserve opened.

There was a range of other components to the project. Giant clams had been overharvested, so areas were seeded and training was given on management. An erosion control programme was set up to plant trees along an eroding road and replant mangroves. A two-day training course on making artificial reefs resulted in two reefs being made inside the reserve and a further 19 made outside for collecting aquarium fish.

Within 2 years of the start of the project there was an increase in fish catch. The Haribon Foundation withdrew in 1994, leaving the local community and the elected management group to maintain the project. The success of the project has been a result of the clear tangible benefits, the acceptance that the community has complete rights over the resource and the collaboration of the local community and local organisations, for example in checking for illegal fishing. The success of this project has inspired many others.

An LTK member discussing the sanctuary with other members of the community (photo: Patrick Christi).

identified a number of issues such as the importance of analysing socio-economic factors to examine if the proposed project is feasible, means that subsequent projects can learn from their experience. The Bismark-Ramu Project, in Papua New Guinea, built explicitly upon the lessons learnt and resulted in the traditional owners designating over 120 000 hectares of land to special conservation status (Ellis 1999). However, in general, the experience of previous projects is greatly underused.

Most aid agencies have their own terminology and approaches for project planning, although these are normally variants of the following methods. If asking for funding from an agency it is important to use their jargon and methodology. Most project cycles include the following stages:

14.4.1 Identification

Projects are often identified following a perceived problem with a species or wildlife community considered to be of high priority. Proposals for integrated conservation development plans usually arise once priority conservation areas have been identified and a conflict has been discovered. Further research is carried out to consider whether it is possible that an integrated conservation development project may help. This stage usually involves considering what could be done, who will be involved and who might finance it. Typically, many more projects are identified as potential projects than are carried out.

14.4.2 Planning

Poorly planned projects are often ineffective or inefficient. Rigorous project planning has the advantage of helping overcome three of the major reasons for failure of development projects: a tendency to concentrate on the project (e.g. creating a fish farm) while losing sight of the goal (conserving biodiversity or reducing poverty), not considering carefully the reasons why the project might fail and not having a sensible process for making decisions (Eggers 1998).

For planning it is necessary to understand the existing institutions, culture and constraints. It is usually easier to modify existing use systems than try and establish a completely unfamiliar pattern (Gardner *et al*. 1999). Complex integrated conservation development project plans are usually ignored and need instead to have the flexibility for people to respond (Wells *et al*. 1999).

The ZOPP (zielorientierte projektplanung = objectives-orientated project planning) methodology (GTZ 1997) was first introduced in 1983, has been very influential and its principles have been widely used. In the ZOPP methodology the creation of the logframe (see Section 14.5.6) drives the whole planning process.

The ZOPP methodology is as follows.

1 Participation analysis. This stage decides who should be involved (see Section 14.5.3) and whose interests should be given priority.

2 Problem analysis. A problem tree (see Section 14.5.4) is used to review the major problems, their causes and consequences.

3 Objectives analysis. An objective tree (see Section 14.5.4) is created to analyse what can be done to remove the problems.

4 Discussion of alternatives. Decisions are made, by using options analysis (see Section 14.5.5), about which approach is to be adopted or even whether it is better to do nothing.

5 Project planning matrix. Decides upon and describes necessary activities, the logic linking these activities to the goals and the necessary monitoring. This is all done using logical framework analysis (see Section 14.5.6).

It is then necessary to identify and allocate tasks (see Section 14.5.8 for methods).

Most other development agencies use similar planning structures and similar components. The Margoluis & Salafsky (1998) approach to planning (see Box 14.1) is also similar but uses very different means of presentation and a different terminology.

Planning is often completed at a workshop which should involve those that will carry out the work and should be multidisciplinary. Such a collaborative production is likely to result in both a better plan and greater commitment to the project. Chapter 7 describes planning principles in more detail.

14.4.3 Appraisal

The project is reviewed to decide whether to proceed, based on the likelihood of success, the costs, social implications and whether the project is compatible with existing institutional structures. If applying for outside funding this is usually a formal process. If funding is already available it is still sensible to carry out a full appraisal to examine the quality of the plan and decide whether the investment is sensible. The project design may be altered as a result of appraisal. In

BOX 14.1

The conceptual model approach to planning projects

Margoluis & Salafsky (1998) give a detailed account of a planning process which may well become widely used by conservationists. Many of their ideas are already in the development literature, for example their conceptual models are really just a different way of formulating problem trees and objective trees (see Section 14.5.4). The main difference is that most of the development agencies use objective trees to devise logframes (Section 14.5.6) while Margoluis and Salafsky simply add further layers (objectives, actions and monitoring) directly onto the conceptual model. I suspect that many will find their approach considerably clearer. However, the terminology and details of Margoluis and Salafsky differ from the methods usually used by funding agencies, yet it is they that often dictate the presentation used.

1 Decide upon the conservation target

This should describe the general conservation aim of the project. For example 'maintain Fictitious Reef at its current boundaries and retain associated species' or 'restore populations of species Y'. If it is a conservation project then it should start with a clear conservation aim and the necessary development work, if any, will follow.

2 Collate all available relevant information

This requires collating published information, reports, government records and local knowledge. It includes ecological data on habitats, species and threats as well as social and economic information (see Section 14.5.1).

3 Develop a preliminary conceptual model

A conceptual model describes the relationships between factors and how these impact on the conservation problem (Fig. 14.3) and is thus very similar to the problem tree (Fig. 14.5).

Write the conservation target (e.g. 'maintain area of coral and fish populations at Fictitious Reef') to the right

of a blackboard or large sheet of paper. Add all the direct threats (e.g. dynamite fishing, tourist boat anchors, sediment deposition) and join with arrows to the conservation target (see Fig. 14.3). Creating a model is often easiest by writing the threats on cards or adhesive memo notes so that they can be altered. Then add all the indirect threats (e.g. deforestation), which are those that underlie or lead to direct threats, and link them to the threats (sediment deposition). Next add contributing factors which impact on the process but are not threats.

As well as being useful for planning, the conceptual model can also be useful for explaining to others the logic of the approach. It is then likely to be useful to take a break and review the model, perhaps clarify issues, collect further data, review more literature and then return to refine the model.

Once the model is agreed it is useful to convert it into a written text. This is best achieved by describing the conservation target and then working towards the left through the model describing the stages, the links and the evidence for the links.

4 Consult fully with the local community over threats

The advantage of waiting until this stage for full consultation is that the project management group is reasonably informed, does not waste time in irrelevant discussions and can give a more confident image. The risks of delaying full consultation are that the group may already feel committed to their analysis so that the consultation is only cosmetic, the community may feel less responsibility for the project and the community does not gain the full experience of developing a project. Ask a range of individuals or groups to state what they consider to be the threats and compile a list. If stages 2 and 3 produced some additional threats then add these to the list. The community should then be fully

(Continued)

(*Continued*)

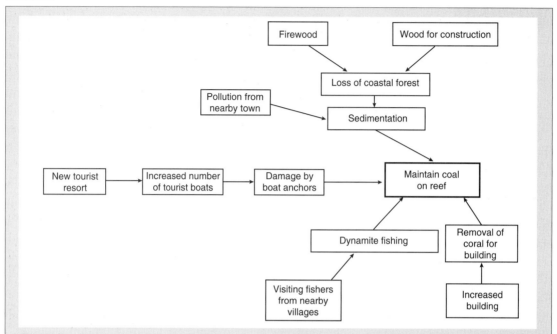

Fig. 14.3 A conceptual model of the threats to the coral on Fictitious Reef. This lists the target in the right-hand box. The direct threats are identified, such as sedimentation. Indirect threats such as felling coastal forest are then added.

consulted on which threats they consider most serious, for example using matrix ranking (see Section 14.5.2). Revise the conceptual model including the community's views on threats. Either the model can be presented to the community or they may be asked to build a new model. Once the model is agreed it is useful to convert it into a written text.

5 Decide upon priorities for action

List the threats and rank according to criteria such as: perceived importance by community (from stage 4), size of area affected, intensity of impact, urgency of the problem, political feasibility, social practicality and ability of your group to tackle the threat. This table is then used to determine the priorities for action. A total score is a useful guide but should not be used as an absolute measure. For example, a threat that is politically impossible to deal with but considered most important by all other criteria may achieve the highest overall score or, as in Table 14.1, the community may consider sedimentation to be unimportant but

the working group, having examined the literature and other sites, is convinced of its importance. Discrepancies between the importance perceived by the community and the conclusions of the group should be discussed with the community and resolved.

6 Agree upon objectives

Objectives should be specific, measurable, time limited and practical. The 'ends objectives' should be able to determine if the project is a success in terms of the conservation target. Thus the ends objective might be to 'restore the population of species Y to 600 individuals by 2005'. These are added to the conceptual model.

7 Decide upon strategies

By reference to the conceptual model, consider the broad policies that are likely to reduce the major threats. Thus these might be to ban the use of anchors on the reef, establish permanent moorings and charge tourists for using these. The broad policies are usually considered

(*Continued*)

(*Continued*)

Table 14.1 Direct ranking of the threats to the Fictitious Marine Reserve. Criteria are ranked between 7 (most important) and 1 (least). The final order is ranked between A (most important) and G (least).

Threat	Community perceived importance	Area	Intensity	Urgency	Political feasibility	Social practicality	Organisational ability	Total score	Final rank
Dynamiting	7	3	6	7	5	6	6	40	A
Sedimentation	1	7	5	5	3	3	5	29	B
Boat anchors	4	2	4	4	7	7	7	35	C
Coral removal for construction	6	1	7	6	2	4	1	27	D
Poisoning	5	4	3	2	1	2	3	20	F
Tourist spearfishing	3	5	2	3	6	5	2	26	E
Overfishing	2	6	1	1	4	1	4	19	G

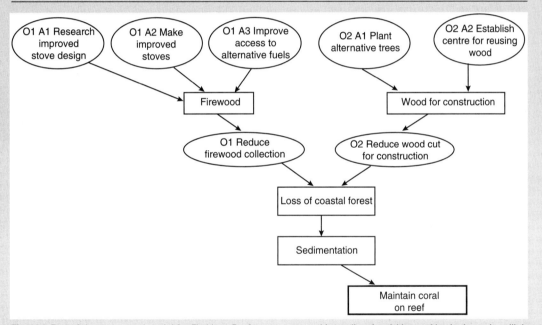

Fig. 14.4 Part of the conceptual model for Fictitious Reef with objectives added and the activities needed to achieve the objectives then added. This conceptual model is then converted into a list of activities, making it clear who will do what and when. The monitoring of the objectives should also be agreed. This can also be added to the model.

in conjunction with producing specific, measurable, time limited and practical objectives ('means objectives') such as 'build at least three moorings by the end of the first year' or 'ensure 90% of tourist boats use the permanent anchors within 6 months of the moorings being completed'. The means objectives are then drawn into the conceptual model (Fig. 14.4).

8 List actions

These are the precise activities to achieve the strategy. The timing, cost and duration need to be calculated. In a

(*Continued*)

(*Continued*)

table, write down what will be done, how it will be done, when it will be done, who will do it, where it will be done and comments. The actions should all lead to means objectives and are drawn in the conceptual model.

9 Agree on monitoring programme

Monitoring (see Section 14.4.5) of the ends objectives could include fish abundance surveys or data from fishers. Monitoring means objectives could include the width of mangrove along ten transect lines, the percentage of tourists that use the permanent anchors in monthly surveys or the number of dynamiting incidents reported by the fishing community. These can be tabulated under each objective. Under each give the indicator that will be monitored, how it will be monitored (including the methods and tasks to be done), when monitoring will take place, who will do it, where it will be open and finally any comments giving further details. It can be useful to prepare a Gantt chart just like Table 14.4 that summarises when each task will be done and by whom. The tasks to be done can be listed under each objective.

10 Run the project and review the successes and failures and modify

See Section 14.4.5.

---

many organisations initial appraisal of an outline plan may be followed by a stage in which detailed plans are created, followed by a stage of final approval.

14.4.4 Implementation

This often starts once the major parties agree on the plan and they may sign a formal agreement.

14.4.5 Monitoring

It is useful to distinguish between monitoring and evaluation. Monitoring involves analysing the current situation in order to improve the existing programme. Evaluation is the process of analysing the entire completed project to consider the successes and failures in order to improve future projects.

Monitoring can be time consuming and if not carefully planned may not be useful. Monitoring is described in detail in Chapter 4. Monitoring can be used to:

• Determine if the planned activities have taken place (e.g. education officer employed).
• Determine if these activities have resulted in the expected outputs (e.g. education poster produced and distributed).
• Determine if the outputs have had the expected consequences (e.g. reduction in hunting protected

species). This is sometimes called 'purposes', 'intermediate objectives' or 'mean objectives'.
• Determine if the main conservation goal is being achieved (e.g. increased abundance of endangered primate). This is also sometimes called the 'ends objective' or 'wider objective'.
• Determine if the project has other negative impacts (e.g. has hunting increased on other species).
• Determine if other factors have influenced the situation (e.g. forest fires reduced primate habitat, crop failure which increased demand for meat).
• Answer research questions (e.g. was the programme more effective in villages where the programme targeted children or the village elders?).
• Provide information to convince others, such as policy makers. This will usually be already covered in the other uses.

All monitoring is usually directed at one of these functions. A sensible approach is to list what needs to be monitored, for example by listing the objectives (from a logframe, see Section 14.5.6) or conceptual model (Box 14.1) and then listing what needs to be known to understand whether each is being achieved, and if not why not. It is easy to create an overelaborate monitoring scheme that takes up too high a proportion of the time or is abandoned. The procedure in Section 5.2 could be used to plan the monitoring programme.

It is often impossible or impractical to make the desired measurement and it is thus usually necessary to use indicators as surrogates for the desired measure. For example, ideally it would be useful to know the size of a sea turtle population, the extent of removal of turtle eggs and number of nests accidentally destroyed by tourists on the beach but these may all be impractical to quantify. Indicators could then be the number of turtles visiting the beach on a standard number of nights, the proportion of mapped nests that are dug up during the period of incubation and the numbers of beach umbrellas used by tourists on the critical areas of beach while the turtle eggs are present. Indicators need to be measurable, precise, consistent and sensitive to actual changes. In each case it is necessary to devise precise monitoring practices and write them down so they can be followed precisely.

Monitoring needs to be fully planned and tasks allocated (for example using the methods in Section 14.5.8). Specify exactly how the data will be collected, when, by whom and how it will be analysed. There is an increasing use of participatory monitoring (see Section 14.5.2) as it is not only often more cost effective but especially as it leads to direct feedback to those who can act on the results. It is probably best to gradually build up the participatory monitoring with increasing experience.

Monitoring should be an integral part of the decision-making process. It is usually useful to have regular opportunities to consider progress and problems, for example with 3-monthly reviews at which data are presented and notes are taken of the conclusions. The collected notes from such regular meetings also make it easier to produce an annual report.

Systematic analysis of experience

Reporting and monitoring usually focus on goals and finances. Field workers rarely have the opportunity, time or mechanism for sharing their experiences, yet it is usually they that have gained the practical knowledge and learnt lessons as to what works and what does not. Systematic analysis of experience (SANE) is a simple method of learning from projects (IUCN International Assessment Team 1997).

1 **Tell the story**. One staff member describes the experience of the project as a story while a facilitator records it on a flip chart. Discussion is encouraged to refine, dispute and correct the story. Gaps in the knowledge and disagreements are documented.

2 **Identify turning points**. Analyse the story to determine changes and why these occurred.

3 **Identify phases of experience**. Intervals between turning points can be called phases and it is useful to name these after distinguishing features.

4 **Phase analysis**. Within each phase analyse the main issues such as objectives, hypotheses, activities, participants, methods, successes and failures.

5 **Analysis**. Compare phases to identify changes and the causes and consequences of changes. Identify trends and evolution of ideas and hypotheses.

6 **Lessons learnt**. It should be straightforward to synthesise the lessons learnt in terms of what should or should not have been done.

7 **Communication**. Record and circulate to those that would benefit.

14.4.6 Completion

This is the end of the provision of external funding or support; for most projects the expectation is that the project will continue and will be self-sustaining. This is often only likely to be successful if capacity building (see Section 14.6) has been an integral part of the project.

14.4.7 Evaluation

As well as monitoring the project as it progresses and modifying accordingly (see Secion 14.4.5) it is useful to evaluate the success of the project once it has ended. The actual progress is compared with the plans and the decisions and actions that were taken are reviewed and ideally made available to others. The lessons learnt from the successes and failings of this project can then be used to improve subsequent projects.

A problem is that everyone is looking for success. The community is likely to want continued funding, the project team wants a reputation of doing their jobs well, while the donors wish to show that they have spent their money wisely. Rigorous objective monitoring is thus important.

14.5 Basic methods for integrated conservation development projects

14.5.1 Key questions

It seems obvious that it is necessary to understand why a conservation problem is occurring, yet many integrated conservation development projects begin with a very limited understanding of the root social, economic, cultural or political causes of the threats. It is also important to understand the lifestyle, problems, differences and political realities of the community and the structure of institutions. It is useful to know the environmental and social conditions at the start of a project 'baseline condition' to see how they change. Some key questions are:

• **What resources are present?** What is the variation in soil, climate, water and natural resources across the area?

• **Why are people here?** What is their history? What are the various ways in which people make a living? Why does this vary between areas and groups? Which food and other products are for the use of the community and which generate money from outside?

• **What are the religious and cultural practices?** Does this vary between groups and does it have social or economic consequences?

• **What are the major social divisions?** What internal divisions are there? Who are the advantaged and why? Who are the disadvantaged and why? Are the interests of the various groups similar or contradictory?

• **Who owns what?** What are their constraints on use?

• **Who holds the power?** Who makes the decisions? What are the political divisions within the community? What is the role of external individuals and institutions?

• **What are the social conditions and how does this vary between groups?** Is the health, nutritional status, literacy, education, hygiene and sanitation satisfactory for all? If not why not? Which of these contribute to continuing poverty?

• **What are the causes of poverty?** What do the poor consider to be the causes of their poverty? What else might contribute to their poverty? What prevents them from escaping poverty?

• **What is the use of species or areas of conservation importance?** Is there conflict with the conservation interest? Which groups or individuals are involved and to what extent? Where, when and why are the species or areas used? Is this done for profit or personal use? Are there alternatives?

• **What changes have there been?** What major social, economic and ecological changes have there been, say over the last 20 years, and why? Has there been deforestation, erosion, drought or declines in exploited populations and what have been the social and economic consequences?

• **What services are available and used?** This should include what other projects are working in the area, what are these doing and who uses them.

It is obviously usually unacceptable, and unlikely to be useful, to arrive and immediately ask direct questions about income, poverty and power, especially as outsiders will often be treated with distrust. One suggestion for making questioning acceptable is that by examining the past it is easier to contemplate future changes (Fuglesang 1982). This could involve asking: how did your ancestors do it? Why did they do it that way? What are you doing now? Why are you doing it like that? What is the current situation? How can it be improved?

14.5.2 Participatory research and monitoring techniques

The traditional means of formal surveys and questionnaires of social and economic conditions are often now accompanied by more informal techniques which encourage participation, so that everyone learns and the responsibility for any planned changes is shared. Rapid rural appraisal uses a range of interactive methods to gain insight and knowledge from local people. Participatory rural appraisal is a process by which people unravel and analyse their own situation and perhaps even plan their future (Chambers 1997). Participatory techniques should compliment conventional analyses such as questionnaires and surveys of land use.

The following are some of the most useful techniques (Mikkelsen 1995, Jackson & Ingles 1998, Margoluis & Salafsky 1998, Guijt 1999).

Building rapport. Not only is building rapport and trust good manners but it reduces suspicion, making contact easier. Techniques include first meeting with village leaders and officials, starting with those that are more likely to be approachable such as shopkeepers and village health workers, explaining the reasons of the visit, choosing times and locations that are convenient for local people and ensuring that the men in the village understand the motives for wishing to talk to the women.

Cross-checking. This is also known as triangulation and is achieved by using different sources or direct observation to check the accuracy of information. This should be a component of any of the following methods.

Semi-structured interviews. These do not use a fixed series of questions but usually have a series of topics that will be used for probing the interviewee so that they may respond more freely. This leaves the opportunity for the interviewer to follow issues raised during the interview so that the results are often more illuminating than if following a strict questionnaire. It also makes it easier to find ways of approaching delicate issues and the relaxed nature of the discussion makes it more likely that sensitive information will be revealed. It is important that the topics are clearly thought out to avoid the exercise simply becoming a pleasant chat. Questions are open ended such as 'where do you hunt?' rather than 'do you hunt in the protected area?' The responses are written down. Opinions should not be given by the interviewer as these may affect the responses. After each interview the topic list may be modified to include new issues raised. Semi-structured interviews cannot be analysed as rigorously as questionnaires.

Key informants. These are individuals who are particularly knowledgeable about a particular issue or subject. For example this might be the local leaders or experienced hunters. Semi-structured interviews of key informants can be a very useful way of gaining an understanding of some important issues. However, their opinions may not be representative.

Focus groups. This involves assembling a group of individuals with similar backgrounds. A process very similar to a semi-structured interview is carried out, again using a topic list. This is a quick and effective method of assessing a wide range of opinions. However, the presence of others may bias the responses and this material is more difficult to interpret than a group of individual interviews.

Semi-structured walks. These involve walking with key informants around a range of sites, such as various wetland areas, to determine their uses, who uses them and how the habitat has changed. The methods are very similar to those of semi-structured interviews.

Mapping. This is the creation of a map of the area, using the medium of choice of the participants (e.g. pen and paper, scratches in the sand, sticks and pebbles). The map is used to clarify an issue of interest such as land ownership, land use or environmental problems. A paper copy is made and kept. Maps can be used for participatory monitoring by either preparing a new map or documenting changes by comparing with the previous map.

Transects. These are similar to semi-structured walks but used to provide more quantifiable information on land use. A route is taken to document land use or environmental problems. The information is usually stored as a cross-section of the route with sketches or notes plotting the distribution of the features of interest. One option is that information on different sites (e.g. on sizes, ownership and uses of different forest blocks) may also be tabulated on the sheet. The same transect can be walked at intervals to monitor changes, in which case the precise route needs careful documentation. It can be useful to take the previous transect results into the field to make it easier to determine changes.

Ranking. A common approach is matrix ranking, a method of creating a table of the preferences of individuals. All the conditions, items or perceptions are listed down the side of a table, such as the uses of a wetland and each individual states which use is most important, then which is next most important until all are ranked. The individual's ranking is recorded along with their name and other data such as that person's age, sex and occupation. The mean score for each use is then calculated. This data can also be analysed to see if there are obvious divisions in opinion. Another means of matrix ranking is to allocate scores, for example by placing different numbers of stones alongside each option. A further possibility is to analyse each option according to a set of criteria, for example when comparing different tree species the criteria may be growth rate, wood quality, ease of growing and fruit value. Thus each tree is allocated a column in the matrix and each criteria a row and each individual or group provides a ranking or score for each combination. As well as the actual ranking scores, much useful information is gained from discussions that ensue during the ranking exercise.

14.5.3 Stakeholder analysis

This is also known as participator analysis. This entails identifying the groups and individuals involved ('stakeholders') and what affects them (Table 14.2). Stakeholder analysis is a useful way of identifying and presenting this information (MacArthur 1997).

1 Determine all the individuals, groups and institutions whose lives may be affected by the conservation or development, who hold influence or who may be capable of affecting the project. This should

include both those who support the project and those who oppose it. It can be useful to divide them into *primary stakeholders*: the intended beneficiaries of the project; *secondary stakeholders*: those involved in the working of the project, including suppliers and users of products but who are not the intended beneficiaries; and *external stakeholders*: those interested in the outcome but not directly involved, such as government departments.

2 For each stakeholder, list the interests that relate to the objectives of the project or the problems it addresses. This is most easily done by considering all the links to the project in terms of expectations, benefits, investment, conflicting interests and attitudes to other stakeholders.

3 Consider whether the project is likely to have a positive (+), negative (−), neutral (0) or uncertain (?) impact on each interest. Alternatively scores can be used from −5 (highly detrimental) to +5 (highly beneficial).

4 Decide which stakeholders should be given priority (1 = highest, 5 = lowest) in meeting these interests while bearing in mind their influence and power. Thus the outcome of the analysis is a table with columns listing stakeholder, interests (can be more then one per stakeholder), impact (given to each interest) and priority (of each stakeholder).

14.5.4 Problem trees and objective trees

The idea behind planning trees (GTZ 1997) is to understand the processes that are causing the

Table 14.2 Part of a stakeholder analysis for Fictitious Marine Reserve. Types: 1 = primary, 2 = secondary, E = external.

Stakeholder	Type	Interest in the project	Potential impact	Priority
Local fishing community	1	Sustained fish catches	+3	1
		Profits from permanent anchors from tourists	+2	
External dynamite fishers	1	Dynamite fishing	−5	1
Construction companies	2	Use of coral for building	−5	4
Tourist operators	2	(a) Improved coral for divers	+2	3
		(b) Destruction of coral by use of anchors	−2	
Local government	E	(a) Facilitate institutional changes	+3	3
		(b) Permitting officers to collaborate in project	+4	

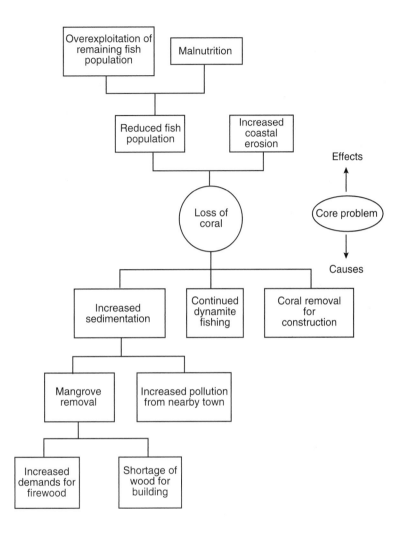

Fig. 14.5 Part of a problem tree of the threats to the coral on Fictitious Reef.

problems, what can be done to improve the situation and how this can be monitored.

The first step is to list all the problems as negative conditions (e.g. overgrazing, shortage of firewood). This is achieved by brainstorming (see Section 8.5.1) in which participants suggest problems each of which is written on a card stuck on the wall or placed on a table. After the brainstorming similar ideas are grouped together and any problems that cannot be tackled by the project are removed.

The problem tree (see Fig. 14.5) is easiest to create by selecting the 'core problem' that the participants

agree is of the greatest importance. The direct effects of this core problem are arranged in a line above while the direct causes are arranged in a line below. Those problems that are direct causes or effects of those in this row are added in a second row. This process continues until the network is considered to be the best possible analysis of the available information.

The objectives tree is created by rewriting the problems as positive conditions to be achieved in the future (see Fig 14.6). The linking within the tree should be checked to see that it follows logically so that each stage causes the next one.

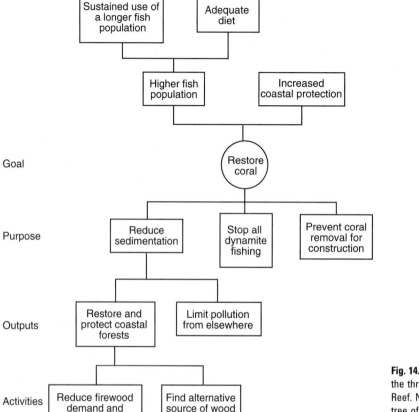

Goal

Purpose

Outputs

Activities

Fig. 14.6 Part of an objectives tree of the threats to the coral on Fictitious Reef. Note that this is the problem tree of Figure 14.5 rewritten as objectives.

The different levels of the objectives tree can then be classified as goal, purpose, outputs and activities as used in the logical framework (see Section 14.5.6).

14.5.5 Options analysis

The analysis of the problems and objectives may produce a range of approaches but these may be incompatible or choices may have to be made based on cost or time. The function of this process is to review the consequences of all options for action, including doing nothing. These options can be taken from the steps of the objective tree (Fig. 14.6). One approach is to list the options in rows of a table and label columns with those criteria considered most important. The criteria could include: expected con-

servation benefit, cost, social risks, likelihood of success, whether funding is available, available personnel, experience with the methodology entailed and development benefits to priority groups. Each option can then be scored on each criteria and this information used to select a strategy.

14.5.6 Logical framework analysis

Logical frameworks, usually abbreviated to logframes and sometimes called project frameworks, are a classical way of planning, managing and evaluating projects (Cusworth & Franks 1993, Mikkelsen 1995, GTZ 1997). The aim is to encourage a clear analysis of the project design so that its internal logic can be checked. They are sometimes completed as a check to see that the logic is correct but are probably more useful as a fundamental part of the planning process.

Many projects fail and it is usually easy to find someone or something else to blame. 'We provided the trees but the villagers failed to look after them.', 'The trees died due to unusually dry weather.', 'The plantation was a great success but the villagers continued to fell the forest.', 'The local government did not provide the promised forestry officer.' It is thus frequently said that the project was good and well planned but failed due to bad luck or a lack of support.

Logical framework analysis makes it much harder to shirk responsibility for failure as such assumptions are made explicit. The plan should consider whether villagers are likely to look after the trees, the probability of dry weather, whether a plantation is likely to reduce demand for wood and the likelihood of government support. If a project fails because of problems that could have been foreseen but were not, then the project has been badly planned. If the success of a project depends upon an assumption that is unlikely to be true (a 'killer assumption') then either the project should not continue or the project should be restructured to avoid this assumption. By making risks explicit it is possible to evaluate them and decide whether to proceed with the project or allocate the money and effort elsewhere.

Logframes divide the project into four stages, from the activities (what the project actually does, such as provide trees and advice) to the goal (such as maintaining the existing area of forest). The aim of using logframes is to find the most effective means of converting activities (labour, investment, training) into outputs (tangible physical or institutional structures) so as to achieve project purposes (objectives that are beneficial to conservation) and thus accomplish the project goal (the larger conservation objective). The simple stepwise logic of this process is that if the logic of the logframe is correct then if the activities are carried out as stated and the assumptions are true, the outputs must happen. If the outputs result and the stated assumptions are true the purpose must occur. If the purpose occurs and the stated assumptions are true then the goal must be achieved. It thus forces planners to consider why the project is being carried out rather than simply creating a programme that produces the outputs, regardless of whether these help towards the goals (MacArthur 1993).

The beauty of a logframe is that a single 4×4 table reveals the logic, assumptions and the necessary monitoring. Alongside each of the four stages of goal, purpose, outputs and activities the indicators are listed which, if achieved, show that the stage has occurred to the expected extent. Finally the table lists the methods by which these indicators will be determined. The simple logic means that flaws in the project design are usually readily revealed. Furthermore, if the project fails then the monitoring should reveal whether it is due to internal failure (poor planning, for example because the logic was faulty or an assumption was omitted) or external failure (one of the identified risks did occur).

The standard logical framework (Table 14.3) usually comprises four rows and four columns.

The four rows usually are:

Goal. The main conservation objective. Often known as 'wider objective'.

Purpose. What the project sets out to achieve. Also known as 'immediate objective'. A single objective is often best as projects with multiple purposes may lose direction.

Outputs. The specific achievements which the project will try to guarantee. Also known as 'results' or 'deliverables'.

Activities. What is needed to be done to achieve the outputs. Some versions of the logframe use 'inputs', what needs to be provided.

The four columns are:

Narrative summary. A description of the logic linking each stage to the next. This stage can be completed early in the planning process, for example at the identification stage of the project cycle (Section 14.4.1).

Indicators. Objectively verified measures which determine whether the narrative summary is as planned. Section 14.4.5 describes the use of indicators in more detail. Indicators should be precise and quantifiable with time limits and the minimum number necessary to verify the narrative summary.

Table 14.3 The logical framework approach.

	Narrative summary	Indicators	Means of verification	Assumptions and risks
Goal	What are the major conservation issues that the project is expected to resolve? / *e.g. Reverse loss of forest due to burning*	Which are the key aspects of biodiversity that can be monitored? / *Reduce area of forest burnt to less than 10 ha per year within 2 yrs of starting project*	How, when and by whom will the conservation objectives be assessed or is the data already routinely collected? / *Fire service records / Annual habitat survey*	What are the assumptions that if true ensure overall long-term continuation towards the goal? / *No droughts*
Purpose	What are the immediate objectives? / *Replacement of wild honey collection with honey from hives*	What shows whether immediate objectives are being achieved? / *Over 95% of hives in operation / Less than 2 incidents of wild honey collecting per month within a year*	How can these indicators be monitored? / *Project records / Field surveys*	What are the assumptions that, if true, ensure that if the purpose is achieved then the goal must also be achieved? / *Fire largely caused accidentally by honey collectors / Fires stop if collecting stops*
Outputs	What will happen that ensures that purposes are achieved? / *150 working hives within one year / 2000 pots of honey sold per month to local town within second year of project*	Which indicators are best for assessing the outputs? / *Number of working hives / Honey sold*	How, when and by whom will the output indicators be monitored? / *Field inspection / Project records*	What are the assumptions that, if true, ensure that if the output is achieved then the purpose must also be achieved? / *People will prefer to use hives rather than collect wild honey / Hives can be maintained*
Activities	What will have to be done to achieve the outputs? / *Provide bee-keeping equipment / Training courses / Help setting up markets*	Specifications of costs/inputs of each activity / *$200 \times \$40 = \$8000$ / $8 \times 5\,days = 40\,days$ / 3 months*	How will the costs and other inputs be recorded? / *Receipts to project office / Project records / Project records*	What are the assumptions that, if true, ensure that if the activities are achieved then the outputs must also be achieved? / *Honey bees can survive in area / Participants will learn and apply bee-keeping skills / Market for honey from hives*

This provides a clear target and allows progress to be measured.

Means of verification. How the indicators can be assessed. It should include the type of data, methodology, timing and who will do it.

Assumptions and risks. Processes over which the project will have no control but are essential if the project is to succeed. Many projects fail due to making incorrect assumptions. If the plan is correct then when the listed assumptions are true the narrative summary has to be true. Thus if the activities occur and the listed assumptions are true then the outputs must also occur. This then applies to every stage until the goal is achieved. It is this simple brutal logic that makes the approach so powerful.

Logical frameworks have the advantage of forcing participants to think in detail about how the project will run and the problems that it may encounter. They should be a simple summary of the plan, not the plan itself. Thus they are often just a page long and should certainly be no more than two to three pages. As monitoring is an integral component it is easier to detect and counter problems at an early stage.

Thus if the project plan is well designed then a logical framework should be easy to prepare. If the framework is difficult to fit then the logic is often poor. Logframes have their critics, for example they can be a format applied without thinking or can result in rigid thinking if not regularly reviewed.

14.5.7 Risk analysis

Risks are anything that could prevent the project from being a success, such as unexpected environmental, economic or social changes. Brainstorm (Section 8.5.1) for risks associated with each stage of the project, review and list. Analyse the likelihood of each occurring from past experience, for example extreme weather or a disease ruining the crop, or experience elsewhere, for example of the risk of a logging company being given permission to log the area. Projects often depend upon policy support from local or national organisations. Risk analysis is also an essential component of logical framework analysis (Section 14.5.6). Carry out further research for those that are critical. Risks should also be reviewed as the project proceeds.

14.5.8 Identifying and allocating tasks

Once the plan has been agreed it is necessary to convert it into a series of tasks with a programme. Thus within the logical framework it is necessary to convert the activities into outputs. Each component of the plan (e.g. build permanent anchors) is then broken into the actual tasks (obtain anchor plans, buy materials, hire workers, borrow boat, construct anchor). Once the main tasks have been agreed it is then necessary to organise them into a logical sequence. Taskboarding involves writing out each key stage on a sticky label or card and placing on a whiteboard or card. For each objective at each stage ask what needs to be complete before this can be started. Join the tasks by arrows which then show the logical flow. If speed is important then critical path analysis can be used which entails finding the most time-consuming chain of tasks 'the critical path' and then adding the other tasks in the manner which least disrupts this critical path.

There are a range of techniques for converting this ordered sequence of tasks into a work programme. The main technique is often called a Gantt chart (although some say that this term should be reserved for charts derived by rigorous analysis) or a bar chart or a timeline. The Gantt chart involves dividing time into periods (hours, days, weeks or months) and allocating the time for each task for each person. Thus Table 14.4 shows the Gantt chart for management tasks. The same process should occur in planning the monitoring programme.

It is then useful to tabulate the tasks that have to be done. The timing, cost and duration of each task is usually calculated. A common approach is to tabulate each task with columns for what will be done, how it will be done, when it will be done, who will do it, where it will be done and comments.

Table 14.4 A Gantt chart to plan the timing of tasks running the Fictitious Reef project.

Management task	Month — Year 1 J F M A M J J A S O N D	Year 2 J F M A M J J A S O N D	Responsibility
Objective O5 A1			
Plant alternative trees			
Research alternative trees	X X X X		Local forestry officer
Locate areas for plantation	X X X		Local forestry officer
Plant trees	X X	X X	Local community
Maintain trees	X X X X X	X X X X X X	Local community
Objective O5 A2			
Establish centre for reusing wood			
Research design for centre	X X X X X X		Project manager
Obtain funding	X X X X X X X X	X X X	Project manager
Build centre		X X X X X X X X	Local community
Buy equipment		X X X X X	Project manager

Table 14.5 A stakeholder participation matrix. For each group involved in the project the level of participation at each stage is identified. For example for a project to regulate access to coral reefs a highly incomplete matrix might look like this. LF = local fishers, DO = tourist diving organisations, LP = local politicians.

	Information	Consultation	Partnership	Control
Identification	LP	LF, DO		
Planning	LP	DO	LF	
Bearing costs	LP, LF		DO	
Implementation	LP	DO		LF
Benefits	LP, DO			LF
Monitoring	LP		LF, DO	
Evaluation	LP	LF, DO		

14.5.9 Stakeholder participation matrix

Although the importance of participation is widely accepted it may range from informing people what will happen to handing over complete control. The nature of participation with a particular group is also likely to vary over the duration of the project. Thus to reduce ambiguity for each group of stakeholders, a matrix can be created to represent the expected participation. One approach is to consider the participation for each stage of the project cycle (see Table 14.5).

14.6 Capacity building

Capacity building is the long-term process of enhancing the capability of an organisation or community to identify and solve its own problems. It is now accepted as a major component of achieving conservation and development goals. It is increasingly realised that aid is more effective if it succeeds in strengthening institutions and policies (World Bank 1998). Capacity building is often essential if decisions are devolved and decentralised and for projects to be likely to persist as long-term viable projects after external support has ended.

Capacity building may apply at many levels from individuals to organisations and sometimes even to government. In most cases projects are run by those with experience in either conservation or development but not both. Those involved in setting up and maintaining projects often require a range of new skills and techniques, such as in planning, management, financial management, agriculture, forestry, socio-economics, law, social anthropology and especially participatory development. Providing support, advice or training can be highly beneficial. As well as asking what skills are required it can also be useful to ask what skills they can teach others, in order to reduce costs, prevent creating a culture of dependency and because teaching really is the best way of learning. A highly successful technique is to identify capable motivated individuals and providing them with the opportunities they need to fulfil their potential.

References

van Aarde, R.J. (1979) Distribution and density of feral cat *Felis catus* on Marion Island. *South African Journal of Antarctic Research* **9**, 14–9.

Adams, D.E. & Anderson, R.C. (1978) The response of a central Oklahoma grassland to burning. *Southwestern Naturalist* **23**, 623–32.

Adams, W.M. (1996) *Future Nature*. London: Earthscan Publications Ltd.

Adams, W.M. (1998) Conservation and development. In: Sutherland, W.J. (ed.) *Conservation Science and Action*. Oxford: Blackwell Science Ltd, pp. 286–315.

Agee, J.K. (1993) *Fire Ecology of Pacific Northwest Forests*. Washington: Island Press.

Akçakaya, H.R. & Burgman, M. (1995) PVA in theory and practice. *Conservation Biology* **9**, 705–7.

Akçakaya, H.R., Burgman, M.A. & Ginzburg, L.R. (1999) *Applied Population Ecology*. Sunderland: Sinauer.

Akeroyd, J. (1994) *Seeds of Destruction*? London: Plantlife, Natural History Museum.

Akeroyd, J. & Wyse Jackson, P. (1995) *A Handbook for Botanic Gardens on the Introduction of Plants to the Wild*. Richmond: Botanic Gardens Conservation International.

Alaric, V. (ed.) (1994) *Remote Sensing and GIS in Ecosystem Management*. Washington: Island Press.

Allee, W.C. (1931) *Animal Aggregations, a Study in General Sociology*. Chicago: University of Chicago Press.

Alpert, P. (1996) Integrated conservation and development projects: examples from Africa. *BioScience* **46**, 845–55.

Altig, R.R. (1980) A convenient killing agent for amphibians. *Herpetological Review* **11**, 35.

Alton, S. (1998) The Millennium seed bank project. *British Wildlife* **9**, 272–7.

Andersen, A.N. (1995) Measuring more of biodiversity: genus richness as a surrogate for species richness in Australian ant faunas. *Biological Conservation* **73**, 39–43.

Anderson, D.R., Burnham, K.P. & White, G.C. (1985) Problems in estimating age-specific survival from recovery data of birds ringed as young. *Journal of Animal Ecology* **54**, 89–98.

Anderson, G.J. & Symon, D.E. (1989) Functional dioecy and andromonoecy in Solanum. *Evolution* **43**, 204–19.

Andrews, J. (1995) Waterbodies. In: Sutherland, W.J. & Hill, D.A (eds) *Managing Habitats for Conservation*. Cambridge: Cambridge University Press, pp. 121–48.

Anon. (1997) Logging threatens rare white bears. *New Scientist* **153**, 2075.

Antman, E.M., Lau, J., Kupelnick, B., Mosteller, F. & Chalmers, T.C. (1992) A comparison of results of meta-analyses of randomised control trails and recommendations of clinical experts. *Journal of the American Medical Association* **268**, 240–8.

Apfelbaum, S.I., Bader, B.J., Faessler, F. & Mahler, D. (1997) Obtaining and processing seeds. In: Packard, S. & Mutel, C.F. (eds) *The Tallgrass Restoration Handbook*. Washington: Island Press, pp. 99–126.

Aplet, G.H. & Keeton, W.S. (1999) Application of historical range of variability concepts to biodiversity conservation. In: Baydeck, R.K., Campa III, H. & Haufler, J.B. (eds) *Practical Approaches to the Conservation of Biological Diversity*. Washington: Island Press.

Ash, H.J., Gemmell, R.P. & Bradshaw, A.D. (1994) The introduction of native plant species on industrial waste heaps: a test of immigration and other factors affecting primary succession. *Journal of Applied Ecology* **32**, 74–84.

Ausden, M. (1996a) Invertebrates. In: Sutherland, W.J. (ed.) *Ecological Census Techniques*. Cambridge: Cambridge University Press, pp. 138–77.

Ausden, M., Sutherland, W.J. & James, R. (in press) The effects of flooding lowland wet grassland on soil macroinvertebrate prey of wading birds. *Journal of Applied Ecology*.

Ausden, M. & Treweek, J. (1995) Grasslands. In: Sutherland, W.J. & Hill, D.A. (eds) *Managing Habitats for Conservation*. Cambridge: Cambridge University Press, pp. 197–229.

Avery, M., Gibbons, D.W., Porter, R. *et al.* (1995) Revising the British Red Data List for birds: the biological basis of UK conservation priorities. *Ibis* **137**, S209–13.

Avice, J.C. (1992) Molecular population structure and the biogeographical history of a regional fauna: a case history with lessons for conservation biology. *Oikos* **63**, 62–76.

Avice, J.C. (1994) *Molecular Markers, Natural History and Evolution*. London: Chapman & Hall.

Avice, J.C. & Hamrick, J.L. (1996) *Conservation Genetics*. New York: Chapman & Hall.

Baines, C. & Smart, J. (1991) *A Guide to Habitat Creation*. Chichester: Packard Publications Ltd.

Baker, C.S. & Palumbi, S.R. (1994) Which whales are hunted? A molecular genetic approach to monitoring whaling. *Science* **265**, 1538–9.

Balinsky, B.I. (1970) *An Introduction to Embryology*. Philadelphia: W.B. Saunders.

Ballou, J.D. & Lacy, R.C. (1995) Identifying genetically important individuals for management of genetic variation in pedigreed populations. In: Ballou, J.D., Gilpin, M. & Foose, T.J. (eds) *Population Management for Survival and Recovery*. New York: Columbia University Press, pp. 76–111.

Balls, H.R., Moss, B. & Irvine, K. (1989) The loss of submerged plants with eutrophication. 1: Experimental design, water chemistry, aquatic plant and phytoplankton biomass in experiments carried out in ponds in the Norfolk Broadland. *Freshwater Biology* **22**, 71–87.

Balmford, A. (1999) (Less and less) great expectations. *Oryx* **33**, 87–8.

Balmford, A., Green, M.J.B. & Murray, M.G. (1996a) Using higher taxon richness as a surrogate for species richness: I. Regional Test. *Proceedings of the Royal Society Series B* **263**, 1267–74.

Balmford, A., Jayasuriya, A.H.M. & Green, M.J.B. (1996b) Using higher-taxon richness as a surrogate for species richness: 2. Local applications. *Proceedings of the Royal Society Series B* **263**, 1571–5.

Balmford, A., Mace, G.M. & Leader-Williams, N. (1997) Redesigning the ark: setting priorities for captive breeding. *Conservation Biology* **11**, 593–4.

Balon, E.K. (1975) Terminology of intervals in fish development. *Journal of Fisheries Research Board, Canada* **32**, 1663–70.

Barnett, A. & Dutton, J. (1995) *Expedition Field Techniques. Small Mammals (Excluding Bats)*. London: Royal Geographical Society.

Barratt, E.M., Deaville, R., Burland, T.M. *et al.* (1997) DNA answers the call of the pipistrelle bat. *Nature* **387**, 138–9.

Baskin, C.C. & Baskin, J.M. (1988) Germination ecophysiology of herbaceous plant species in a temperate region. *American Journal of Botany* **75**, 286–305.

Baskin, C.C. & Baskin, J.M. (1998) *Seeds: Ecology, Biogeography and Evolution of Dormancy and Germination*. San Diego: Academic Press.

Baskin, J.M. & Baskin, C.C. (1981) Temperature relations of seed germination and ecological implications in *Galinsoga parviflora* and *G. quadriradiata*. *Bartonia* **48**, 12–18.

Bayliss, P. (1989) Population dynamics of Magpie Geese in relation to rainfall and density: implications for harvest models in a fluctuating environment. *Journal of Applied Ecology* **26**, 913–24.

Beck, B.B., Rapaport, L.G. & Wilson, A.C. (1994) Reintroduction of captive-born animals. In: Olney, P.J.S., Mace, G.M. & Feistner, A.T.C. (eds) *Creative Conservation: Interactive Management of Wild and Captive Animals*. London: Chapman & Hall, pp. 265–86.

Beddington, J.R. & May, R.M. (1977) Fluctuating natural populations in a naturally fluctuating environment. *Science* **197**, 463–5.

Begon, M. (1979) *Investigating Animal Abundance*. London: Edward Arnold.

Beintema, A. (1992) Mayfield moet: oegeningen in het berekenen van uitkomstsucces. *Limosa* **65**, 155–62.

Beissinger, S.R. & Westphal, M.I. (1998) On the use of demographic models of population viability in endangered species recovery. *Journal of Wildlife Management* **62**, 821–41.

Bell, B.D. (1994) Translocation of fluttering shearwaters: developing a method to re-establish seabird populations. In: Serena, M. (ed.) *Reintroduction Biology of Australian and New Zealand Fauna*. Chipping Norton: Surrey Beatty, pp. 143–5.

Benstead, P., Drake, M., José, P. *et al.* (1997) *The Wet Grassland Guide*. Sandy: Royal Society for the Protection of Birds.

Benton, T.G. & Grant, A. (1999) Elasticity analysis as an important tool in evolutionary and population ecology. *Trends in Ecology and Evolution* **14**, 467–71.

Berger, J. (1990) Persistence of different-sized populations: an empirical assessment of rapid extinctions in big horn sheep. *Conservation Biology* **4**, 91–8.

Berger, L., Speare, R., Daszak, P. *et al.* (1998) Chytridiomycosis causes amphibian mortality associated with population declines in the rain forests of Australia and Central America. *Proceedings of the National Academy of Sciences, USA* **95**, 9031–6.

Beverton, R.J.H. & Holt, S.J. (1957) *On the Dynamics of Exploited Fish Populations*. London: H.M. Stationery Office.

Bibby, C.J. (1995) A global view of priorities for bird conservation: a summary. *Ibis* **137**, S247–8.

Bibby, C.J. (1998) Selecting areas for conservation. In: Sutherland, W.J. (ed.) *Conservation Science and Action*. Oxford: Blackwell Science Ltd, pp. 176–201.

Bibby, C.J. (1999) Making the most of birds as environmental indicators. *Ostrich* **70**, 81–8.

Bibby C.J. & Etheridge, B. (1993) Status of the hen harrier *Circus cyaneus* in Scotland 1988–89. *Bird Study* **40**, 1–11.

Bibby, C.J., Burgess, N.D. & Hill, D.A. (1992b) *Bird Census Techniques*. London: Academic Press.

Bibby, C.J., Collar, N.J., Crosby, M.J. *et al.* (1992a) *Putting Biodiversity on the Map: Priority Areas for Global Conservation*. Cambridge: International Council for Bird Preservation.

Bibby, C.J., Jones, M. & Marsden, S. (1998) *Expedition Field Techniques. Bird Surveys*. London: Royal Geographical Society.

Bird, C. (1995) Communal lands, communal problems. *Focus* **16**, 7–8.

Birkhead, T.R., Veiga, J.P. & Fletcher, F. (1995) Sperm competition and unhatched eggs in the House Sparrow. *Journal of Avian Biology* **26**, 343–5.

Biswell, H.H. (1989) *Prescribed Burning in Californian Wildlands Vegetation Management*. Berkeley: University of California Press.

Black, J.M. (1995) The Nene *Branta sandvicensis* Recovery Initiative: research against extinction. *Ibis* **137**, S147–52.

Blanchard, K.A. (1995) Reversing population declines in seabirds on the north shore of the Gulf of St Lawrence,

Canada. In: Jacobson, S.K. (ed.) *Conserving Wildlife. International Education and Communication Approaches*. New York: Columbia University Press, pp. 51–63.

Bland, M. (1998). *Communicating Out of a Crisis*. London: Macmillan.

Blangy, S. & Wood, M.E. (1993) Developing and implementing ecotourism guidelines for wildlands and neighbouring communities. In: Lindberg, K. & Hawkins, D.E. (eds) *Ecotourism: A Guide for Planners and Managers*. North Bennington, VT: The Ecotourism Society, pp. 32–49.

Block, W.M., Brennan, L.A. & Gutierrez, R.J. (1987) Evaluation of guild-indicator species for use in resource management. *Environmental Management* **11**, 265–9.

Blomberg, S. & Shine, R. (1996) Reptiles. In: Sutherland, W.J. (ed.) *Ecological Census Techniques*. Cambridge: Cambridge University Press, pp. 218–21.

Blower, J.G., Cook, L.M. & Bishop, J.A. (1981) *Estimating the Size of Animal Populations*. London: Allen & Unwin.

Bolton, M.P. & Specht, R.L. (1983) *A method for selecting conservation reserves*. Australian National Parks & Wildlife Service Occasional Paper, No. 8.

Bond, W.J. & van Wilgen, B.W. (1996) *Fire and Plants*. London: Chapman & Hall.

Boo, E. (1993) Ecotourism planning for protected areas. In: Lindberg, K. & Hawkins, D.E. (eds) *Ecotourism: A Guide for Planners and Managers*. North Bennington, VT: The Ecotourism Society, pp. 15–31.

Boone, J.L. & Laurie, E.A. (1999) Effects of marking *Uta stansburiana* with xylene-based paint. *Herpetological Review* **30**, 33–4.

Bormann, F.H. (1953) The statistical efficiency of sample plot, size and shape in forest ecology. *Ecology* **53**, 474–87.

Bowen, B.W. & Avise, J.C. (1996) Conservation genetics of marine turtles. In: Avise, J.C. & Hamrick, J.L. (eds) *Conservation Genetics*. New York: Chapman & Hall, pp. 190–237.

Bowles, M.L. & Whelan, C.J. (1994) Conceptual issues in restoration ecology. In: Bowles, M.L. & Whelan, C.J. (eds) *Restoration of Endangered Species*. Cambridge: Cambridge University Press, pp. 1–7.

Boyce, M.S. (1992) Population viability analysis. *Annual Review of Ecology and Systematics* **23**, 481–506.

Bradshaw, A.D. (1983) The reconstruction of ecosystems. *Journal of Applied Ecology* **20**, 1–17.

Bragg, O.M., Hulme, P.D., Ingham, H.A.P., Johnston, J.P. & Wilson, A.A. (1994) A maximum–minimum recorder for shallow water tables, developed for ecohydrological studies on mires. *Journal of Applied Ecology* **31**, 581–92.

Braithwaite, R.W. (1991) Aboriginal fire regimes of monsoonal Australia in the nineteenth century. *Search* **22**, 247–9.

Branden, K.L., Pollard, D.A. & Reimers, H.A. (1994) A review of recent artificial reef developments in Australia. *Bulletin of Marine Science* **55**, 982–94.

Bratton, W. (1998) *Turnaround. How American Top Cop Reversed the Crime Epidemic*. New York: Random House.

Braus, J.A. & Wood, D. (1993) *Environmental Education in the Schools*. Washington: North American Association for Environmental Education.

Brayshaw, T.C.C. (1996) *Plant Collecting for the Amateur*. Victoria: Royal British Columbia Museum.

Brickell, C. (ed.) (1996) *The Royal Horticultural Society Encyclopaedia of Garden Plants*. London: Dorling Kindersley.

Britten, H.B. (1996) Meta-analysis of the association between multilocus heterozygosity and fitness. *Evolution* **50**, 2158–64.

Brocklesby, M.A., Oji, B.A. & Fon, T.C. (1999) Developing participatory forest management: the user group analysis on Mount Cameroon. In: Doolan, S. (ed.) *African Rainforests and the Conservation of Biodiversity*. London: Earthwatch, pp. 51–61.

Brook, B.W. & Kikkawa, J. (1998) Examining threats faced by island birds: a population viability analysis on the Capricorn silvereye using long-term data. *Journal of Applied Ecology* **33**, 491–503.

Brooks, S. & Stoneman, R. (1997) *Conserving Bogs. The Management Handbook*. Edinburgh: The Stationery Office.

Brown, J.C. & Stoddart, D.M. (1977) Killing mammals and general post-mortem methods. *Mammal Review* **7**, 63–94.

Brown, J.H. (1995) *Macroecology*. Chicago: University of Chicago Press.

Brown, J.H. & Kodric Brown, A. (1977) Turnover rates in insular biogeography: effect of immigration on extinction. *Ecology* **58**, 445–9.

Brown, W.M., George, M.J. & Wilson, A.C. (1979) Rapid evolution of animal mitochondrial DNA. *Proceedings of the National Academy of Sciences* **76**, 1967–71.

Brownie, C., Anderson, D.R., Burham, K.P. & Robson, D.S. (1985) *Statistical Inference from Band Recovery Data – a Handbook*, 2nd edn. Washington, DC: US Fish and Wildlife Service Resources Publication 159.

Bruna, E.M. (1999) Seed germination in rainforest fragments. *Nature* **402**, 139.

Bryson, J.M. (1995) *Strategic Planning for Public and Non-Profit Organisations*. New York: Prentice-Hall.

Buckland, S.T., Anderson, D.R., Burnham, K.P. & Laake, J.L. (1993) *Distance Sampling – Estimating Abundance of Biological Populations*. London: Chapman & Hall.

Buckley, J. & Inns, H. (1998) Field identification, aging and sexing. In: Gent, A.H. & Gibson, S.D. (eds) *Herpetofauna Workers' Manual*. Peterborough: Joint Nature Conservation Committee, pp. 15–32.

Buhat, D.Y. (1994) Community-based coral reef and fishery management, San Salvador Island, Philippines. In: White, A.T., Hale, L.Z., Renard, Y. & Cortesi, L. (eds) *Collaborative and Community-based Management of Coral Reefs*. West Hartford: Kumarian Press, pp. 33–50.

Bullock, D. & North, S. (1984) Round Island in 1982. *Oryx* **18**, 36–41.

Bullock, J. (1996) Plants. In: Sutherland, W.J. (ed.) *Ecological Census Techniques*. Cambridge: Cambridge University Press, pp. 111–38.

Bunch, R. (1982) *Two Ears of Corn: A Guide to People-Centred Agricultural Improvement*. Oklahoma: World Neighbors.

Burkey, S. (1993) *People First: A Guide to Self-reliant, Participatory Rural Development*. New York: Zed Books.

Burnham, K.P. & Overton, W.S. (1979) Robust estimation of population size when capture probabilities vary amongst animals. *Ecology* **60**, 927–36.

Burnham, K.P., Anderson, D.R. & Laake, J.L. (1980) Estimation of density from line transect sampling of biological populations. *Wildlife Monographs* **72**, 1–200.

Butchart, S.H.M., Seddon, N. & Ekstrom, J.M.M. (1999) Polyandry and competition for territories in Bronze-winged Jacanas. *Journal of Animal Ecology* **68**, 928–39.

Butler, P.J. (1995) Marketing the conservation message: using parrots to promote protection and pride in the Caribbean. In: Jacobson, S.K. (ed.) *Conserving Wildlife*. International Education and Communication Approaches. New York: Columbia University Press, pp. 87–102.

Cadbury, J. & Lambton, S. (1996) The importance of RSPB nature reserves for amphibians and non-marine reptiles. *RSPB Conservation Review* **10**, 82–9.

Caithness, T.A. & Williams, G.R. (1971) Protecting birds from poison baits. *New Zealand Journal of Agriculture* **122**, 1–4.

Caldecott, J.C. (1996) *Designing Conservation Projects*. Cambridge: Cambridge University Press.

Calder, I.R. (1976) The measurement of water losses from a forested area using a 'natural' lysimeter. *Journal of Hydrology* **30**, 311–25.

Cambell, A. (1999) Mission, vision and strategic development. In Crainer, S. (ed.) *Financial Times Book of Management*. London: Financial Times, pp. 134–44.

Camper, J.D. & Dixon, J.K. (1988) *Evolution of a microchip marking system for amphibians and reptiles*. Research Publication 7100-159. Austin: Texas Parks and Wildlife Department.

Canter, L.W. (1996) *Environmental Impact Assessment*. New York: McGraw-Hill.

Carbyn, L.N. (1979) Wolf-howling as a technique for ecosystem interpretation in national parks. In: Klinghammer, E. (ed.) *The Behaviour and Ecology of Wolves*. New York: Garland, pp. 458–70.

Caro, T. (2000) Controversy over behaviour and genetics in cheetah conservation. In: Gosling, M. & Sutherland, W.J. (eds) *Behaviour and Conservation*. Cambridge: Cambridge University Press, pp. 221–37.

Carter, A.L. & McNeilly, T. (1975) Effects of increased humidity on pollen tube growth and seed set following self-pollination in Brussels sprout (*Brassica oleraceae* var *gemmifera*). *Euphytica* **24**, 805–13.

Castro, I., Alley, J.C., Empson, R.A. & Minot, E.O. (1994) Translocation of Hihi or Stitchbird *Notiomystis cincto* to Kapiti Island, New Zealand: transfer techniques and comparison of release strategies. In: Serena, M. (ed.) *Reintroduction Biology of Australian and New Zealand Fauna*. Chipping Norton: Surrey Beatty, pp. 113–20.

Caswell, H. (1978) A general formula for the sensitivity of population growth rate to changes in life history parameters. *Theoretical Population Biology* **14**, 215–30.

Caswell, H. (1989) *Matrix Population Models: Construction, Analysis and Interpretation*. Sunderland: Sinauer.

Caughley, G. (1974) Bias in aerial surveys. *Journal of Wildlife Management* **38**, 921–33.

Caughley, G. (1977) *Analysis of Vertebrate Populations*. London: Wiley.

Caughley, G. (1994) Directions in conservation biology. *Journal of Animal Ecology* **63**, 215–44.

Caughley, G. & Gunn, A. (1995) *Conservation Biology in Theory and Practice*. Oxford: Blackwell Science Ltd.

Caughley, G. & Sinclair, A.R.E. (1994) *Wildlife Ecology and Management*. Oxford: Blackwell Scientific Publications Ltd.

Chambers, R. (1993) *Challenging the Professions: Frontiers for Rural Development*. London: Intermediate Technology Publication.

Chambers, R. (1997) *Whose Reality Counts?* London: Intermediate Technology Publications.

Chase, A. (1987) *Playing God in Yellowstone*. Orlando: Harcourt Brace.

Chen, X. (1993) Comparisons of inbreeding and outbreeding in hermaphroditic *Arianta arbustorum* (L.) (land snail). *Heredity* **71**, 456–61.

Chin, H.F. & Roberts, E.H. (1980) *Recalcitrant Crop Seeds*. Kuala Lumpar: Tropical Press.

Chou, L.M. (1997) Artificial reefs of southeast Asia – Do they enhance or degrade the marine environment? *Environmental Monitoring and Assessment* **44**, 45–52.

Christensen, N.L., Agee, J.K., Brussard, P.F. *et al.* (1989) Interpreting the Yellowstone fires of 1988. *Biosciences* **39**, 678–85.

Clark, S. & Edwards, A.J. (1994) Use of artificial reef structure to rehabilitate reef flats degraded by coral mining in the Maldives. *Bulletin of Marine Science* **55**, 724–44.

Clark, T.W. (1997) *Averting Extinction. Reconstructing Endangered Species Recovery*. New Haven: Yale University Press.

Clark, T.W., Reading, R.P. & Clarke, A.L. (1994a) Introduction. In: Clark, T.W., Reading, R.P. & Clarke, A.L. (eds) *Endangered Species Recovery*. Washington: Island Press, pp. 3–17.

Clark, T.W., Reading, R.P. & Clarke, A.L. (eds) (1994b) *Endangered Species Recovery*. Washington: Island Press.

Clarke, C.W. (1973) The economics of overexploitation. *Science* **181**, 630–4.

Clarke, C.W. (1981) Bioeconomics. In: May, R.M. (ed.) *Theoretical Ecology*, 2nd edn. Oxford: Blackwell Scientific Publications Ltd, pp. 387–418.

Clarke, C.W. (1990) *Mathematical Bioeconomics: The Optimal Management of Renewable Resources*. New York: Wiley Interscience.

Clavijo, I.E. & Donaldson, P.L. (1994) Spawning behaviour in the Labrid, *Halichoeres bivittatus*, on artificial and natural substrates in Onslow Bay, North Carolina, with notes on early life history. *Bulletin of Marine Science* **55**, 383–7.

Clout, M.N. & Craig, J.L. (1995) The conservation of critically endangered flightless birds in New Zealand. *Ibis* **137**, S181–90.

Cohen, J. (1988) *Statistical Power Analysis for the Behavioural Sciences.* New Jersey: Lawrence Erlbaum Associates.

Collar, N.J. (1996) Species concepts and conservation: a reply to Hazevoet. *Bird Conservation International* **6**, 197–200.

Collar, N.J. (1997) Taxonomy and conservation: chicken and egg. *Bulletin British Ornithologists' Club* **117**, 122–36.

Collar, N.J., Crosby, M.J. & Stattersfield, A.J. (1994) *Birds to Watch 2: The World List of Threatened Birds.* Cambridge: Birdlife International.

Condit, R., Hubbell, S.P., La Frankie, J.V. *et al.* (1996) Species-area and species-individual relationship for a tropical tree. *Journal of Ecology* **84**, 549–62.

Constantine, D.G. (1988) Health precautions for bat researchers. In: Kunz, T.H. (ed.) *Ecological and Behavioural Methods for the Study of Bats.* Washington: Smithsonian Institution Press, pp. 491–528.

Cooper, C., Fletcher, J., Gilbert, D. & Wanhill, S. (1998) *Tourism Principles and Practice.* Harlow: Longman.

Cooper, J.E., Needham, J.R. & Applebee, K. (1988) Clinical and pathological studies on the Mauritius Pink Pigeon. *Ibis* **130**, 57–64.

Costanza, R., d'Arge, R., de Groot, R. *et al.* (1997) The value of the world's ecosystem services and natural capital. *Nature* **387**, 253–60.

Côté, I.M. & Sutherland, W.J. (1997) Removing predators to protect bird populations: does it work? *Conservation Biology* **11**, 395–405.

Coulthard, N. (1996) Conservation in the community. *World Birdwatch* **18**, (4) 12–15.

Courchamp, F., Grenfell, B. & Clutton-Brock, T.H. (1999) Inverse density dependence and the Allee effect. *Trends in Ecology and Evolution* **14**, 405–10.

Courtois, R. & Jolicoeur, H. (1993) The use of Schaefer's and Fox's surplus-yield models to estimate optimal moose harvest and hunting effort. *Alces* **29**, 149–62.

Covell, D.G., Uman, G.C. & Manning, P.R. (1985) Information needs in office practice: are they being met? *Annals of Internal Medicine* **103**, 596–9.

Cowan, D.P., Reynolds, J.C. & Gill, E.L. (2000) Manipulating predatory behaviour through conditioned taste aversion: can it help endangered species? In: Gosling, M. & Sutherland, W.J. (eds) *Behaviour and Conservation.* Cambridge: Cambridge University Press, pp. 281–99.

Cowie, N.R., Sutherland, W.J., Ditlhago, M.K.M. & James, R. (1993) The effects of conservation management of reed beds. II The flora and litter disappearances. *Journal of Applied Ecology* **29**, 277–84.

Cracraft, J. (1983) Species concepts and speciation analysis. *Current Ornithology* **1**, 159–87.

Cracraft, J. (1992) The species of the birds-of-paradise (Paradisaeidae): applying the phylogenetic species concept to a complex pattern of diversification. *Cladistics* **8**, 1–43.

Cracraft, J., Feinstein, J., Vaughn, J. & Helm-Bychowski, H. (1998) Sorting out tigers (*Panthera tigris*), mitochondrial sequences, nuclear inserts, systematics and conservation genetics. *Animal Conservation* **1**, 139–50.

Craig, J.L. (1994) Meta-populations: is management as flexible as nature? In: Olney, P.J.S., Mace, G.M. & Feistner, A.T.C. (eds) *Creative Conservation: Interactive Management of Wild and Captive Animals.* London: Chapman & Hall, pp. 50–66.

Crockford, N.J., Green, R., Rocamora, G. *et al.* (1996) Action plan for the corncrake (*Crex crex*) in Europe. In: Heredia, B., Rose, L. & Painter, M. (eds) *Globally Threatened Birds in Europe.* Strasbourg: Council of Europe, pp. 205–43.

Crockford, N.J., Williams, G. & Fox, T.C. (1993) Action plans for birds in the UK. In: Andrews, J. & Carter, S.P. (eds) *Britain's Birds in 1990–91.* Thetford: British Trust for Ornithology, pp. 44–51.

Cropper, S.C. (1993) *Management of Endangered Plants.* Melbourne: Commonwealth Scientific and Industrial Research Organization.

Cropper, S.C., Calder, D.M. & Tonkinson, D. (1989) *Thelymitra epipactoides* F. Muell. (Orchidaceae): the morphology, biology and conservation of an endangered species. *Proceedings of the Royal Society of Victoria* **101**, 89–101.

Crosby, M.J. (1994) Mapping the distributions of restricted-range birds to identify global conservation priorities. In: Miller, R.I. (ed.) *Mapping the Diversity of Nature.* London: Chapman & Hall, pp. 145–54.

Crouse, D., Crowder, L. & Caswell, H. (1987) A stage-based population model for loggerhead sea turtles and implications for conservation. *Ecology* **68**, 1412–23.

Croxall, J.P. & Rothery, P. (1991) Population regulation of seabirds. Implications of their demography for conservation. In: Perrins, C.M., Lebreton, J.-D. & Hirons, G.J.M. (eds) *Bird Population Studies: Relevance to Conservation.* Oxford: Oxford University Press, pp. 272–96.

Croxall, J.P., Rothery, P., Pickering, S.P.C. & Prince, P.A. (1990) Reproductive performance, recruitment and survival of wandering albatrosses *Diomedia exulans* at Bird Island, South Georgia. *Journal of Animal Ecology* **59**, 775–96.

Crump, M.L. & Scott, N.J., Jr. (1994) Visual encounter surveys. In: Heyer, W.R., Donnelly, M.A., McDiarmid, R.W., Hayek, L.C. & Foster, M.S. (eds) *Measuring and Monitoring Biological Diversity. Standard Method for Amphibians.* Washington: Smithsonian Institution Press, pp. 84–92.

Cummings, S.L. (1994) Colonisation of a nearshore artificial reef at Boca Raton (Palm Beach County), Florida. *Bulletin of Marine Science* **55**, 1193–215.

Curio, E. (1993) Proximate and developmental aspects of antipredator behaviour. *Advances in the Study of Behaviour* **22**, 135–238.

Curnutt, J., Lockwood, J., Luh, H.-K., Nott, P. & Russell, G. (1994) Hotspots and species diversity. *Nature* **367**, 326–7.

Cusworth, J.W. & Franks, T.R. (eds) (1993) *Managing Projects in Developing Countries*. Harlow: Longman.

Danna, G., Badalamenti, F., Gristina, M. & Pipitone, C. (1994) Influence of artificial reefs on coastal nekton assemblages of the Gulf of Castellammarine (northwest Sicily). *Bulletin of Marine Science* **55**, 418–33.

Darwin, C.R. (1842) *The Structure and Distribution of Coral Reefs*. London: Smith, Elder & Co.

Davis, S.D., Heywood, V.H. & Hamilton, A.C. (eds) (1994) *Centres of Plant Diversity: A Guide and Strategy for their Conservation*, Vol. 1: *Europe, Africa, South West Asia and the Middle East*. Cambridge: International Union for the Conservation of Nature and Natural Resources Publications.

Davis, S.D., Heywood, V.H. & Hamilton, A.C. (eds) (1995) *Centres of Plant Diversity: A Guide and Strategy for their Conservation*, Vol. 2: *Asia, Australasia and the Pacific*. Cambridge: International Union for the Conservation of Nature and Natural Resources Publications.

Davis, S.D., Heywood, V.H., Herrera-MacBryde, O., Villa-Lobos, J. & Hamilton, A.C. (eds) (1997) *Centres of Plant Diversity: A Guide and Strategy for their Conservation*, Vol. 3: *The Americas*. *Cambridge*: International Union for the Conservation of Nature and Natural Resources Publications.

Davis, T.J. (ed.) (1994) *The Ramsar Convention Manual: A Guide to the Convention on Wetlands of International Importance especially as Waterfowl Habitat*. (French translation: Le Manuel de la Convention de Ramsar). Gland: Ramsar Convention Bureau.

Day, M. (1998) *Environmental Action: A Citizen's Guide*. London: Pluto Press.

DeMauro, M.M. (1995) Development and implementation of a recovery program for the federal threatened Lakeside Daisy (*Hymenoxys acualis* var *glabra*). In: Bowles, M.L. & Whelan, C.J. (eds) *Restoration of Endangered Species*. Cambridge: Cambridge University Press, pp. 298–321.

Diamond, J. (1997) *Guns, Germs and Steel: A Short History of Everybody for the Last 13 000 Years*. London: Jonathan Cape.

Dinesen, L., Lehmberg, T., Svendsen, J.O., Hansen, L.A. & Fjeldså, J. (1994) A new genus and species of perdicine bird (Phasianidae, Perdicini) from Tanzania: a relief form with Indo-Malaya affinities. *Ibis* **136**, 2–11.

Ditlhago, M.K.M., James, R., Laurence, B.R. & Sutherland, W.J. (1993) The effects of conservation management of reed beds: 1 The invertebrates. *Journal of Applied Ecology* **29**, 265–76.

Dobson, A.P. (1996) *Conservation and Biodiversity*. New York: Scientific American.

Duke, G. (1994) Mountains, forests and pheasants. *World Birdwatch* **16** (1) 10–13.

Dung, V.V., Giao, P.M., Chinh, N.N., Tuoc, D., Arctander, P. & Mackinnon, J. (1993) Discovery of a new bovid from Vietnam. *Nature* **363**, 443–5.

Durdin, C.J. (1993) Little terns and kestrels. *British Wildlife* **3**, 194–5.

East, M.L. & Perrins, C.M. (1988) Effect of nestboxes on breeding populations of birds in broadleaved temperate woodlands. *Ibis* **130**, 393–401.

Echt, D.S., Liebson, P.R. & Mitchell, B. (1991) Mortality and morbidity in patients receiving encainide, flecainide, or placebo: the Cardiac Arrhythmia Suppression Trial. *New England Journal of Medicine* **324**, 781–8.

Eggers, H.W. (1998) Project cycle management revisited. *The Courier* **169**, 69–72.

Eisner, T., Lubchenko, J., Wilson, E.O., Wilcove, D.S. & Bean, M.J. (1995) Building a scientifically sound policy for protecting endangered species. *Science* **268**, 1231–2.

Ellegren, H. & Sheldon, B.C. (1997) New tools for sex identification and the study of sex allocation in birds. *Trends in Ecology and Evolution* **12**, 255–9.

Ellis, J.-A. (1999) *Race for the Rainforest II. Applying the Lessons Learned from Lak to the Bismark-Ramu Integrated Conservation and Development Initiative in Papua New Guinea*. Papua New Guinea: United Nations Development Programme.

Ellis, R.H., Hong, T.D. & Roberts, E.H. (1985) *Handbook of Seed Technology for Seedbanks. II Principles and Methodology*. Rome: International Plant Genetics Resource Institute.

Ellstrand, N.C. (1992) Gene flow by pollen: implications for plant conservation genetics. *Oikos* **63**, 77–86.

Elmberg, J. (1989) Knee-tagging – a new marking technique for anurans. *Amphibia-Reptilia* **10**, 101–4.

Emlen, S.T. (1968) A technique for marking anuran populations for behavioural studies. *Herpetologica* **24**, 172–3.

English, J. & English, B. (1996) *Police Training Manual*. Maidenhead: McGraw-Hill.

Erlinge, S. (1983) Demography and dynamics of a stoat *Mustela erminea* population in a diverse community of vertebrates. *Journal of Animal Ecology* **52**, 705–26.

Evans, B. (1999) Wild words, wild lands. *Wild Earth* **9**, 9–11.

Evans, C. & Lambton, S. (1992) Red data and nationally scarce plants on RSPB reserves. *RSPB Conservation Review* **6**, 57–61.

Fa, J., Juste, E.J., Perez del Val, J. & Satroviejo, J. (1995) Impact of market hunting on mammal species in Equatorial Guinea. *Conservation Biology* **9**, 1107–15.

Faith, D.P. & Walker, P.A. (1996) Environmental biodiversity: on the best-possible use of surrogate data for assessing the relative biodiversity of sets of areas. *Biodiversity and Conservation* **5**, 399–415.

Falk, D.A., Millar, C.I. & Olwell, M. (eds) (1996) *Restoring Diversity Strategies for Reintroduction of Endangered Plants*. Washington: Island Press.

Fasola, M., Barbieri, F. & Canova, L. (1993) Test of an electronic individual tag for newts. *Herpetological Journal* **3**, 149–50.

Fellers, G.M. & Drost, C.A. (1994) Sampling with artificial cover. In: Heyer, W.R., Donnelly, M.A., McDiarmid, R.W.,

References

Hayek, L.-A.C. & Foster, M.S. (eds) *Measuring and Monitoring Biological Diversity: Standard Methods for Amphibians.* Washington & London: Smithsonian Institution Press, pp. 146–50.

Fennell, D.A. (1999) *Ecotourism.* London: Routledge.

Fenster, C.B. & Dudash, M.R. (1994) Genetic considerations for plant population restoration and conservation. In: Bowles, M.L. & Whelan, C.J. (eds) *Restoration of Endangered Species.* Cambridge: Cambridge University Press, pp. 34–62.

Fitch, H.S. (1992) Methods of sampling snake populations and their relative success. *Herpetological Review* **23**, 17–19.

Fitzgibbon, C.D., Mogaka, H. & Fanshaw, J.H. (1995) Subsistence hunting in Arabuko-Sokoke forest, Kenya, and its effects on mammal populations. *Conservation Biology* **9**, 1116–26.

Forman, L. & Bridson, D. (eds) (1992) *The Herbarium Handbook.* London: Kew.

Francis, C.N., Anthony, E.L.P., Brunton, J.A. & Kunz, T.H. (1994) Lactation in male fruit bats. *Nature* **367**, 691–2.

Francis, J.R.D. & Minton, P. (1984) *Civil Engineering Hydraulics.* London: Arnold.

Frankham, R. (1995) Conservation genetics. *Annual Review of Genetics* **29**, 305–27.

Frankham, R. (1998) Inbreeding and extinction: island populations. *Conservation Biology* **12**, 665–75.

Frankham, R. & Loebel, D.A. (1992) Modelling problems in conservation genetics using captive *Drosophila* populations: rapid genetic adaptation to captivity. *Zoo Biology* **1**, 333–42.

Friedland, D.J. (ed.) (1998) *Evidence-based Medicine: A Framework for Clinical Practice.* Stamford: Appleton & Lange.

Friend, M. (ed.) (1987) *Field Guide to Wildlife Diseases.* Washington, DC: US Department of the Interior Fish and Wildlife Series.

Froehlich, J.W., Thorington, R.W., Jr & Otis, J.S. (1981) The demography of howler monkeys (*Alouatta palliata*) on Barro Colorado Island, Panama. *International Journal of Primatology* **2**, 207–36.

Fuglesang, A. (1982) *About Understanding: Ideas and Observations on Cross-Cultural Communication.* Uppsala: Dag Hammarskjöld Foundation.

Galbraith, M.P. & Hayson, C.R. (1995) Tiriti Matangi Island, New Zealand: public participation in species translocation to an open sanctuary. In: Serena, M. (ed.) *Reintroduction Biology of Australian and New Zealand Fauna.* Chipping Norton: Surrey Beatty, pp. 149–54.

Galster, S. (1999) Global Survival Network. http://www.5tigers.org/YOT/YOTgalster.htm

Galster, S.R. & Eliot, K.V. (1999) Roaring back: anti-poaching strategies for the Russian Far East and the comeback of the Amur tiger. In: Seidensticker, J. Christie, S. & Jackson, P. (eds) *Riding the Tiger: Tiger Conservation in Human-dominated Landscapes.* Cambridge: Cambridge University Press, pp. 230–9.

Gardner, A., Anders, S., Asanga, C. & Jeremiah, N. (1999) Community forest management at the Kilum-Ijum Forest Project: lessons learnt so far. In: Doolan, S. (ed.) *African Rainforests and the Conservation of Biodiversity.* London: Earthwatch, pp. 62–7.

Gaston, K.J. (1996) Biodiversity-congruence. *Process in Physical Geography* **20**, 105–12.

Gaston, K.J. (1996) Species richness: measure and measurement. In: Gaston, K.J. (ed.) *Biodiversity: a Biology of Numbers and Difference.* Oxford: Blackwells Science Ltd.

Gauld, I.D. (1999) Inventory and monitoring of biodiversity: a taxonomist's perspective. In: Doolan, S. (ed.) *African Rainforests and the Conservation of Biodiversity.* Oxford: Earthwatch, pp. 1–9.

Gault, C. (1997) *A Moving Story. Species and Community Translocation in the UK: a Review of Policy, Principle, Planning and Practice.* Godalming: World Wide Fund for Nature.

Gibbons, D.W., Hill, D.A. & Sutherland, W.J. (1996) Birds. In: Sutherland, W.J. (ed.) *Ecological Census Techniques.* Cambridge: Cambridge University Press, pp. 227–59.

Gibbons, D.W., Reid, J.B. & Chapman, R.A. (1993) *The New Atlas of Breeding Birds in Britain and Ireland: 1988–1991.* London: Poyser.

Gibson, J. & Warren, L. (1995) Legislative requirements. In: Gubay, S. (ed.) *Marine Protected Areas.* London: Chapman & Hall, pp. 32–60.

Gilbert, O.L. (1995) Urban commons: colourful alternatives. *Enact* **4**, (4) 10–11.

Gilbert, O.L. & Anderson, P. (1998) *Habitat Creation and Repair.* Oxford: Oxford University Press.

Gilman, K. (1994) *Hydrology and Wetland Conservation.* Chichester: Wiley.

Gilpin, M. & Soulé, M. (1986) Minimum viable populations: processes of species extinction. In: Soulé, M. (ed.) *Conservation Biology.* Sunderland: Sinauer, pp. 19–34.

Gilpin, M.E. (1987) Spatial structure and population vulnerability. In: Soulé, M. (ed.) *Viable Populations for Conservation.* Cambridge: Cambridge University press, pp. 125–39.

Ginsberg, J.R., Alexander, K.A., Cleveland, S.L. *et al.* (1997) Some techniques for studying Wild Dogs. In: Woodruffe, R., Ginsberg, J. & Macdonald, D. (eds) *The African Wild Dog.* Gland and Cambridge: International Union for the Conservation of Nature and Natural Resources, pp. 139–46.

Ginsberg, J.R. & Milner-Gulland, E.J. (1993) Sex-biased harvesting and population dynamics: implications for conservation and sustainable use. *Conservation Biology* **8**, 157–66.

Glowka, L., Burhenne-Guilmin, F. & Synge, H. (1994) *A Guide to the Convention on Biological Diversity.* Gland and Cambridge: International Union for the Conservation of Nature and Natural Resources.

Godfray, H.C.J. & Crawley, M.J. (1998) Introductions. In: Sutherland, W.J. (ed.) *Conservation Science and Action.* Oxford: Blackwell Science Ltd, pp. 39–65.

Goff, F.G., Dawson, G.A. & Rochow, J.J. (1982) Site examination for threatened and endangered plant species. *Environmental Management* **6**, 307–16.

Golterman, H.L., Clyme, R.S. & Ohnstad, M.A.M. (1978) *Methods for Physical and Chemical Analysis of Fresh Waters.* Oxford: Blackwell Scientific Publications.

Goodwin, H. (1931) Studies in the ecology of Wicken Fen – 1 The ground water level of the fen. *Journal of Ecology* **19**, 449–73.

Gosling, L.M. & Baker, S.M. (1987) Planning and monitoring an attempt to eradicate Coypu from Britain. *Symposia of the Zoological Society of London* **58**, 99–113.

Götmark, F., Åhlund, M. & Eriksson, M.O.G. (1986) Are indices reliable for assessing conservation value of natural areas? An avian case study. *Biological Conservation* **38**, 55–73.

Gough, M.W. & Marrs, R.H. (1990) A comparison of soil fertility between seminatural and agricultural plant communities: implications for the creation of species-rich grassland on abandoned agricultural land. *Biological Conservation* **5**, 83–6.

Gray, D.H. (1986) Uses and misuses of strategic planning. *Harvard Business Review* (January/February) 87–97.

Green, R.E. (1995) Diagnosing causes of bird population decline. *Ibis* **137**, S47–55.

Green, R.E., Pienkowski, M.W. & Love, J.A. (1996) Long-term viability of the reintroduced population of the white-tailed eagle *Haliaeetus albicilla* in Scotland. *Journal of Applied Ecology* **33**, 357–68.

Greenwood, A. (1996) Veterinary support for *in situ* avian conservation programmes. *Bird Conservation International* **6**, 285–92.

Greenwood, J.J.D. (1993) Statistical power. *Animal Behaviour* **46**, 1011.

Greenwood, J.J.D. (1996) Basic techniques. In: Sutherland, W.J. (ed.) *Ecological Census Techniques.* Cambridge: Cambridge University Press, pp. 11–110.

Griffiths, R. & Tiwari, B. (1995) Sex of the last Spix macaw. *Nature* **375**, 454.

Griffiths, R., Daan, S. & Dijkstra, C. (1996) Sex identification in birds using CHD genes. *Proceedings of the Royal Society Series B* **263**, 1251–6.

Groombridge, B. (1992) *Global Biodiversity: Status of the Earth's Living Resources.* London: Chapman & Hall.

GTZ (1997) *ZOPP: An Introduction to the Method.* Frankfurt: Deutsche Gesellschaft für Technische Zusammenarbeit.

Guijt, I. (1999) *Participatory Monitoring and Impact Assessment of Sustainable Agriculture Initiatives.* London: International Institute for Environment and Development.

Gunn, A. (1999) Caribou and Muskox harvesting in the Northwest Territories. In: Milner-Gulland, E.J. & Mace, R. *Conservation of Biological Resources.* Oxford: Blackwell Science Ltd, pp. 314–30.

Hadfield, M.G. (1986) Extinction in Hawaiian achatinelline snails. *Malacologia* **27**, 67–81.

Hagström, T. (1973) Identification of newt specimens (*Urodela triturus*) by recording the belly pattern and a description of photographic equipment for such registration. *British Journal of Herpetology* **4**, 321–6.

Haig, S.M., Belthoff, J.R. & Allen, D.H. (1993) Population viability analysis for a small population of red-cockaded woodpeckers and an evaluation of enhancement strategies. *Conservation Biology* **7**, 289–301.

Hall, P. & Selinger, B. (1986) Statistical significance: balancing evidence against doubt. *Australian Journal of Statistics* **28**, 354–70.

Halliday, T.R. (1996) Amphibians. In: Sutherland, W.J. (ed.) *Ecological Census Techniques.* Cambridge: Cambridge University Press, pp. 205–17.

Halliday, T.R. (1998) A declining amphibian conundrum. *Nature* **394**, 418–9.

Halliday, T.R. & Verrell, P.A. (1988) Body size and age in amphibians and reptiles. *Journal of Herpetology* **22**, 253–65.

Hammond, C.O. (1983) *The Dragonflies of Great Britain and Ireland.* Colchester: Harley Books.

Hammond, P.S. & Thompson, P.M. (1991) Minimum estimate of the number of Bottlenose Dolphins (*Tursiops truncatus*) in the Moray Firth. *Biological Conservation* **56**, 79–88.

Hanlin, R.T. (1972) Preservation of fungi by freeze-drying. *Bulletin of the Torrey Botanical Club* **99**, 23–7.

Hanski, I. (1999) *Metapopulation Ecology.* Oxford: Oxford University Press.

Hanski, I. & Gilpin, M.E. (eds) (1997) *Metapopulation Biology.* San Diego: Academic Press.

Harcourt, A.H. (1995) PV estimates: theory and practice for a wild gorilla population. *Conservation Biology* **9**, 134–42.

Hardin, G. (1968) The tragedy of the commons. *Science* **162**, 1243–8.

Harrington, J.E. (1970) Seed and pollen storage for conservation of plant gene resources. In: Fraenkel, O.H. & Bennett, E. (eds) *Genetic Resources in Plants.* Oxford: Blackwell Scientific Publications, pp. 501–21.

Harris, R.B., Metzgar, L.H. & Bevin, C.D. (1986) *GAPPS: Generalised Animal Population Project System. Version 3.0. User's Manual.* Montana Co-operative Wildlife Research Unit. Missoula: University of Montana.

Harris, S., Cresswell, W.J., Forde, P.G. *et al.* (1990) Home-range analysis using radio-tracking data – a review of problems and techniques particularly as applied to the study of mammals. *Mammal Review* **20**, 97–123.

Harris, T.M. (1958) Forest fire in the Mesozoic. *Journal of Ecology* **46**, 447–57.

Harrison, J.A. & Martinez, P. (1995) Measurement and mapping of avian diversity in southern Africa: implications for conservation planning. *Ibis* **137**, 410–7.

Harrison, J.A., Allan, D.G., Underhill, L.G. *et al.* (1997) *The Atlas of Southern African Birds.* Rhodes Gift: Birdlife South Africa.

Harrison, S. (1994) Metapopulation and conservation. In: Edwards, P.J., Webb, N.R. & May, R.M. (eds) *Large-scale*

Ecology and Conservation. Oxford: Blackwell Science Ltd, pp. 111–28.

Harrison, S., Stahl, A. & Doak, D. (1993) Spatial models and spotted owls: exploring some biological issues behind recent events. *Conservation Biology* **7**, 950–3.

Harrop, D.O. & Nixon, J.A. (1999) *Environmental Assessment in Practice*. London: Routledge.

Hart, J. (1996) *Storm over Mono*. Berkeley: University of California Press.

Hart, P.J. & Pitcher, A.J. (1969) Field trials of fish marking using a jet inoculator. *Journal of Fish Biology* **1**, 383–5.

Harvey, H.W. (1960) *The Chemistry and Fertility of Seawaters*. Cambridge: Cambridge University Press.

Hazevoet, C.J. (1995) *The Birds of the Cape Verde Islands*. British Ornithologists' Union Check-list No 13. Tring: British Ornithologists' Union.

Hazevoet, C.J. (1996) Conservation and species lists: taxonomic neglect promotes the extinction of endemic birds, as exemplified by taxa from eastern Atlantic islands. *Bird Conservation International* **6**, 181–96.

Heath, M.F. & Evans, M.I. (2000) *Important Bird Areas in Europe: Priority Sites for Conservation*. Cambridge: Birdlife International.

Hedrick, P.W. & Miller, P.S. (1992) Conservation genetics: techniques and fundamentals. *Ecological Applications* **2**, 30–46.

Heezik, Y. van., Seddon, P.J. & Maloney, R.F. (1998) Helping reintroduced Houbara bustards avoid predation: effective anti-predator training and the predictive value of pre-release behaviour. *Animal Conservation* **2**, 155–64.

Heil, G.W. & Diemont, W.H. (1983) Raised nutrient levels change heathland into grassland. *Vegetatio* **53**, 113–20.

Heller, R. (1997) *In Search of European Excellence*. London: HarperCollins Publishers Ltd.

Henderson-Sellers, B. & Martland, H.R. (1987) *Decaying Lakes: the Origins and Control of Cultural Eutrophication*. Chichester: Wiley.

Heyer, W.R., Donnelly, M.A., McDiarmid, R.W., Hayek, L.-A.C. & Foster, M.S. (eds) (1994) *Measuring and Monitoring Biological Diversity: Standard Methods for Amphibians*. Washington & London: Smithsonian Institution Press.

Hibberd, B.G. (1989) *Urban Forestry Practice*. Forestry Commission Handbook No 5. London: HMSO.

Higuchi, R.G., Bowman, B., Freiberger, M., Ryder, O.A. & Wilson, A.C. (1984) DNA sequence from the quagga, an extinct member of the horse family. *Nature* **312**, 282–4.

Hilborn, R. & Walters, C.J. (1992) *Quantitative Stock Assessment: Choice, Dynamics and Uncertainty*. London: Chapman & Hall.

Hirons, G., Goldsmith, B. & Thomas, G. (1995) Site management planning. In: Sutherland, W.J. & Hill, D.A. (eds) *Managing Habitats for Conservation*. Cambridge: Cambridge University Press, pp. 22–41.

Hobbs, R.J. & Gimingham, C.H. (1984) Studies on fire in Scottish heathland communities. I. Fire characteristics. *Journal of Ecology* **72**, 223–40.

Hoeh, W.R., Blakley, K.H. & Brown, W.M. (1991) Heteroplasmy suggests limited biparental inheritance of *Mytilus* mitochondrial DNA. *Science* **251**, 1488–90.

Hong, T.D., Linington, S. & Ellis, R.H. (1996) *Seed Storage Behavior: A Compendium*. Rome: International Plant Genetics Resources Institute.

Hopkinson, P. (1999) *Evaluating reserve networks*. PhD thesis, Imperial College, University of London.

Höss, M., Kohn, M., Pääbo, S., Knauer, F. & Schröder, W. (1992) Excrement analysis by PCR. *Nature* **359**, 199.

Howard, P., Davenport, T. & Kigenyi, F.W. (1997) Planning conservation areas in Uganda's natural forests. *Oryx* **31**, 253–64.

Howard, P.C., Viskanic, P., Davenport, T.R.B. *et al.* (1998) Complementarity and the use of indicator groups for reserve selection in Uganda. *Nature* **394**, 472–5.

del Hoyo, J., Elliot, A. & Sargatal, J. (eds) (1992) *Handbook of the Birds of the World*, Vol 1. Barcelona: Lynx Edicions.

Hudson, M. (1995) *Managing without Profit*. London: Penguin.

Huff, D.E. & Varley, J.D. (1999) Natural regulation in Yellowstone National Park's northern range. *Ecological Applications* **9**, 17–29.

Hugenholz, P. & Pace, N.R. (1996) Identifying microbial diversity in the natural environment: a molecular phylogenetic approach. *Tibtech* **14**, 190–7.

Hunt, C. (1997) Cooperative management: geese rebound in Alaska. *Arctic Bulletin* **3**, 18–19.

Hunter, M.L. (1996) *Fundamentals of Conservation Biology*. Oxford: Blackwell Science Ltd.

Hurlbert, S.H. (1984) Pseudoreplication and the design of ecological field experiments. *Ecological Monographs* **54**, 187–211.

Hurst, J. (1998) *Education Projects*. London: Royal Geographical Society.

ICBP (1992) *Putting Biodiversity on the Map*. Cambridge: International Council for Bird Preservation.

Innes, J., Warburton, B., Williams, D., Speed, H. & Bradfield, P. (1995) Largescale poisoning of ship rats (*Rattus rattus*) in indigenous forests of North Island, New Zealand. *New Zealand Journal of Ecology* **19**, 5–17.

Iriondo, J.M. & Pérez, C. (1999) Propagation from seeds and seed preservation. In Bowles, B.G. (ed.) *A Colour Atlas of Plant Propagation and Conservation*. London: Manson Publishing Ltd, pp. 46–57.

Irvine, K., Moss, B. & Balls, H.R. (1989) The loss of submerged plants with eutrophication 11 Relationships with fish and zooplankton in a set of experimental ponds and conclusions. *Freshwater Biology* **22**, 89–107.

IUCN (1994) *IUCN Red List Categories*. Cambridge: International Union for the Conservation of Nature and Natural Resources, The World Conservation Union.

IUCN International Assessment Team (1997) *An Approach to Assessing Progress Towards Sustainability.* Cambridge: International Union for the Conservation of Nature and Natural Resources, The World Conservation Union.

IUCN (1998) *Guidelines for Re-introductions.* Gland: International Union for the Conservation of Nature and Natural Resources, The World Conservation Union.

Jackson, W.J. & Ingles, A.W. (1998) *Participatory Techniques for Community Forestry.* Gland: International Union for the Conservation of Nature and Natural Resources, The World Conservation Union.

Jacobs, J. (1974) Quantitative measurement of food selection. *Oecologia* **14**, 413–7.

Jacobson, S.K. & McDuff, M.D. (1998) Conservation education. In: Sutherland, W.J. (ed.) *Conservation Science and Action.* Cambridge: Cambridge University Press, pp. 237–55.

Jacobson, S.K. & Padua, S.M. (1995) A systems model for conservation education in parks: examples from Malaysia and Brazil. In: Jacobson, S.K. (ed.) *Conserving Wildlife. International Education and Communication Approaches.* New York: Columbia University Press, pp. 3–15.

James, A.N., Gaston, K.J. & Balmford, A. (1999) Balancing the earth's accounts. *Nature* **401**, 323–4.

Jarvinen, O. & Vaisenen, R.A. (1975) Estimating relative densities of breeding birds by the line transect method. *Oikos* **26**, 316–22.

Jefferies, A.J., Wilson, V. & Thein, S.L. (1985a) Hypervariable 'minisatellite' regions in human DNA. *Nature* **314**, 67–73.

Jefferies, A.J., Wilson, V. & Thein, S.L. (1985b) Individual-specific 'fingerprints' of human DNA. *Nature* **316**, 76–9.

Jennings, S., Kaiser, M.J. & Reynolds, J.D. (2000) *Marine Fisheries Ecology.* Oxford: Blackwell Science Ltd.

Jennings, S., Reynolds, J.D. & Mills, S.C. (1998) Life history correlates of responses to fisheries exploitation. *Proceedings of the Royal Society London Series B* **265**, 333–9.

Jennings, T.J. (1979) A simple technique for the production of reference slides in the study of herbivore diets by faecal analysis. *Journal of Zoology* **188**, 256–98.

Jensen, A.C., Collins, K.J., Lockwood, A.P.D., Mallinson, J.J. & Turnpenny, W.H. (1994) Colonisation and fishery potential of a coal-ash artificial reef, Poole Bay, United Kingdom. *Bulletin of Marine Science* **55**, 1263–76.

Jiménez, J.A., Hughes, K.A., Alaks, G., Graham, L. & Lacy, R.C. (1994) An experimental study of inbreeding depression in a natural habitat. *Science* **266**, 271–3.

Johannes, R.E. (1998) The case for data-less marine resource management: examples from tropical nearshore fin fisheries. *Trends in Ecology and Evolution* **13**, 243–6.

Johnson, D.H. (1979) Estimating nest success: The Mayfield method and an alternative. *Auk* **96**, 651–61.

Johnson, D.H. (1996) Population analysis. In: Bookhout, T.H. (ed.) *Research and Management Techniques for Wildlife and Habitats.* Maryland: The Wildlife Society, pp. 419–44.

Johnson, E.A. (1992) *Fire and Vegetation Dynamics.* Cambridge: Cambridge University Press.

Jolly, G.M. (1965) Explicit estimates from capture–recapture data with both death and immigration – stochastic model. *Biometrika* **52**, 225–47.

Joly, P. & Miaud, C. (1989) Tattooing as an individual marking scheme in urodeles. *Alytes* **8**, 11–16.

Jones, A.T. & Evans, P.R. (1984) A comparison of the growth and morphology of native and commercially obtained continental European *Crataegus monogyna* Jacq. (Hawthorn) at an upland site. *Watsonia* **20**, 97–103.

Jones, C.G. & Duffy, K. (1993) Conservation management of the Echo Parakeet. *Dodo* **29**, 126–48.

Jones, C.G., Heck, W., Lewis, R.E. *et al.* (1995a) The restoration of the Mauritius kestrel *Falco punctatus* population. *Ibis* **137**, S173–80.

Jones, G.H., Trueman, I.C. & Millett, P. (1995b) The use of hay strewing to create species-rich grassland. (1) General principles and hay strewing versus seed mixes. *Land Contamination and Reclamation* **3**, 104–7.

Jones, J.C. & Reynolds, J.D. (1996) Environmental variables. In: Sutherland, W.J. (ed.) *Ecological Census Techniques.* Cambridge: Cambridge University Press, pp. 281–316.

Jones, M. (1998) Study design. In: Bibby, C.J., Jones, M. & Marsden, S. (eds) *Expedition Field Techniques. Bird Surveys.* London: Royal Geographical Society, pp. 15–34.

Jones, P.H. (1994) Photomonitoring on sites of wildlife interest in Wales. *British Wildlife* **6**, 23–7.

Jules, C. & Leal Filho, W. (1998) Environmental education in the Caribbean region with emphasis on protected areas in St Lucia. In: Leal Filho, W., de Carvalho, C.A.R. & Hale, W.H.G. (eds) *Environmental Education in Protected Areas.* Carnforth: Parthenon Publishing Group Ltd, pp. 121–37.

Kanouse, D., Killich, D. & Kahan, J.P. (1995) Dissemination of effectiveness and outcomes research. *Health Policy* **34**, 167–92.

Karp, A., Isaac, P.G. & Ingram, D.S. (eds) (1998) *Molecular Tools for Screening Biodiversity.* London: Chapman & Hall.

Kay, C.E. (1997) Ungulate herbivory, willows, and political ecology in Yellowstone. *Journal of Range Management* **50**, 130–45.

Kearns, C.A. & Inouye, D.W. (1993) *Techniques for Pollination Biologists.* Colorado: University Press of Colorado.

Keegan, D.W., Coblentz, B.E. & Winchell, C.S. (1994) Feral goat eradication on San Clemente Island, California. *Wildlife Society Bulletin* **22**, 56–61.

Keller, L.F., Arcese, P., Smith, J.N.M., Hochachka, W.M. & Stearns, S.C. (1994) Selection against inbred song sparrows during a natural population bottleneck. *Nature* **372**, 356–7.

Kelling, G.L. & Catherine M.C. (1998) *Fixing Broken Windows: Restoring Order and Reducing Crime in Our Communities.* New York: Free Press.

Kent, M. & Coker, P. (1992) *Vegetation Description and Analysis: a Practical Approach.* Chichester: Wiley.

Kenwood, R. (2000) *Wildlife Radio Tracking.* London: Academic Press.

Ketelhöhn, W. (1997) Toolboxes are out: thinking is in. In: Bickerstaffe, G. (ed.) *Mastering Management*. London: Financial Times, pp. 639–43.

Kincaid, H.L. (1976) Inbreeding in rainbow trout (*Salmo gairdneri*). *Journal of Fisheries Research Board, Canada* **105**, 273–80.

King, M. (1995) *Fisheries Biology: Assessment and Management*. Oxford: Fishing News Books.

Kingdon, J. (1990) *Island Africa*. London: Collins.

Kirkwood, J.K., Bennett, P.M., Jepson, P.D. *et al.* (1997) Entanglement in fishing gear and other causes of death in cetaceans stranded on the coasts of England and Wales. *Veterinary Record* **141**, 94–8.

Kleiman, D.G., Stanley Price, M.R. & Beck, B.B. (1994) Criteria for reintroductions. In: Olney, P.J.S., Mace, G.M. & Feistner, A.T.C. (eds) *Creative Conservation Interactive Management of Wild and Captive Animals*. London: Chapman & Hall, pp. 287–303.

Kline, V.M. & Howell, E.A. (1987) Prairies. In: Jordan III, W.R., Gilpin, M.E. & Aber, J.D. (eds) *Restoration Ecology*. Cambridge: Cambridge University Press, pp. 75–83.

Klötzli, F. (1987) Disturbance in transplanted grasslands and wetlands. In: van Andel, J., Bakker, J.P. & Snaydon, R.W. (eds) *Disturbance in Grasslands*. Dordrecht: Junk, pp. 79–96.

Knudsen, J.W. (1972) *Collecting and Preserving Plants and Animals*. New York: Harper & Row.

Komdeur, J. (1992) Importance of habitat saturation and habitat quality for evolution of cooperative breeding in the Seychelles warbler. *Nature* **358**, 493–5.

Komdeur, J. (1996) Breeding of the Seychelles magpie robin *Copsychus sechellarum* and implications for conservation. *Ibis* **138**, 485–98.

Komdeur, J., Huffstadt, A., Prast, W. *et al.* (1995) Transfer experiments of Seychelles warblers to new islands – changes in dispersal and helping behaviour. *Animal Behaviour* **48**, 695–708.

Koslin, I.L. (1944) Macro- and microscopic methods of detecting fertility in unhatched hen's eggs. *Poultry Science* **23**, 266–9.

Kraemer, H.C. & Thiemann, S. (1987) *How Many Subjects?* London: Sage Publications Ltd.

Krapovickas, S. & Lyons de Perez, J.A. (1997) Swainson's hawk in Argentina. *World Birdwatch* **19**, 12–15.

Krebs, C.J. (1999) *Ecological Methodology*. New York: Harper Collins.

de Kroon, H., Plaiser, A., van Groendael, J. & Caswell, H. (1986) Elasticity: the relative contribution of demographic parameters to population growth rate. *Ecology* **67**, 1427–31.

Kumar, A. & Wright, B. (1999) Combating tiger poaching and illegal wildlife trade in India. In: Seidensticker, J., Christie, S. & Jackson, P. (eds) *Riding the Tiger: Tiger Conservation in Human-dominated Landscapes*. Cambridge: Cambridge University Press, pp. 243–51.

Kunz, T.H. (1996) Methods of marking bats. In: Wilson, D.E., Cole, F.R., Nichols, J.D., Rudran, R. & Foster, M.S. (eds) *Measuring and Monitoring Biological Diversity, Standard Methods for Mammals*. Washington & London: Smithsonian University Press, pp. 304–10.

Kunz, T.H., Wemmer, C. & Hayssen, V. (1996) Sex, age and reproductive condition of mammals. In: Wilson, D.E., Cole, F.R., Nichols, J.D., Rudran, R. & Foster, M.S. (eds) *Measuring and Monitoring Biological Diversity, Standard Methods for Mammals*. Washington & London: Smithsonian University Press, pp. 279–90.

Laake, J.L., Buckland, S.T., Anderson, D.R. & Burnham, K.P. (1994) *DISTANCE Users Guide Version 2.1*. Fort Collins: Colorado Cooperative Fish and Wildlife Research Unit, Colorado State University.

Lachance, S. & Mangan, P. (1990) Performance of domestic, hybrid and wild strains of brook trout *Salvelinus fortinalis* after stocking: the impact of intra and inter-specific competition. *Canadian Journal of Fisheries and Aquatic Sciences* **47**, 2278–84.

Lacy, R.C. (1987) Loss of genetic diversity from managed populations: interacting effects of drift, mutation, immigration, selection and population subdivision. *Conservation Biology* **1**, 143–58.

Lamberson, R.H., McKelvey, K.S., Noon, B.R. & Voss, C. (1992) A dynamic analysis of northern spotted owl viability in a fragmented forest landscape. *Conservation Biology* **6**, 505–12.

Lande, R. (1988) Genetics and demography in biological conservation. *Science* **241**, 1455–60.

Lande, R., Engen, S. & Saether, B. (1994) Optimal harvesting, economic discounting and extinction risk in fluctuating environments. *Nature* **372**, 88–90.

Larson, P., Freudenberger, M. & Wyckoff-Baird, B. (1997) *Lessons from the Field: A Review of World Wildlife Fund's Experience with Integrated Conservation and Development Projects*. Gland: World Wildlife Fund.

Latter, B.D.H. & Mulley, J.C. (1995) Genetic adaptation to captivity and inbreeding depression in small laboratory populations of *Drosophila melanogaster*. *Genetics* **139**, 287–97.

Laurance, W.F. (1991) Edge effects in tropical forest fragments: application of a model for the design of nature reserves. *Biological Conservation* **57**, 205–19.

Lawrence, W.J.C. (1968) *Plant Breeding*. London: Edward Arnold.

Lawton, J.H., Prendergast, J.R. & Eversham, B.C. (1994) The number and spatial distributions of species: analyses of British data. In: Forey, P.L., Humphries, C.J. & Vane-Wright, R.I. (eds) *Systematics and Conservation Evaluation*. Oxford: Clarendon Press, pp. 177–95.

Lebreton, J.-D., Burnham, K.P., Clobert, J. & Anderson, D.R. (1992) Modeling survival and testing biological hypotheses using marked animals. Case studies and recent advances. *Ecological Monographs* **62**, 67–118.

Leduc, N., Douglas, G.C., Monnier, M. & Connolly, V. (1990) Pollination in vitro: effects on the growth of pollen tubes, seeds set and gametophytic self-incompatability in *Trifolium pratense* L. and *T. repens* L. *Theoretical and Applied Genetics* **80**, 657–64.

Lens, L., Galbusera, P., Brooks, T., Waiyaki, E. & Schenck, T. (1998) Highly skewed sex ratios in the critically endangered Tiata Thrush as revealed by CHD genes. *Biodiversity and Conservation* **7**, 869–73.

Leopold, L.B., Clarke, F.E., Kanshaw, B.B. & Balsley, J.R. (1971) *A Procedure for Evaluating Environmental Impact.* Washington, DC: US Geological Survey Circular No 654, US Geological Survey.

Lewis, C. (ed.) (1996) *Managing Conflict in Protected Areas.* Gland: International Union for the Conservation of Nature and Natural Resources, The World Conservation Union.

Lichtman, P. (1998) The politics of wildfire: lessons from Yellowstone. *Journal of Forestry* **96**, 4–9.

Lincoln, F.C. (1930) Calculating waterfowl abundance on the basis of banding returns. *US Department of Agriculture Circular* **118**, 1–4.

Lincoln, R.J. & Sheals, J.G. (1979) *Invertebrate Animals, Collection and Preservation.* London: British Museum (Natural History).

Lindberg, K. & Hawkins, D.E. (eds) (1993) *Ecotourism: A Guide for Planners and Managers.* North Bennington, VT: The Ecotourism Society.

Lindenmayer, D.B., Burgman, M.A., Akçakaya, H.R., Lacy, R.C. & Possingham, H.P. (1995) A review of the genetic computer programs ALEX, RAMAS/space and VORTEX for modelling the viability of wildlife populations. *Biological Modelling* **82**, 161–74.

Lindenmayer, D.B., Clark, T.W., Lacey, R.C. & Thomas, V.C. (1993) Population viability analysis as a tool in wildlife conservation policy: with reference to Australia. *Environmental Management* **17**, 745–58.

Linderman, P.V. (1990) Closed and open model estimates of abundance and tests of model assumptions for two populations of the turtle *Chrysemys picta. Journal of Herpetology* **24**, 78–81.

Litvaitis, J.A., Sherburne, J.A. & Bissonett, J.A. (1986) Bobcat habitat use and home range size in relation to prey density. *Journal of Wildlife Management* **50**, 110–17.

Lloyd, H., Cahill, A., Jones, M. & Marsden, S. (1998) Estimating bird densities using distance sampling. In: Bibby, C.J., Jones, M. & Marsden, S. (eds) *Expedition Field Techniques. Bird Surveys.* London: Royal Geographical Society, pp. 35–52.

Lombard, A.J. (1995) The problems with multispecies conservation: do hotspots, ideal reserves and existing reserves coincide? *South African Journal of Zoology* **30**, 145–63.

Loucks, O.L. (1994) Art and insight in remnant native ecosystems. In: Baldwin, A.D., Jr., Luce, J.D. & Pletsch, C. (eds) *Beyond Preservation: Restoring and Inventing Landscapes.* Minneapolis: University of Minnesota Press, pp. 127–35.

Ludwig, D. (1999) Is it meaningful to estimate a probability of extinction? *Ecology* **80**, 298–310.

Ludwig, D., Hilborn, R. & Walters, C. (1993) Uncertainty, resource exploitation and conservation: lessons from history. *Science* **260**, 17–36.

Lynch, M., Conery, J. and Burger, R.C. (1995) Mutation accumulation and the extinction of small populations. *American Naturalist* **146**, 489–518.

MacArthur, J.D. (1993) *The Logical Framework. A Tool for the Management of Project Planning and Evaluation.* New Series Discussion Papers No 42. Development and Project Planning Centre, University of Bradford.

MacArthur, J. (1997) Stakeholder analysis in project planning: origins, applications and refinements of the method. *Project Appraisal* **12**, 251–65.

McCallum, H. (1996) Immunocontraception for wildlife population control. *Trends in Ecology and Evolution* **11**, 491–3.

McCallum, H. & Dobson, A.C. (1995) Detecting disease and parasite threats to endangered species and ecosystems. *Trends in Ecology and Evolution* **10**, 190–4.

McCallum, R. & Sekhran, N. (1997) *Race for the Rainforest: Evaluating Lessons from an Integrated Conservation and Development 'Experiment' in New Ireland, Papau New Guinea.* Papau New Guinea: United Nations Development Programme.

McCollough, D.R. (1996) Spatially structured populations and harvest theory. *Journal of Wildlife Management* **60**, 1–9.

McDiarmid, R.W. (1994) Preparing amphibians as scientific specimens. In: Heyer, W.R., Donnelly, M.A., McDiarmid, R.W., Hayek, L.-A.C. & Foster M.S. (eds) *Measuring and Monitoring Biological Diversity. Standard Methods for Amphibians.* Washington: Smithsonian Institution Press, pp. 289–99.

McNeely, J.A. (1995) Partnerships for conservation: an introduction. In: McNeely, J.A. (ed.) *Expanding Partnerships in Conservation.* Washington: Island Press, pp. 1–12.

Mace, G.M., Smith, T.B., Bruford, M.W. & Wayne, R.K. (1996) An overview of the issues. In: Smith, T.B. & Wayne, R.K. (eds) *Molecular Genetic Approaches in Conservation.* New York: Oxford University Press, pp. 3–21.

MacKinnon, J. & De Wulf, R. (1994) Designing protected areas for giant pandas in China. In: Miller, R.I. (ed.) *Mapping the Diversity of Nature.* London: Chapman & Hall, pp. 127–42.

MacKinnon, J. & Phillips, K. (1993) *A Field Guide to the Birds of Sumatra, Java and Bali.* Oxford: Oxford University Press.

Madsen, M., Nielsen, B.O., Holter, P. *et al.* (1990) Treating cattle with Ivermectin and effects on the fauna and decomposition of cow pats. *Journal of Applied Ecology* **27**, 1–15.

Madsen, T., Shine, R., Olsson, M. & Wittzell, H. (1999) Restoration of an inbred adder population. *Nature* **402**, 34–5.

Madsen, T., Stille, B. & Shine, R. (1996). Inbreeding depression in an isolated population of adders *Vipera berus. Biological Conservation* **75**, 113–18.

Magurran, A.E. (1988) *Ecological Diversity and its Measurement*. London: Croom Helm.

Mallory, F.F., Elliot, J.R. & Brooks, R.J. (1981) Change in body size in fluctuating populations of the collared lemming: age and photoperiod influences. *Canadian Journal of Zoology* **59**, 174–82.

Marchant, J.H., Hudson, R., Carter, S.P. & Whittington, P. (1990) *Population Trends in British Breeding Birds*. Tring: British Trust for Ornithology.

Margoluis, R. & Salafsky, N. (1998) *Measures of Success: Designing, Managing, and Monitoring Conservation and Development Projects*. Washington: Island Press.

Margules, C.M. (1986) Conservation evaluation in practice. In: Usher, C.M.B. (ed.) *Wildlife Conservation*. London: Chapman & Hall, pp. 297–314.

Marren, P. (1999) *Britain's Rare Flowers*. London: Poyser.

Martin, G.J. (1995) *Ethnobotany*. London: Chapman & Hall.

Mayer, S.S. & Charlesworth, D. (1991) Cryptic dioecy in flowering plants. *Trends in Ecology and Evolution* **6**, 320–5.

Mayfield, H.F. (1961) Nesting success calculated from exposure. *Wilson Bulletin* **73**, 255–61.

Mayfield, H.F. (1975) Suggestions for analysing nest success. *Wilson Bulletin* **87**, 456–66.

Mayr, E. (1942) *Systematics and the Origin of Species*. Cambridge: Harvard University Press.

Meffe, G.K. & Carroll, C.R. (1994) *Principles of Conservation Biology*. Sunderland: Sinauer.

Merton, D. (1987) Eradication of rabbits from Round Island, Mauritius: a conservation success story. *Dodo* **24**, 19–43.

Merton, D.V., Atkinson, A.E., Strahm, W. *et al.* (1989) *A Management Plan for the Restoration of Round Island Mauritius*. Jersey: Jersey Wildlife Preservation Trust.

Messenger, T., Birks, J. & Jefferies, D. (1997) What is the status of the pine marten in England and Wales? *British Wildlife* **8**, 273–9.

Metcalfe, C.R. (1960) *Anatomy of the Monocotyledons. I. Graminae*. Oxford: Clarendon Press.

Metcalfe, S.C. (1995) Communities, parks and regional planning: a co-management strategy based on the Zimbabwean experience. In: McNeely, J.A. (ed.) *Expanding Partnerships in Conservation*. Washington: Island Press, pp. 270–9.

Mikkelsen, B. (1995) *Methods for Development Work and Research*. New Delhi: Sage Publications Ltd.

Milinski, M. (1997) How to avoid the seven deadly sins in the study of behaviour. *Advances in the Study of Behaviour* **26**, 159–80.

Miller, B. & Mullette, R.J. (1985) Rehabilitation of an endangered Australian bird: the Lord Howe Island Woodhen *Tricholimnas sylvestris* (Sclater). *Biological Conservation* **34**, 55–95.

Miller, B., Biggins, D., Hanebury, L. & Vargas, A. (1994) Reintroduction of the black-footed ferret (*Mustela nigripes*). In: Olney, P.J.S., Mace, G.M. & Feistner, A.T.C. (eds) *Creative Conservation. Interactive Management of Wild and Captive Animals*. London: Chapman & Hall, pp. 456–64.

Mills, L.S. & Smouse, P.E. (1994) Demographic consequences of inbreeding in remnant populations. *American Naturalist* **144**, 412–31.

Mills, L.S., Doak, D.F. & Wisdom, M.J. (1999) Reliability of conservation actions based on elasticity analysis of matrix models. *Conservation Biology* **13**, 815.

Mills, L.S., Hayes, S.G., Baldwin, C. *et al.* (1996) Factors leading to different viability predictions for a grizzly bear data set. *Conservation Biology* **3**, 863–73.

Millsap, B.A., Gore, J.A., Runde, D.E. & Cerulean, S.I. (1990) Setting priorities for the conservation of fish and wildlife species in Florida. *Wildlife Monographs*, **111**, 1–57.

Milner-Gulland, E.J. & Leader-Williams, N. (1992) A model of incentives for the illegal exploitation of black rhinos and elephants: poaching pays in Luangwa Valley, Zambia. *Journal of Applied Ecology* **29**, 388–401.

Minns, C.K. & Hurley, D.A. (1988) Effects of net length and set time on fish catches in gill nets. *North American Journal of Fisheries Management* **8**, 216–23.

Mintzberg, H. (1994) *The Rise and Fall of Strategic Planning*. New York: Prentice-Hall.

Mintzberg, H., Ahlstrand, B. & Lampel, J. (1998) *Strategy Safari*. Hemel Hempstead: Prentice-Hall.

Mitchell, M.K. & Stapp, W.B. (1997) *Field Manual for Water Quality Monitoring. An Environmental Education Program for Schools*. Dubuque: Kendal/Hunt.

Møller, A.P. (1987) Egg predation as a selective factor for nest design: an experiment. *Oikos* **50**, 91–4.

Monaghan, P. & Duncan, W.N. (1979) Variation in plumage characteristics of known age herring gulls. *British Birds* **72**, 100–3.

Morganroth, J., Bigger, J.T., Jr. & Anderson, J.L. (1990) Treatment of ventricular arrhythmia by United States cardiologists: a survey before the Cardiac Arrhythmia Suppression Trial results were available. *American Journal of Cardiology* **65**, 40–8.

Morin, P.A., Moore, J.J. & Woodruff, D.S. (1992) Identification of chimpanzee subspecies with DNA from hair and allele-specific probes. *Proceeding of the Royal Society B* **249**, 293–7.

Moritz, C. (1994) Defining 'Evolutionary Significant Units' for conservation. *Trends in Ecology and Evolution* **9**, 373–5.

Moritz, C. (1995) Uses of molecular phylogenies for conservation. *Philosophical Transactions of the Royal Society London Series B* **349**, 113–18.

Morton, P.A. & Murphy, M.J. (1995) Comprehensive approaches for saving bats. In: Jacobson, S.K. (ed.) *Conserving Wildlife. International Education and Communication Approaches*. New York: Columbia University Press, pp. 103–18.

Mosquera, I., Côté, I.M., Jennings, S. & Reynolds, J.D. (in press) Conservation benefits of marine reserves for fish populations. *Animal Conservation*.

Moss, B., Balls, H., Irvine, K. & Stansfield, J. (1986) Restoration of two lowland lakes by isolation from nutrient-rich water. *Journal of Applied Ecology* **23**, 391–414.

Moss, B., Madgewick, J. & Phillips, G. (1996) *A Guide to the Restoration of Nutrient-enriched Shallow Lakes*. Norwich: Broads Authority.

Moss, B.C. (1988) *Ecology of Freshwaters*. Oxford: Blackwell Science Ltd.

Moss, R., Picozzi, N., Summers, R.W. & Baines, D. (2000) Capercaillies *Tetrao urogallis* in Scotland – demography of a declining population. *Ibis* **142**, 259–67.

Mueller-Dombois, D. & Ellenberg, H. (1974) *Aims and Methods of Vegetation Ecology*. New York: Wiley.

Mullarney, K., Svensson, L., Zetterström, D. & Grant, P.J. (1999) *Collins Bird Guide*. London: Collins.

Murombedzi, J.C. (1999) Devolution and stewardship in Zimbabwe's CAMPFIRE programme. *Journal of International Development* **11**, 287–94.

Murton, R.K., Westwood, N.J. & Isaacson, A.J. (1974) A study of wood-pigeon shooting: the exploitation of a natural animal population. *Journal of Applied Ecology* **11**, 61–81.

Myers, N. (1988) Threatened biotas: 'hotspots' in tropical forests. *Environmentalist* **8**, 1–20.

Myers, N. (1990) The biodiversity challenge: expanded hotspots analysis. *Environmentalist* **10**, 243–56.

National Research Council Committee on Non-human primates (1981) *Techniques for the Study of Primate Population Ecology*. Washington: National Academy Press.

Nelson, J.R. (1987) Rare plant surveys: techniques for impact assessment. In: Elias, T.S. (ed.) *Conservation and Management of Rare and Endangered Plants*. California: California Native Plant Society, pp. 156–66.

Newman, D. & Pilson, D. (1997) Increased probability of extinction due to decreased effective population size: experimental populations of *Clarkia pulchela*. *Evolution* **51**, 354–62.

Nielsen, L.A. & Johnson, D.L. (1983) (eds) *Fisheries Techniques*. Bethesda, MD: American Fisheries Society.

Nilsson, S.G. & Nilsson, I.N. (1976) Valuation of South Swedish wetlands for conservation with the proposal of a new method for valuation of wetlands as breeding habitats for birds. *Fauna och Flora* **71**, 136–44.

Nishikawa, K.C. & Service, P.M. (1988) A fluorescent marking technique for individual recognition of terrestrial salamanders. *Journal of Herpetology* **22**, 351–3.

Noss, R.F. (1987) Corridors in real landscapes: a reply to Simberloff and Cox. *Conservation Biology* **1**, 159–64.

Noss, R.F. (1992) The Wildlands Project land conservation strategy. *Wild Earth*, Special Issue 10–25.

Noss, R.F. & Cooperrider, A.Y. (1994) *Saving Nature's Legacy*. Washington: Island Press.

Nygrén, T. (1987) The history of moose in Finland. *Swedish Wildlife Research Supplement* **1**, 49–54.

Nygrén, T. & Pasonen, M. (1993) The moose population (*Alces alces* L.) and methods of moose management in Finland 1975–89. *Finnish Game Research* **48**, 46–53.

Oates, J.F. (1995) The dangers of conservation by rural development – a case study from the forests of Nigeria. *Oryx* **29**, 115–22.

Oglesby, R.T. (1969) Effects of controlled nutrient dilution on the eutrophication of a lake. In: National Academy of Sciences (eds) *Eutrophication: Causes, Consequences and Correctives*. Washington, DC: pp. 483–93.

Olindo, P. (1991) The old man on nature tourism. In: Whelan, T. (ed.) *Nature Tourism*. Washington: Island Press, pp. 23–38.

Omar, R.M.N.R., Kean, C.E., Wagiman, S. *et al.* (1994) Design and construction of artificial reefs in Malaysia. *Bulletin of Marine Science* **55**, 1050–61.

O'Neill, P., Singh, M.B. & Knox, R.B. (1988) Cell biology of the stigma of *Brassica campestris* in relation to CO_2 effects on self-pollination. *Journal of Cell Science* **89**, 541–9.

Oren, U. & Benayahu, Y. (1997) Transplantation of juvenile corals: a new approach for enhancing colonisation of artificial reefs. *Marine Biology* **127**, 499–505.

O'Toole, R. (1991) Recreation fees and the Yellowstone Forests. In: Heiter, R.B. & Boyce, M.S. (eds) *The Greater Yellowstone Ecosystem*. Yale: Yale University Press, pp. 41–8.

Otto, S.P. & Whitlock, M.C. (1997) The probability of fixation in populations of changing size. *Genetics* **146**, 723–33.

Overseas Development Administration (1995) *A Guide to Social Analysis for Projects in Developing Countries*. London: HMSO.

Paine, R.T. (1974) Intertidal community structure: experimental studies on the relationship between a dominant competitor and its principal predator. *Oecologia* **15**, 93–120.

Palmerzwahlen, M.L. & Aseltine, D.A. (1994) Successional development of the turf community on a quarry rock artificial reef. *Bulletin of Marine Science* **55**, 902–23.

Parker, D.M. (1995) *Habitat Creation – a Critical Guide*. Peterborough: English Nature.

Parsons, T.R., Maita, Y. & Lalli, C.M. (1984) *A Manual of Chemical and Biological Methods for Seawater Analysis*. Oxford: Pergamon Press.

Passmore, V. & Carruthers, N. (1995) *South African Frogs – a Complete Guide*. Johannesburg: Southern Book Publishers.

Paton, P.W.C. (1994) The effect of edge on avian nest success: how strong is the evidence? *Conservation Biology* **8**, 17–26.

Paxinos, E., McIntosh, C., Ralls, K. & Fleischer, R. (1997) A non-invasive method for distinguishing among canid species: amplification and enzyme restriction of DNA from dung. *Molecular Ecology* **6**, 483–6.

Payne, N.F. (1992) *Techniques for Wildlife Habitat Management of Wetlands*. New York: McGraw-Hill.

Peepre, J. (1998) On wilderness values and wild rivers. A Canadian perspective. *Wild Earth* **8**, (4) 19–22.

Peiperi, M. (1997) Does empowerment deliver the goods? In: Bickerstaffe, G. (ed.) *Mastering Management*. London: Financial Times, pp. 283–7.

Peres, C.A (2000) Evaluating the impact and sustainability of subsistence hunting at multiple Amazonian forest sites. In:

Robinson, J.G. & Bennett, E.L. (eds) *Evaluating the Sustainability of Hunting in Tropical Forests*. New York: Columbia University Press, pp. 31–57.

Perrow, M.R., Côté, I.M. & Evans, M. (1996) Fish. In: Sutherland, W.J. (ed.) *Ecological Census Techniques*. Cambridge: Cambridge University Press, pp.178–204.

Peterken, G.F. (1996) *Natural Woodland: Ecology and Conservation in Northern Temperate Regions*. Cambridge: Cambridge University Press.

Peters, C.M., Gentry, A.H. & Mendelsohn, R.O. (1989) Valuation of an Amazonian rainforest. *Nature* **339**, 655–6.

Petersen, C.G.J. (1896) The yearly immigration of young plaice into Limfjord from the German Sea. *Report Danish Biological Station* **6**, 1–48.

Petersen, R.T. (1934) *A Field Guide to the Birds*. Boston: Houghton Mifflin.

Petokas, P.J. & Alexander, M.M. (1979) A new trap for basking turtles. *Herpetological Review* **10**, 90.

Piersma, T. (1998) Phenotypic flexibility during migration: physiological optimisation of organ sizes contingent on the risks and rewards of refuelling and flight. *Journal of Avian Biology* **92**, 511–20.

Pimm, S.L. (1987) The snake that ate Guam. *Trends in Ecology and Evolution* **2**, 293–5.

Pimm, S.L. (1992) *Balance of Nature*. Chicago: Chicago University Press.

Pletscher, D.H. (1995) Age and sex criteria with special reference to Indian Wildlife. In: Berwick, S.H. & Saharia, V.B. (eds) *The Development of International Principles and Practices of Wildlife Research*. Delhi: Oxford University Press, pp. 107–21.

Pollock, K.H., Nichols, J.D., Brownie, C. & Hines, J.E. (1990) Statistical inference from capture–recapture experiments. *Wildlife Monographs* **107**.

Pomeroy, D. & Tengecho, B. (1986) Studies of birds in a semi-arid area of Kenya III – the use of 'timed-species-counts' for studying regional avifaunas. *Journal of Tropical Ecology* **2**, 231–47.

Posey, D. (1990) Intellectual property rights: what is the position of ethnobiology? *Journal of Ethnobiology* **10**, 93–8.

Potts, G.R. (1986) *The Partridge: Pesticides, Predation and Conservation*. London: Collins.

Potts, G.R. & Aebischer, N.J. (1995) Population dynamics of the grey partridge *Perdix perdix* 1793–1993: monitoring, modelling and management. *Ibis* **137**, S29–37.

Prater, A.J. (1979) Trends in accuracy of counting birds. *Bird Study* **26**, 198–200.

Prendergast, J.R., Quinn, R.M., Lawton, J.H., Eversham, B.C. & Gibbons, D.W. (1994) Rare species, the coincidence of diversity hotspots and conservation strategies. *Nature* **365**, 335–7.

Pressey, R.L., Johnson, I.R. & Wilson, P.D. (1994) Shades of irreplaceability: towards a measure of the contribution of sites to a reservation goal. *Biodiversity and Conservation* **3**, 242–62.

Primack, R. (1998) *Essentials of Conservation Biology*. Sunderland: Sinauer.

Primmer, C.R., Møller, A.P. & Ellegren, H. (1996) A wide-range survey of cross-species microsatellite amplification in birds. *Molecular Ecology* **5**, 365–78.

Proctor, M., Yeo, P. & Lack, A. (1996) *The Natural History of Pollination*. London: HarperCollins.

Pye-Smith, C. & Feyerabend, G.B. (1994) *The Wealth of Communities: Stories of Success in Local Environmental Management*. London: Earthscan Publications Ltd.

Quinn III T.J. & Deriso, R.B. (1999). *Quantitative Fish Dynamics*. New York: Oxford University Press.

Racey, P.A. & Swift, S.M. (1986) The residual effects of remedial timber treatment on bats. *Biological Conservation* **35**, 205–14.

Rackham, O. (1998) Implications of historical ecology for conservation. In: Sutherland, W.J. (ed.) *Conservation Science and Action*. Oxford: Blackwell Science Ltd, pp. 152–75.

Rackham, O. & Moody, J. (1997) *The Making of the Cretan Landscape*. Manchester: Manchester University Press.

Rao, N.K., Roberts, E.H. & Ellis, R.H. (1987) Loss of viability in lettuce seeds and the accumulation of chromosomal damage under different storage conditions. *Annals of Botany* **60**, 85–96.

Reading, C.J. (1997) A proposed standard method for surveying reptiles on dry lowland heath. *Journal of Applied Ecology* **34**, 1057–69.

Reading, C.J. & Davies, J.L. (1996) Predation by grass snakes (*Natrix natrix* L.) at a site in southern England. *Journal of Zoology* **239**, 73–82.

Reaka-Kundla, M.L., Wilson, D.E. & Wilson, E.O. (eds) (1997) *Biodiversity II*. Washington: Joseph Henry Press.

Rebelo, A.G. & Siegfried, W.R. (1992) Where should nature reserves be located in the Cape Floristic Region, South Africa? Models for the spatial configuration of a reserve network aimed at maximising the protection of floral diversity. *Conservation Biology* **6**, 243–52.

Recher, H.F. & Clark, S.S. (1974) A biological survey of Lord Howe Island with recommendations for the conservation of the island's wildlife. *Biological Conservation* **6**, 263–73.

Reed, J.M. (1996) Using statistical probability to increase confidence of inferring species extinction. *Conservation Biology* **10**, 1283–95.

Rich, T.C.G. & Smith, P.A. (1996) Botanical recording, distribution maps and species frequencies. *Watsonia* **21**, 155–67.

Rich, T.C.G., Donovan, P., Harmes, P. *et al.* (1996) *Flora of Ashdown Forest*. East Grinstead: Sussex Botanical Recording Society.

Richards, A.J. (1999) *Plant Breeding Systems*. London: Chapman & Hall.

Richer, W.E. (1975) Competition and interpretation of biological statistics of fish populations. *Bulletin of the Fisheries Research Board of Canada* **191**, 1–382.

Ridgeway, B., McCabe, M., Bailey, J., Saunders, R. & Sadler, B. (1996) *Environmental Impact Assessment Training Course*

Resource Manual. Nairobi: United Nations Environment Programme.

Roberts, E.H. (1992) Physiological aspects of *ex situ* seed conservation. In: Kapoor-Vijay, P. & White, J. (eds) *Conservation Biology: A Training Manual for Biological Diversity and Genetic Resources*. London: Commonwealth Secretariat, pp. 171–7.

Robertson, A.W., Kelly, D., Ladley, J.J. & Sparrow, A.D. (1999) Effects of pollinator loss on endemic New Zealand mistletoes (Loranthaceae). *Conservation Biology* **13**, 499–508.

Robertson, B.C., Minot, E.O. & Lambert, D.M. (1999) Molecular sexing of individual kakapo *Strigops habroptilus* Aves from faeces. *Molecular Ecology* **8**, 1347–50.

Robertson, C.J.R. & Nunn, G.B. (1997) Towards a new taxonomy for albatrosses. In: Robertson, G. & Gales, R. (eds) *Albatross Biology and Conservation*. Chipping Norton: Surrey Beatty, pp. 13–19.

Robertson, P. & Liley, D. (1998) Assessment of sites – measurement of species richness and diversity. In: Bibby, C., Jones, M. & Marsden, S. (eds). *Bird Surveying and Conservation*. London: Royal Geographical Society, pp. 76–98.

Robertson, P.A. (1993) *The management of artificial coastal lagoons in relation to invertebrates and avocets* Recurvirostra avosetta *(L).* PhD thesis, University of East Anglia.

Robinson, J.G. & Redford, K.H. (1991) Sustainable harvest of neotropical forest mammals. In: Robinson, J.G. & Redford, K.H. (eds) *Neotropical Wildlife Use and Conservation*. Chicago: University of Chicago Press, pp. 415–29.

Robinson, J.G. & Redford, K.H. (1994) Measuring the sustainability of hunting in tropical forests. *Oryx* **28**, 249–56.

Roffe, T.J., Friend, M. & Locke, L.N. (1996) Evaluation of causes of wildlife mortality. In: Bookhout, T.A. (ed.) *Research and Management Techniques for Wildlife and Habitats*. Lawrence, KS: The Wildlife Society, pp. 324–48.

Royal Society for the Protection of Birds, English Nature & Institute for Terrestrial Ecology (1997) *The Wet Grassland Guide*. Sandy: Royal Society for the Protection of Birds.

Royal Society Study Group (1992) *Risk: Analysis, Perception and Management*. London: The Royal Society.

Rozhnov, V.V. (1993) Extinction of the European mink: ecological catastrophe or a natural process. *Lutreola* **1**, 10–16.

Saccheri, I., Kuussaari, M., Kankare, M., Vikman, P., Fortelius, W. & Hanski, I. (1998) Inbreeding and extinction in a butterfly metapopulation. *Nature* **392**, 491–4.

Sackett, D.L., Haynes, R.B., Taylor, D.W. *et al.* (1977) Clinical determinants of the decision to treat primary hypertension. *Clinical Research* **24**, 648.

Sackett, D.L., Richardson, W.S., Rosenberg, W. & Haynes, R.B. (1998) *Evidence-based Medicine. How to Practice and Teach EBM*. Edinburgh: Churchill Livingstone.

Samuel, M.D. & Fuller, M.R. (1994) Wildlife radiotelemetry. In: Bookhout T.A. (ed.) *Research and Management Techniques for Wildlife and Habitats*. Bethesda: The Wildlife Society, pp. 370–418.

Savage, M. & Swetman, T.W. (1990) Early 19th-century fire decline following sheep pasturing in a Navajo Ponderosa Pine Forest. *Ecology* **71**, 2374–8.

Savidge, J.A. (1987) Extinction of an island forest avifauna by an introduced snake. *Ecology* **68**, 660–8.

Scarlet, C.G., Flake, L.D. & Willis, D.W. (1996) *Introduction to Wildlife and Fisheries*. New York: W.H. Freeman.

Schaefer, M.B. (1954) Some aspects of the dynamics of populations important to the management of commercial marine fisheries. *Inter-American Tropical Tuna Commission Bulletin* **1**, 27–54.

Schaffer, M.L. (1981) Minimum population size for species conservation. *Bioscience* **31**, 131–4.

Schaffer, M.L. (1987) Minimum viable populations: coping with uncertainty. In: Soulé, M. (ed.) *Viable Populations for Conservation*. Cambridge: Cambridge University Press, pp. 69–86.

Schaller, G.A. & Rabinowitz, A. (1995) The saola or spindle-horn bovid *Pseudoryx nghetinhensis* in Laos. *Oryx* **29**, 107–14.

Schaller, G.A. & Vrba, E.S. (1996) Description of the giant muntjac (*Megamuntiacus vuquangensis*) in Laos. *Journal of Mammology* **77**, 675–83.

Schaller, G.B. (1993) *The Last Panda*. Chicago: Chicago University Press.

Scheepers, J.L. & Venzke, K.A.E. (1995) Attempts to reintroduce African wild dogs *Lycaon pictus* into Etosha National Park, Namibia. *South African Journal of Wildlife Research* **25**, 138–40.

Schmidt, K. (1996) Rare habitats vie for protection. *Science* **274**, 916–18.

Schriver, P., Bøgestrand, J., Jeppesen, E. & Søndergaard, M. (1995) Impact of submerged macrophytes on fish–zooplankton–phytoplankton interactions: large-scale enclosure experiments in a shallow eutrophic lake. *Freshwater Biology* **33**, 255–70.

Schullery, P.D. (1997) *Searching for Yellowstone: Ecology and Wonder in the Last Wilderness*. Boston: Houghton-Mifflin.

Slessman, M.A., Lowry, P.P., II & Lloyd, D.G. (1990) Functional dioecism in the New Caledonian endemic *Polyscias pancheri* (Araliaceae). *Biotropica* **22**, 133–9.

Scott, P. (1998) *From Conflict to Collaboration. People and Forests at Mount Elgon, Uganda*. Gland: International Union for the Conservation of Nature and Natural Resources, The World Conservation Union.

Seber, G.A.F. (1965) A note on the multiple-recapture census. *Biometrika* **52**, 249–59.

Seber, G.A.F. (1982) *Estimation of Animal Abundance*. London: Griffin.

Shah, P. & Shah, M.K. (1995) Participatory methods: precipitating or avoiding conflict? *PLA Notes* No 24. 48–51.

Sharrock, J.T.R. (1976) *The Atlas of Breeding Birds in Britain and Ireland*. Berkhamstead: Poyser.

Shaw, E.M. (1994) *Hydrology in Practice*. London: Chapman & Hall.

Shaw, G. & Williams, A. (1994) *Critical Issues in Tourism.* Oxford: Blackwell Science Ltd.

Shepard, D. & McNeely, J. (1998) Education and protected areas: a perspective from IUCN. In: Leal Filho, W., de Carvalho, C.A.R. & Hale, W.H.G. (eds) *Environmental Education in Protected Areas.* Carnforth: Parthenon Publishing Group Ltd, pp. 139–47.

Sherden, W.A. (1998) *The Fortune Sellers: The Big Business of Buying and Selling Predictions.* New York: Wiley.

Shorrocks, G. (1997) The success of DNA fingerprinting in wildlife law enforcement. *RSPB Conservation Review* **11**, 96–100.

Short, J., Bradshaw, S.D., Giles, J., Prince, R.I.T. & Wilson, G.R. (1992) Reintroduction of macropods (Marsupialia: Macropodoidae) in Australia – a review. *Biological Conservation* **62**, 189–204.

Sibley, C.G. & Monroe, B.L. (1990) *Distribution and Taxonomy of Birds of the World.* New Haven: Yale University Press.

Simberloff, D. (1998) Small and declining populations. In: W.J. Sutherland (ed.) *Conservation Science and Action.* Oxford: Blackwell Science Ltd, pp. 116–34.

Simberloff, D. & Cox, J. (1987) Consequences and costs of conservation corridors. *Conservation Biology* **1**, 63–71.

Simberloff, D., Farr, J.A., Cox, J. & Mehlman, D.W. (1992) Movement corridors: conservation bargains or poor investments? *Conservation Biology* **6**, 493–504.

Sinclair, A.R.E. (1989) Population regulation in animals. In: Cherrett, J.M. (ed.) *Ecological Concepts.* Oxford: Blackwell Scientific Publications, pp. 197–241.

Sinclair, I., Hockey, P. & Tarboton, W. (1999) *Birds of Southern Africa.* Cape Town: Struik.

Slade, N.A., Gomulkiewicz, R. & Alexander, H.M. (1998) Alternatives to Robinson and Redford's method of assessing overharvest from incomplete demographic data. *Conservation Biology* **12**, 148–55.

Slatis, M.N. (1960) An analysis of inbreeding in the European bison. *Genetics* **45**, 275–87.

Smith, P.G.R. & Theberge, J.B. (1986) A review of criteria for evaluating natural areas. *Environmental Management* **10**, 715–34.

Smith, T.B. & Wayne, R.K. (eds) (1996) *Molecular Genetic Approaches in Conservation.* Oxford: Oxford University Press.

Snow, D.W. (1997) Should the biological be superseded by the phylogenetic species concept? *Bulletin British Ornithologists Club* **117**, 110–21.

Snyder, N.F.R. (1994) The Californian Condor recovery program: problems in organisation and execution. In: Clark, T.W., Reading, R.P., & Clark, A.L. (eds) *Endangered Species Recovery.* Washington: Island Press, pp. 183–204.

Soberón, J.M. & Llorente, J.B. (1993) The use of species accumulation functions for the prediction of species richness. *Conservation Biology* **7**, 480–8.

Soderquist, T.R. (1994) The importance of hypothesis testing in reintroduction biology: examples from the reintroduction of the carnivorous marsupial *Phascogale tapoutafa.* In: M. Serena (ed.) *Reintroduction Biology of Australian and New Zealand Fauna.* Chipping Norton: Surrey Beatty, pp. 159–64.

Söderström, L. (1988) The occurrence of epixylic bryophyte and lichen species in an old natural and a managed forest stand in north-east Sweden. *Biological Conservation* **45**, 169–78.

Søndergaard, M., Jeppesen, E. & Berg, S. (1997) Pike (*Esox lucius* L.) stocking as a biomanipulation tool. 2. Effects on lower trophic levels in Lake Lyng, Denmark. *Hydrobiologia* **342**, 319–25.

Soulé, M. (1992) A vision for the meantime. *Wild Earth,* Special Issue 10–25.

Southwood, T.R.E. (1978) *Ecological Methods with Particular Reference to the Study of Insect Populations.* London: Chapman & Hall.

Splettstoesser, J. & Folks, M.C. (1994) Environmental guidelines for Antarctica. *Annals of Tourism Research* **21**, 231–44.

Squires, N.R.W., Hagger, R.J. & Elliot, J.G. (1979) A one-pass seeder for introducing grasses, legumes and fodder crops into swards. *Journal of Agricultural Engineering Research* **24**, 199–208.

Stapp, W.B., Cromwell, M.M. & Wals, A. (1995) The Global Rivers Environmental Education Network. In: Jacobson, S.K. (ed.) *Conserving Wildlife. International Education and Communication Approaches.* New York: Colombia University Press, pp. 177–97.

Stattersfield, A., Crosby, M., Long, A. & Wege, D. (1998) *Endemic Bird Areas of the World: Priorities for Biodiversity Conservation.* Cambridge: BirdLife International.

Stephens, P.A. & Sutherland, W.J. (1999) Consequences of the Allee effect for ecology and conservation. *Trends in Ecology and Evolution* **14**, 401–5.

Stephens, P.A., Sutherland, W.J. & Freckleton, R.F. (1999) What is the Allee effect? *Oikos* **87**, 185–90.

Stevensen, M.J., Bullock, J.M. & Ward, L.K. (1995) Recreating semi-natural communities: effect of sowing rate on establishment of calcareous grassland. *Restoration Ecology* **3**, 279–89.

Stewart, D.R.M. (1967) Analysis of plant epidermis in faeces: a technique for studying the food preferences of grazing herbivores. *Journal of Applied Ecology* **4**, 83–111.

Stokes, M.A. & Smiley, T.L. (1996) *An Introduction to Tree-ring Dating.* Tucson: University of Arizona Press.

Surridge, A.K., Timmins, R.J., Hewitt, G.M. & Bell, D.J. (1999) Striped rabbits in Southeast Asia. *Nature* **400**, 726.

Sutherland, W.J. (1987) Random and deterministic components of variance in mating success. In: Bradbury, J.W. & Andersson, M.B. (eds) *Sexual Selection: Testing the Alternatives.* Chichester: Wiley, pp. 207–19.

Sutherland, W.J. (1990) The great pigeonhole in the sky. *New Scientist* **1720**, 73–4.

Sutherland, W.J. (1994) How to save the corncrake. *Nature* **372**, 223.

Sutherland, W.J. (1996a) Mammals. In: Sutherland, W.J. (ed.) *Ecological Census Techniques*. Cambridge: Cambridge University Press, pp. 260–80.

Sutherland, W.J. (1996b) Predicting the consequences of habitat loss for migratory populations. *Proceedings of the Royal Society* **263**, 1325–7.

Sutherland, W.J. (1996c) *From Individual Behaviour to Population Ecology*. Oxford: Oxford University Press.

Sutherland, W.J. (1998a) *Conservation Science and Action*. Oxford: Blackwell Science.

Sutherland, W.J. (1998b) The effect of change in habitat quality on populations of migratory species. *Journal of Applied Ecology* **35**, 418–21.

Sutherland, W.J. (1998c) Managing species and habitats. In: W.J. Sutherland (ed.) *Conservation Science and Action*. Oxford: Blackwell Science Ltd, pp. 203–19.

Sutherland, W.J. & Crockford, N.J (1993) Factors affecting the feeding distribution of Red-breasted Geese *Branta ruficollis* wintering in Romania. *Biological Conservation* **63**, 61–5.

Sutherland, W.J. & Walton, D. (1990) The changes in morphology and demography of *Iris pseudacorus* at different heights on a saltmarsh. *Functional Ecology* **4**, 661–5.

Svensson, L. (1992) *Identification Guide to European Passerines*. Stockholm: Naturhistoriska Riksmuseet.

Szmidt, A.E., Alden, T. & Hallgren, J.-E. (1987) Paternal inheritance of chloroplast DNA in *Larix*. *Plant Molecular Biology* **9**, 59–64.

Taberlet, P. (1996) The use of mitochondrial DNA control region sequencing in conservation genetics. In: Smith, T.B. & Wayne, R.K. (eds) *Molecular Genetic Approaches in Conservation*. New York: Oxford University Press, pp. 125–42.

Taberlet, P. & Bouvet, J. (1991) A single plucked hair as a source of DNA for bird genetic studies. *Auk* **108**, 959–60.

Taberlet, P. & Luikart, G. (1999) Non-invasive genetic sampling and individual recognition. *Biological Journal of the Linnean Society* **68**, 41–55.

Taberlet, P., Camarra, J.-J., Griffin, S. *et al.* (1997) Noninvasive genetic tracking of the endangered Pyrenean brown bear population. *Molecular Ecology* **6**, 869–76.

Taberlet, P., Griffin, S., Goosens, B. *et al.* (1996) Reliable genotyping of samples with very low DNA quantities using PCR. *Nucleic Acids Research* **26**, 3189–94.

Tasker, M.L., Hope Jones, P., Dixon, T. & Blake, B.F. (1984) Counting seabirds at sea from ships: a review of methods employed and a suggestion for a standardised approach. *Auk* **101**, 567–77.

Templeton, A.R. & Read, B. (1983) The elimination of inbreeding depression in a captive herd of Speke's gazelle. In: Schonewald-Cox, C.M., Chambers, S.M., MacBryde, B. & Thomas, L. (eds) *Genetics and Conservation: A Reference for Managing Wild Animal and Plant Populations*. Menlo Park: Benjamin-Cummins, pp. 241–61.

Templeton, A.R. & Read, B. (1984) Factors eliminating inbreeding depression in a captive herd of Speke's gazelle (*Gazella spekei*). *Zoo Biology* **3**, 177–99.

Therival, R., Wilson, E., Thompson, S., Heaney, D. & Pritchard, D. (1992) *Strategic Environmental Assessment*. London: Earthscan Publications Ltd.

Thomas, C.D. & Hanski, I. (1997) Butterfly metapopulations. In: Hanski, I. & Gilpin, M.E. (eds) *Metapopulation Biology*. London: Academic Press, pp. 359–86.

Thomas, C.D. & Jones, T.M. (1993) Partial recovery of a skipper butterfly (*Hesperia comma*) from population refuges: lessons for conservation in a fragmented habitat. *Journal of Animal Ecology* **62**, 563–7.

Thomas, C.D., Baguette, M. & Lewis, O.T. (2000) Butterfly movement and conservation in patchy landscapes. In: Gosling, M. & Sutherland, W.J. (eds) *Behaviour and Conservation*. Cambridge: Cambridge University Press, pp. 85–104.

Thomas, D. (in press) Devolution of decision making: lessons from community forest management at the Kilum-Ijum Forest project, Cameroon. In Jeffery, R. & Vira, B. (eds) *Co-operation and Conflict in Natural Resource Management: Lessons from Case Studies*. London: Macmillan.

Thomas, D.C. (1977) Metachromatic staining of dental cementum for mammalian age determination. *Journal of Wildlife Management* **41**, 207–10.

Thomas, J.A. (1991) Rare species conservation: case studies of European butterflies. In: Spellerberg, I.F., Goldsmith, F.B. & Morris, M.G. (eds) *The Scientific Management of Temperate Communities for Conservation*. Oxford: Blackwell Scientific Publications, pp. 144–97.

Thomas, J.A., Thomas, C.D., Simcox, D.J. & Clarke, R.T. (1986) Ecology and declining status of the silver-spotted skipper butterfly (*Hesperia comma*) in Britain. *Journal of Applied Ecology* **23**, 365–80.

Thompson, C.F. & Neill, A.J. (1991) House wrens do not prefer clean nestboxes. *Animal Behaviour* **42**, 1022–4.

Thompson, C.F. & Neill, A.J. (1993) Statistical power and accepting the null hypothesis. *Animal Behaviour* **46**, 1022–4.

Thompson, K. & Grime, J.P. (1983) A comparative study of germination responses to diurnally fluctuating temperatures. *Journal of Applied Ecology* **20**, 141–56.

Thompson, P.A. (1974) The use of seed banks for conservation of populations of species and ecotypes. *Biological Conservation* **6**, 15–19.

Thompson, W.L., White, G.C. & Gowan, C. (1998) *Monitoring Vertebrate Populations*. San Diego: Academic Press.

Thornton, I. (1971) *Darwin's Islands: A Natural History of the Galapagos*. New York: Natural History Press.

Tobias, D. & Mendelsohn, R. (1991) Valuing ecotourism in a tropical rain forest. *Ambio* **20**, 91–3.

Tomiałojc, L. (1991) Characteristics of old growth in the Białowieza Forest, Poland. *Natural Areas Journal* **11**, 7–18.

Treweek, J. (1996) Ecology and environmental impact assessment. *Journal of Applied Ecology* **33**, 191–9.

Treweek, J. (1999) *Ecological Impact Assessment*. Oxford: Blackwell Science Ltd.

Trivers, R.L. (1972) Parental investment and sexual selection. In: Campbell, B. (ed.) *Sexual Selection and the Descent of Man 1871–1971*. Chicago: Aldine, pp. 136–79.

Tuljapurkar, S. (1990) *Population dynamics in variable environments*. Lecture Notes in Biomathematics 85. New York: Springer-Verlag.

Turpie, J.K. (1995) Prioritising South African estuaries for conservation: a practical example using waterbirds. *Biological Conservation* **74**, 175–85.

Tuttle, M.D. (1990) Return to Thailand. *Bats* **8** (3) 6–11.

Uhl, C. & Kaufman, J.B. (1990) Deforestation, fire susceptibility and potential tree responses to fire in the eastern Amazon. *Ecology* **71**, 437–49.

Urban, E.K., Fry, C.H. & Keith, S. (eds) (1997) *Birds of Africa*. Vol. 5. London: Academic Press.

Usher, M.B. (1986) Wildlife conservation evaluation: attributes, criteria and values. In: Usher, M.B. (ed.) *Wildlife Conservation Evaluation*. London: Chapman & Hall, pp. 3–44.

Valutis, L.L. & Marzluff, J.M. (1999) The appropriateness of puppet-reared birds for reintroduction. *Conservation Biology* **13**, 584–91.

Vane-Wright, R.I., Humphries, C.J. & Williams, P.H.C . (1991) What to protect? Systematics and the agony of choice. *Biological Conservation* **55**, 235–54.

Vera, F.W.M. (1997) *Metaphors for the Wilderness: Oak, Hazel, Cattle and Horse*. The Hague: Ministry of Agriculture, Nature Management and Fisheries.

Vickers, W.T. (1991) Hunting yields and game composition over ten years in an Amazon Indian territory. In: Robinson, J.G. & Redford, K.H. (eds) *Neotropical Wildlife Use and Conservation*. Chicago: Chicago University Press, pp. 58–81.

Vickery, J. (1996) Access. In: Sutherland, W.J. & Hill, D.A. (eds) *Managing Habitats for Conservation*. Cambridge: Cambridge University Press, pp. 42–58.

Walters, C. & Maguire, J.J. (1996) Lessons from stock assessment from the northern cod collapse. *Reviews in Fish Biology and Fisheries* **6**, 125–37.

Walters, C.J. (1986) *Adaptive Management of Renewable Resources*. London: Macmillan.

Warren, K.S. & Mosteller, F. (eds) (1993) Doing more harm than good: the evaluation of health care interventions. *Annals of the New York Academy of Sciences* **703**, 1–341.

Waterman, R. (1994) *The Frontiers of Excellence*. London: Brealey Publishing Ltd.

Watkinson, A.R. (1985) Plant responses to crowding. In: White, J. (ed.) *Studies on Plant Demography: a Festschrift for John L. Harper*. London: Academic Press, pp. 275–89.

Webb, N.R. & Thomas, J.A. (1994) Conserving insect habitats in heathland biotopes: a question of scale. In: Edwards, P.J., May, R.M. & Webb, N.R. (eds) *Large-scale Ecology and Conservation Biology*. Oxford: Blackwell Science Ltd, pp. 129–51.

Welch, E.B. & Patmont, C.R. (1980) Lake restoration by dilution: Moses Lake, Washington. *Water Research* **14**, 1317–25.

Weller, S.G. (1995) The relationship of rarity to plant reproductive biology. In: Bowles, M.L. & Whelan, C.J. (eds) *Restoration of Endangered Species*. Cambridge: Cambridge University Press, pp. 90–117.

Wells, M., Guggenheim, S., Khan, A., Wardojo, W. & Jepson, P. (1999) *Investing in Biodiversity. A Review of Indonesia's Integrated Conservation and Development Projects*. Washington, DC: The World Bank.

Wells, M.P. & Brandon, K.E. (1993) The principles and practice of buffer zones and local participation in biodiversity conservation. *Ambio* **22**, 157–62.

Wells, T.C.E. (1989) The re-creation of grassland habitats. *The Entomologist* **108**, 97–108.

Weston, P.B. & Wells, K.M. (1997) *Criminal Investigation*. New Jersey: Prentice-Hall.

Wetton, J.H. & Parkin, D.T. (1997) A suite of falcon single-locus minisatellite probes: a powerful alternative to DNA fingerprinting. *Molecular Ecology* **6**, 119–28.

Whelan, R.J. (1995) *The Ecology and Fire*. Cambridge: Cambridge University Press.

Whitbread, A. & Jenman, W. (1995) A method of conserving biodiversity in Britain. *British Wildlife* **7**, 84–93.

White, G.C. & Garrott, R.A. (1990) *Analysis of Wildlife Radiotracking Data*. London: Academic Press.

Whitney, G.G. (1996) *From Coastal Wilderness to Fruited Plain: A History of Environmental Change in Temperate North America 1500 – Present*. New York: Cambridge University Press.

Wilberforce, S. (1998) *Legacy Fundraising*. West Malling: Charities Aid Foundation.

Wilkie, D.S & Carpenter, J.F. (1999a) Bushmeat hunting in the Congo Basin: an assessment of impacts and options for mitigation. *Biodiversity and Conservation* **8**, 927–55.

Wilkie, D.S. & Carpenter, J.F. (1999b) The potential role of safari hunting as a source of revenue for protected areas in the Congo Basin. *Oryx* **33**, 339–45.

Wilkinson, T. (1998) *Science Under Siege: The Politicians' War on Nature and Truth*. Boulder: Johnson Press.

Williams, G. (1980) An index for ranking of wildfowl habitats, as applied to eleven sites in west Surrey, England. *Biological Conservation* **18**, 93–9.

Williams, G., Holmes, J. & Kirby, J. (1995) Action plans for United Kingdom rare, threatened and internationally important birds. *Ibis* **137**, S209–13.

Williams, P., Biggs, J., Corfield, A., Fox, G., Walker, D. & Whitfield, M. (1997) Designing new ponds for wildlife. *British Wildlife* **8**, 137–50.

Williams, P., Gibbons, D., Margules, C. *et al.* (1996) A comparison of richness hotspots, rarity hotspots, and complementary areas for conserving diversity of British birds. *Conservation Biology* **10**, 155–74.

Williams, P.H. (1999) Key sites for conservation: area-selection methods for biodiversity. In: Mace, G.M.,

Balmford, A. & Ginsberg, J.R. *Conservation in a Changing World*. Cambridge: Cambridge University Press, pp. 211–49.

Wilson, B., Hammond, P.S. & Thompson, P.M. (1999) Estimating size and assessing trends in a coastal Bottlenose Dolphin population. *Ecological Applications* **9**, 288–300.

Wilson, D.E., Cole, F.R., Nichols, J.D., Rudran, R. & Foster, M.S. (1996) *Measuring and Monitoring Biological Diversity: Standard Methods for Mammals*. Washington & London: Smithsonian Institution Press.

Wilson, I. (1994) Strategic planning isn't dead – it changed. *Long-Range Planning* **27**, 12–24.

Wilson, J.E. & Macdonald, J.W. (1967) Salmonella infection in wild birds. British *Veterinary Journal* **123**, 212–19.

Wishart, G.J. (1987) Regulation of the length of the fertile period in the domestic fowl by numbers of oviductal spermatozoa as reflected by those trapped in laid eggs. *Journal of Reproductive Fertility* **80**, 493–8.

Wood, C. (1995) *Environmental Impact Assessment. A Comparative Review*. Harlow: Longman.

Woodell, S.J. (1975) Five years of a county flora project. *Watsonia* **10**, 265–72.

Woodford, M.H. & Rossiter, P.B. (1994) Disease risks associated with wildlife translocation projects. In: Olney, P.J.S., Mace, G.M. & Feistner, A.T.C. (eds) *Creative Conservation*. London: Chapman & Hall, pp. 178–200.

Woodruffe, R. & Ginsberg, J.R. (2000) Ranging behaviour and extinction in carnivores: how behaviour affects species vulnerability. In: Gosling, M. & Sutherland, W.J. (eds) *Behaviour and Conservation*. Cambridge: Cambridge University Press, pp. 125–140.

World Bank (1990) *World Development Report 1990*. New York: Oxford University Press.

World Bank (1998) *Assessing Aid: What Works, What Doesn't and Why*. New York: Oxford University Press.

Worrell, R. (1992) A comparison between European, continental and British provenances of some British native trees: growth, survival and stem form. *Forestry* **65**, 253–80.

Wright, H.E. & Heinselman, M.L. (1973) The ecological role of fire in natural conifer forests of western and northern North America introduction. *Quaternary Research* **3**, 319–28.

Wright, R., Ray, S., Green, D.R. & Wood, M. (1998) Development of a GIS of the Moray Firth (Scotland, UK) and its application in environmental management. *Science of the Total Environment* **223**, 65–76.

Wright, R.G. (1999) Wildlife management in the national parks: questions in search of answers. *Ecological Applications* **9**, 30–6.

Wynne, G. (1998) Conservation policy and policies. In: Sutherland, W.J. (ed.) *Conservation Science and Action*. Oxford: Blackwell Science Ltd, pp. 256–85.

Yates, T.L., Jones, C. & Cook, J.A. (1996) Preservation of voucher specimens. Standard method for mammals. In: Wilson, D.E., Cole, F.R., Nichols, J.D., Rudron, R. & Foster, M.S. (eds) *Measuring and Monitoring Biological Diversity*. Washington: Smithsonian Institution Press, pp. 265–74.

Yerli, S., Canbolat, A., Brown, L.J. & MacDonald, D.W. (1997) Wire grids protect turtle nests from red fox predation. *Biological Conservation* **82**, 109–11.

Yoccoz, N.G. (1991) Use, overuse, and misuse of significance tests in evolutionary biology and ecology. *Bulletin of Ecological Society of America* **72**, 106–11.

Youell, R. (1998) *Tourism. An Introduction*. Harlow: Addison Wesley Longman.

Zackrisson, O. (1977) Influence of forest fires on the north Swedish boreal forest. *Oikos* **29**, 22–3.

Zar, J.H. (1974) *Biostatistical Analysis*. New Jersey: Prentice-Hall.

Zouros, E., Freeman, K.R., Ball, A.O. & Pogson, G.H. (1992) Direct evidence for extensive parental mitochondrial DNA inheritance in the marine mussel *Mytilus*. *Nature* **359**, 412–4.

Index

Note: Page numbers in *italics* indicate figures and **bold** type refers to tables.

Index